Contents

1 **Introduction: The Quest for Naturbanity** 1
Estienne Rodary, Louise Bruno-Lézy, Julien Dellier, Sylvain Guyot and Frédéric Landy

Part I Actors: Whose Parks?

2 **The History of the Four Parks: Favouring or Protecting from Urban Growth, Different Successive Conservation Policies** 35
Estienne Rodary, Louise Bruno-Lézy, Frédéric Landy, Mayara Morokawa and Janie Swanepoel

3 **Urban National Parks and the Rich: Friends with Benefits** 63
Julien Dellier, Sylvain Guyot, Frédéric Landy and Rafael Soares Gonçalves

4 **The "Poor", the Park and the City: Policies of Social Stigmatisation Rather Than Inclusion** 85
Frédéric Landy, Pauline Texier and Sylvain Guyot

5 **Categorisation of People and Places, Indigenous Peoples and Urban National Parks: Between Eviction, Instrumentality and Empowerment** 109
Nadia Belaidi, Karl-Heinz Gaudry and Frédéric Landy

Part II Beyond Local Challenges: The City-Park Relationships

6 **Urban National Parks and the Making of the Housing Market in Emerging Cities: Places of Exclusiveness, Land of Opportunities** 131
Julien Migozzi, Abdul Shaban and Rafael Soares Gonçalves

7 Why Did Leopards Kill Humans in Mumbai but not in Nairobi? Wildlife Management in and Around Urban National Parks 157
Frédéric Landy, Estienne Rodary and Bernard Calas

8 When Cities Host Parks: When Will Urban Frontiers Become Eco-Frontiers? 181
Sylvain Guyot and Estienne Rodary

9 National Parks in the Urban Arena: The Interplay of Actors for Incomplete Models of Governance 201
Véronique Fourault-Cauët and Jean-Fabien Steck

Part III Circulation of Practices, Discourses and Models

10 A Conservation View: Towards New Urban Protected Area Governance? 219
Louise Bruno-Lézy

11 What Makes Urban National Parks "Urban"? Their Specifics Within the National Systems of Protection 243
Bernard Calas, Frédéric Landy and Theresa Mbatia

12 Environmental Education in Urban National Parks: A Political Tool for the Democratisation Process in Emerging Countries. Studies on Tijuca and Table Mountain National Parks 265
Nadia Belaidi, Rafael Soares Gonçalves and Glaucio Glei Maciel

Part IV To Conclude

13 Conclusion: National Parks Between Urbanisation and Globalisation 287
Frédéric Landy

14 Afterword: Seen Through the Eyes of Researchers, Are Practitioners Partners, Research Objects or Hurdles? 317
Glen Hyman, Frédéric Landy and Louise Bruno-Lézy

Editor and Contributors

About the Editor

Frédéric Landy was the coordinator of the UNPEC project. A professor of social geography at the University of Paris-Nanterre (UMR LAVUE), he is currently the director of the French Institute of Pondicherry. Frédéric is the author of *Feeding India. The Spatial Parameters of Food Grain Policy*, Manohar-CSH, 2009. He has coauthored *Agriculture and food in India*, Manohar-CSH-Quae, 2009, and coedited *Megacity Slums. Social Exclusion, Space and Urban Policies in Brazil and India*, Imperial College Press, 2013 and *Reconfiguring Identities and Building Territories in India and South Africa*, Manohar, 2005. He is presently working on the relationships between agriculture, environment and ethnicity.

Contributors

Nadia Belaidi is a researcher at the CNRS, and a fellow of the eco-anthropology and ethnobiology laboratory (UMR 7206, Paris) of the French National Museum of Natural History. Based in the anthropology of law, Nadia's work focuses on social links and the way human-nature relationships contribute to them. She focuses on the Peace Parks experience in Southern Africa and social-ecology strategies in the global South and North.

Louise Bruno-Lézy heads the directorate of Environmental Projects and Partnerships for the Essonne district (France), which manages natural sites in the suburbs of Paris. Louise holds a master's degree in urban planning and a Ph.D. in geography from the University of Paris-Nanterre. She works on the city-nature relationship. As a founding member of the Libertas Institute for Applied Research, she launched the Urban Protected Areas Network program, which she has been developing since 2009. Louise is a member of the IUCN-WCPA, working as

Deputy Chair of the Urban Conservation Strategies Specialist Group (http://www.iucn-urban.org/).

Bernard Calas lectures African geopolitics and geoeconomics as a professor in the Department of Geography at the University of Bordeaux-Montaigne, France. Bernard is a fellow of "Les Afriques dans le Monde", a research unit of the CNRS and the Institut d'Etudes Politiques, Bordeaux. For 30 years, he has carried out in-depth field research in Eastern Africa. He has been director of the French Institute for Research in Africa, Nairobi. Currently, Bernard coordinates a program on the global value chain of roses, with a personal focus on the Kenyan cut-rose industry.

Julien Dellier is an associate professor in geography at the University of Limoges (UMR Geolab), France. Julien's research focuses mainly on social and critical approaches to urban nature and "edible cities". He has studied in various places in South Africa since 2008 and his work for the UNPEC project concentrated on Cape Town.

Véronique Fourault-Cauët is an assistant professor of geography at the University of Paris-Nanterre (UMR LAVUE), France. After a Ph.D. studying Mediterranean forest landscapes, Véronique's research now focuses on the new relationships between societies and their environment. Thus, from ordinary and unprotected landscapes to urban national parks, the goal of her research is to understand the way people participate—in a political and cultural sense—in the management of their living environment.

Karl-Heinz Gaudry is a senior researcher at the University of Freiburg (Institute for Environmental Social Sciences and Geography). After his doctoral research on the territorial governance of nature conservation regimes, he has gained expertise on Agenda 2030, the Convention of the Conservation for Biological Diversity and the use of green infrastructure for the development of nationally determined contributions as part of the United Nations Framework-Convention on Climate Change. Karl-Heinz is a member of the IUCN-World Commission of Protected Areas, and the Urban Conservation Strategies Specialist Group.

Sylvain Guyot is professor of geography at the University of Bordeaux-Montaigne (UMR Passages), France. Sylvain's research focuses on the political dimensions of nature conservation, with a particular interest in the aesthetics of eco-frontiers.

Glaucio Glei Maciel is an environmental trainer. He is a Ph.D. candidate in social services at the Pontifical Universidade Catholica, Rio de Janeiro, Brazil.

Glen Hyman is a doctoral researcher at the Center for the Sociology of Organizations at Sciences-Po in Paris, and a former Deputy Chair of the IUCN-WCPA Urban Specialist Group. Long interested in the interdependence of nature and cities, Glen has conducted extensive field research on UNESCO Biosphere Reserves near urban areas in Australia, Canada and South Africa. Glen lives in Brazil.

Theresa Mbatia holds a Ph.D. in geography from the University of Bordeaux 3 (France) and the University of Nairobi (Kenya). Theresa works on urban protected areas in Kenya.

Mayara Morokawa is a doctoral researcher in geography and urban planning at the University of Lyon 2 (France). Mayara was awarded a scholarship by CAPES Foundation, Brasilia.

Julien Migozzi is a fellow of the Ecole Normale Supérieure, Lyon, and is a Ph.D. candidate in urban and economic geography at the University of Grenoble Alpes, France. His research focuses on real-estate markets, social stratification and neighbourhood changes in Cape Town, South Africa. Julien is currently a research and teaching assistant at the École Normale Supérieure, Paris.

Estienne Rodary holds a Ph.D. in geography and political science and is a senior researcher at the French Research Institute for Development (IRD). Estienne's research interests focus on biodiversity conservation policies, protected areas, transnational governance, political ecology and the politics of connectivity, in Africa, the Pacific and at the global level. His latest book, *Politique de connectivité: conservation de la nature et extinction de la modernité en Afrique australe* will be published in 2018. Estienne is currently based in Nouméa, New Caledonia, and works on marine conservation policies in the new rush for oceanic resources.

Abdul Shaban is a professor at the School of Development Studies and a deputy director (Tuljapur Campus), at the Tata Institute of Social Sciences, Mumbai. Abdul has authored and edited several books and research papers. His most recent book is *Mega-Urbanization in the Global South: Fast Cities and New Urban Utopias of the Postcolonial State* (Routledge, 2012, coeditor).

Rafael Soares Gonçalves is a lawyer and historian. After a Ph.D. in history (Paris 7 University, France), he took a post-doctorate in anthropology (EHESS, Paris). Rafael is a professor in the Department of Social Service at the Pontifical Universidade Catholica, Rio de Janeiro, Brazil, where he heads the Laboratory of Urban and Socio-Environmental Studies (LEUS/PUC-Rio).

Jean-Fabien Steck is an associate professor in urban geography in the Department of Geography and Planning at the University of Paris-Nanterre (UMR LAVUE), France. The main topics of Jean-Fabien's research, based on fieldwork in Africa, are informality and public space, informal urbanism and planning, and planning and development. Jean-Fabien's contribution to the UNPEC research program was to analyse relationships between urban national parks and urban planning and policies, especially in Nairobi.

Janie Swanepoel is a Ph.D. fellow in the Social Anthropology Department at the University of the Free State (South Africa) and the University of Cologne (Germany). Janie conducted archival and ethnographic research on urban conservation practices in Cape Town, and her current research interests include agrarian change, white identity politics and multispecies ethnography in Namibia.

Pauline Texier is associate professor at the university of Lyon 3, France. He works on slums, informal settlements and urban poverty mostly in Indonesia and Brazil.

List of Figures

Fig. 2.1 Sanjay Gandhi National Park in the urban agglomeration of Mumbai 40
Fig. 2.2 Tijuca National Park in the agglomeration of Rio de Janeiro 44
Fig. 2.3 Nairobi National Park in the agglomeration of Nairobi. 49
Fig. 2.4 The Maasai reserves 51
Fig. 2.5 Table Mountain National Park in the Cape Town agglomeration 54
Fig. 3.1 The house of Karen Blixen built in the 1920s (Karen district, Nairobi) 66
Fig. 3.2 Constantia Valley looking over False Bay towards Handklip in the south-east, by Johannes Schumacher 1776–1777, a watercolour on paper 66
Fig. 3.3 The Cosme Velho district in Rio, mid-nineteenth century at the foot of the Tijuca massif. 67
Fig. 4.1 Location of the *favelas* surrounding the protected areas of Tijuca and other urban parks in Rio de Janeiro 91
Fig. 4.2 Exhibition panel at Tijuca Park Museum 96
Fig. 4.3 Layout and zoning map of Parque Da Cidade *favela* 97
Fig. 4.4 Informal pipe system bringing water from the park (**a**); accumulated waste at the bottom of the Parque Da Cidade district (**b**). 98
Fig. 4.5 A discovery trek on the mining history and endemic plants of the Enchanted Valley (**a**); community kitchen that employs three women, cooking with local produce (**b**). 102
Fig. 5.1 Washing dishes without tap water in Taleopada village (Sanjay Gandhi National Park). On the right, the Warli house has walls built traditionally with bamboos and mud. 119
Fig. 5.2 At dusk, an old Maasai takes his goats and cattle back to his family settlement, one kilometer from NNP 119

Fig. 5.3	KhoiSan Rights Now! Cape Town residents protesting for their Khoisan rights as part of a larger movement against a new development project on the scenic Chapman's Peak drive showcasing parts of Table Mountain National Park	121
Fig. 6.1	Localities around Table Mountain National Park	135
Fig. 6.2	Number of sales in Cape Town (1990–2015)	139
Fig. 6.3	Average prices (in Rs.) of flats (per square foot) in Mumbai	140
Fig. 6.4	Average price of residential flats along the Western Railway line and Western Expressway Corridor (2003 and 2012)	141
Fig. 6.5	Average price of residential flats along the Central Railway line and the Eastern Expressway corridor (2003 and 2012)	141
Fig. 6.6	Housing prices and distance to Table Mountain National Park	143
Fig. 6.7	Housing indexes and distance to Table Mountain National Park	144
Fig. 6.8	Market value of residential flats in Mumbai (Rs./sq. ft)	145
Fig. 6.9	Housing prices around Table Mountain National Park in 2000	146
Fig. 6.10	Housing prices around Table Mountain National Park in 2014	147
Fig. 7.1	Fatal leopard attacks in or around the park 1987–2015	158
Fig. 7.2	Compensations ("consolations") granted by Kenya Wildlife Service to herders in 2012	160
Fig. 8.1	Hybridisation working framework in a context of encounter between the park and the city	188
Fig. 8.2	Behind the gate, a villa with garden: is the Karen district in Nairobi a symbol of hybridisation?	190
Fig. 8.3	Silvermine Reservoir and Picnic Area (TMNP)	193
Fig. 8.4	Hang-gliding from the national park above Barra de Tijuca	193
Fig. 8.5	Green Belt Corridor (Constantia)	194
Fig. 8.6	Environmental education, or simply recreation? School students in Mumbai park	194
Fig. 8.7	Baba in her garden created with the support of an NGO (Gugulethu Township, Cape Town)	195
Fig. 8.8	Property towards Oudekraal (Camps Bay, Cape Town)	195
Fig. 8.9	Hiranandani Gardens (Mumbai North): Despite the name, a derisory effort to greenify a middle class suburb	196
Fig. 11.1	The Kenyan conservationist system	245
Fig. 11.2	A violent landscape gradient, specificity of NNP	249

Fig. 11.3 East Gate in the GreenLine, July 2014 252
Fig. 11.4 The Standard Gauge Railway (SGR) project:
The end of NNP? 260
Fig. 11.5 Advertisement for a cultural event in NNP: culture for all? 262
Fig. 13.1 The four national parks seen at the same scale 305

List of Maps

Map 1.1 Nairobi National Park, Nairobi 22
Map 1.2 Sanjay Gandhi National Park, Mumbai 24
Map 1.3 Tijuca National Park, Rio de Janeiro 26
Map 1.4 Table Mountain National Park, Cape Town 28

List of Tables

Table 1.1 National parks between globalisation and urban integration . . . 11
Table 2.1 Some features of the four sites . . . 36
Table 3.1 Influence of the national parks on the territorial entrenchment in upmarket districts . . . 68
Table 3.2 Evolution of the influence of well-to-do populations on the national parks . . . 76
Table 6.1 Comparison of factory workers' wages and prices of flats in Mumbai . . . 140
Table 6.2 Determinant of flat prices in Mumbai . . . 150
Table 8.1 Frontier dynamics in the four cities . . . 185
Table 8.2 Hybridisation between eco-frontier and urban frontier in the four cities . . . 189
Table 8.3 The different effects of the park/city contact hybridisation: generalisation . . . 192
Table 10.1 Analysis of the governance of the four parks studied by UNPEC in the light of IUCN typology . . . 225
Table 10.2 Evolution of parks in context compared to a national park traditional conservation model . . . 226
Table 11.1 NNP in relation to other major parks in the KWS . . . 246
Table 13.1 Configurations favourable or unfavourable to naturbanity (park-city integration) . . . 309
Table 13.2 National parks between globalization and urban integration . . . 309

List of Boxes

Box 1.1	The UNPEC Project	2
Box 6.1	Contrasted Situations Around Nairobi National Park (Frédéric Landy)	136
Box 6.2	Tijuca National Park and the Rio Housing Market (Rafael Soares Gonçalves)	137
Box 13.1	Urban National Park or National Urban Park? (Monique Poulot)	288

Chapter 1
Introduction: The Quest for Naturbanity

Estienne Rodary, Louise Bruno-Lézy, Julien Dellier, Sylvain Guyot and Frédéric Landy

> The survey found that the leopard density at Sanjay Gandhi National Park [in Mumbai] is 21 in a 100 sq km, which, the report said, was one of the highest density of leopards found anywhere…Breaking previously-held notions of leopards entering buildings along the periphery due to lack of prey inside the national park, the study showed wild prey contributed to 57 per cent of the leopards' diet. However, dogs along the periphery of the park was still found to be the principal prey species.
>
> *Indian Express,* July 1, 2015.

> [In Corcovado, Tijuca National Park,] "the statue of Christ the Redeemer is a global image and in reality is as wonderful. Very efficient elevators and escalators and the cleanest gondola windows ever!".
>
> www.tripadvisor.ca, October 21, 2013.

The two megacities of Mumbai and Rio de Janeiro host a national park. But the quotes above suffice to show how they differ! A good part of this book will be devoted to the explanation of these differences between urban national parks, and the consequences of these in terms of scenarios. The fact that nature conservation policies have been in line with the way societies function for more than a century has nurtured recurring

---

E. Rodary (✉)
French Research Institute for Development (IRD), Nouméa, New Caledonia
e-mail: Estienne.Rodary@ird.fr

L. Bruno-Lézy
Environmental Projects & Partnerships, Essonne, France

J. Dellier
University of Limoges/UMR GEOLAB, Limoges, France

S. Guyot
University of Bordeaux-Montaigne/UMR Passages, Pessac, France

F. Landy
University of Paris Nanterre/UMR LAVUE, Nanterre, France

F. Landy
French Institute of Pondicherry, Pondicherry, India

F. Landy (ed.), *From Urban National Parks to Natured Cities in the Global South*,
https://doi.org/10.1007/978-981-10-8462-1_1

controversies on the links between protection measures on the one hand and exploitation and development objectives on the other. These controversies have been exacerbated particularly in cities: they were for a long time considered as highly artificialised spaces, and for this reason the least likely to meet environmental concerns. The gap between nature protection and urban spaces has been widened to such an extent that conservation professionals and activists have, up until recently, neglected these areas, preferring to focus on rural spaces or even more so on the "wilderness". Nevertheless, things have begun to change and, today, cities are recognised not only as spaces likely to welcome nature conservation policies, but also as spaces able to generate a specific nature and therefore specific policies. From being characterised as marginal spaces in nature management issues, cities are now characterised as potentially original and innovative spaces. Yet, while in theory structuring perspectives are being opened up, actual coordination and/or collaboration practices and policies between conservation and urban public action are still fragile. They are greatly slowed down by spatial and institutional opposition logics inherited from past nature conservation as well as urban expansion and planning policies. This is particularly visible in the large metropolises of so-called "emerging" countries where urban pressures are exacerbated, and in national parks where fairly strict "nature" protection measures are being imposed.

Box 1.1: The UNPEC Project

UNPEC (Urban National Parks in Emerging Countries and Cities), is an interdisciplinary fundamental and applied research project (2012–2016) based in Cape Town, Mumbai, Nairobi and Rio de Janeiro, where it operated in collaboration with the national parks of these cities, i.e. respectively Table Mountain National Park, Sanjay Gandhi National Park, Nairobi National Park and Tijuca National Park. The largest of these parks, Table Mountain, has its continental part 73 times bigger than Central Park in New York. Too often, park and city managements work independently, which given the diversity of actors and their interests, is not without creating tensions and conflicts. UNPEC aimed to understand the implications and consequences of these dynamics.

UNPEC was funded by the French National Research Agency (ANR). It brought together 20 social and environmental science researchers from three continents, working jointly with agencies in charge of national parks, namely South African National Parks (SANParks) (South Africa), the Maharashtra Forest Department (India), Instituto Chico Mendes de Conservação da Biodiversidade (ICMBio) (Brazil) and the Kenya Wildlife Service. The project was coordinated by the University of Paris Nanterre, with the collaboration of the French Institute of Research in Africa (IFRA-Nairobi), the French Institute of South Africa (IFAS), the Tata Institute of Social Sciences (Mumbai), the Pontifical Universidade Catholica in Rio de Janeiro, the Libertas Institute for Applied Research, and the Urban Protected Areas Network. In close collaboration with park and city managers, the research work of the project's multidisciplinary teams on each site has been punctuated by three international conferences on BiodiverCities in Rio de Janeiro (2012), Cape Town (2014) and Marseille (2015).

The four national parks under review are located in countries considered more or less as "emerging" countries. For our purpose, the notion of emergence is defined as a double process characterised by the development of areas, activities, social groups and values drawing closer to those of the countries of the North, and by the fact that emerging countries are still underlain by poverty and socio-spatial injustice to varying degrees, depending on the countries and the extent of their emergence. The notion of emergence in countries, cities and parks is also tackled as a new relationship between parties, institutions and actors potentially leading to new tendencies, including in the field of conservation.

UNPEC offered a critical look at the notions of nature and city, making it possible to revisit this dichotomy in the metropolises of the South, thereby opening up to a critical analysis of the circulation of models regarding approaches to conservation. How can a national park survive in the megacity of an emerging country? *E pur si muove!* And yet these parks exist! They compel us to move from the idea of nature being opposed to humans' territory, to the recognition of intrinsic relations between natural and inhabited spaces. They also compel us to imagine the political interface between central and local powers: although these "national" parks are part of a "local" metropolis, they must face "global" issues: their management is made difficult by the diversity of actors operating at different levels, often bringing divergent representations and meeting contradictory interests, and endowed with unequal powers.

This work offers an overview of the main problems underlying urban space as part of the conservationist field and, symmetrically, nature as an object of urban policies. Not only does it mean to contribute to the theoretical thinking on "cities" and "national parks" as subjects, on their limits, structuring and coexistence, but also to the more pragmatic debates on the formulation of the actual and political means of such coexistence. This approach seems all the more necessary, since the explosion in the past fifteen years or so of academic works on urban biodiversity conservation, does not always come with clear and sufficiently detailed results (Shwartz et al. 2014)—not only as regards species conservation—even though urban managers are developing an increasingly structured discourse around environmental policies.

All the national parks under review have in common the fact that they are situated in or close to large urban areas in more or less emerging countries, although their characteristics are very different: What is common to, for example, the forest of the Sanjay Gandhi National Park (SGNP) in Mumbai, which was managed for many years like a "fortress"—which did not prevent leopards from getting out nor thousands of slum dwellers from entering—on the one hand, and the *fynbos* shrub of the Table Mountain National Park (TMNP) of Cape Town, which is criss-crossed by tar roads as well as numerous residential and economically disparate suburbs on the other hand, which is part of the image of a world-class city enhanced by the "nature" criterion? Contrasts internal to each case also reinforce this diversity, as found with the *favelas* above the plush residences of Rio de Janeiro, or with the "nature" representations of Maasai herders in Nairobi, which contrast with those of well-off and westernised Kenyans. Confronted with such diversity, the objective of this publication is to answer one central question: Whose parks are we talking about? Stuck between biodiversity protection requirements and city needs—whether these are legitimate such as ecosystem services, or more questionable such as urban sprawl—urban national parks are prey to identity crises. What are their objectives? Which actors must define them? With our research entries being so diverse, we can use, at best, the comparative approach, with all these entries leading to the following conclusion: not only should cities take their national parks more into account, in that they represent an incredible asset for local urban management and global competition, but also and reciprocally parks should negotiate and coordinate more with the cities around them: even if they wish to conduct an isolated "fortress" policy, for this they need the cooperation of the urban sphere—otherwise it could lead to disaster, as was the case for a while with SGNP in Mumbai.

'Naturbanity' is a hybrid neologism expressing this need. Neither a city nor a park has a right or the possibility to negate the presence of the other: as will be seen, the principle of "non-exclusivity" summarises the main argument of our book. The fact that parks and cities must and will live together, for better or for worse, must be considered empirically: challenges such as informal settlements encroaching on parks or industrial pollution, may be balanced out by, for example, getting support from the urban "civil society" for funding parks, or by benefitting from the awareness of ecosystem services provided by the park to the city. It also has to be considered philosophically, since it blurs the old "modern" nature/culture dichotomy: animals and human beings can often jump the physical and ideological barriers separating parks from adjacent cities.

Photo 1.1 View from Kanheri Caves towards Borivali, in the middle of Sanjay Gandhi National Park. *Source* Photo by Frédéric Landy

Photo 1.2 Aarey Milk Colony: on the right-hand side in the background is Sanjay Gandhi National Park, dominated by the illegally built Royal Palm Hotel. *Source* Photo by Frédéric Landy

Photo 1.3 The *favela* of Rocinha seen from Tijuca National Park. *Source* Photo by M. Morokawa

Photo 1.4 Panorama from Santa Teresa, with Botafogo district and Tijuca National Park. *Source* Photo by Chensiyuan (https://commons.wikimedia.org/w/index.php?curid=34129978)

Photo 1.5 An urban giraffe in Nairobi National Park. *Source* Photo by Frédéric Landy

Photo 1.6 The slum of Kibera: which possibility of environmental connectivity with the neighbouring Nairobi National Park? *Source* Photo by Frédéric Landy

Photo 1.7 View of Cape Flats and False Bay from Constantia Berg trail, Table Mountain National Park. Note the *fynbos* vegetation in the foreground. *Source* Photo by S. Guyot

Photo 1.8 View of Devil's Peak from Bo-Kaap, an heritage Cape Malay district. New business buildings are being constructed in the background. *Source* Photo by J. Dellier

1.1 Conceiving the Park-City Inclusion in Four Dimensions (Space and Time)

1.1.1 The Non-exclusivity Principle

The issue of national parks is generally considered from the perspective of some opposition between (nature) conservation and (societal) development. The precarious balance between human needs and biodiversity protection still greatly structures biodiversity management in the countries of the South. There is abundant literature on the subject, mostly insisting on the fact that it is necessary to include local populations in conservation policies (Berkes 2007; Brosius et al. 2005; Dellier 2010; Dressler et al. 2010; Ghimire and Pimbert 1997; Kepe 2007; Shackleton et al. 2010). In this binary approach, two positions are possible. One insists on reconciling or even merging biodiversity with anthropised landscapes, i.e. on policies able to articulate nature protection issues with the well-being of populations (Balmford and Cowling 2006; Mathevet 2012). The other points to competition potentially established between nature and exploited or built spaces: in this approach, an economic-type analysis prevails, where the advantages and inconveniences of nature conservation are assessed against the yardstick of a comparison with other non-environmental planning or exploitation options. This economic or even market assessment of conservation, has known these last years a spectacular reinforcement, with the introduction of the notion of ecosystem services (Balvanera et al. 2001; Boisvert et al. 2013; Laurans et al. 2013).

When placed back in an urban context, and one of large metropolises in particular, this dialectic is highly questioned. Cities and add huge after their populations cannot be moved due to biodiversity protection reasons—with the exception of the destruction of hamlets housing populations unfortunately living in spaces that suddenly became protected (as was the case in the Sanjay Gandhi National Park), or squatter camps invading protected areas (again in SGNP). Reciprocally, urban *national* parks cannot be delisted either as a whole because of urban development, whatever the reason, be it economic, social justice or land speculation. Thoughts on the links between urban national parks and their respective cities should be dominated by this non-exclusivity principle, but should not be without ambiguity! Some highlight the fact that cities "host" parks (cf. Chap. 8), while others say the opposite: parks house cities, because nature was there first![1]

1.1.2 The Dialogue Principle

In this context of non-exclusivity, the deep interweaving between a park and its urban environment is not a temporary situation that can be corrected by reinforcing any wilderness or conversely imposing a concrete jungle. On the contrary, the interconnected dimension is a central characteristic of urban parks. In this context of cohabitation, where biodiversity cannot rid itself of infrastructures and humans, and where "development" (the city) cannot rid itself of wild species that, in any case, often contribute to the city's identity, dialogue, as the first step towards participation, becomes the central node of interaction between city and park. This is not only the case within the restricted framework of an institutional instrument for the management of interface, i.e. a set place for meetings and collaboration; dialogue is also essential in a more political approach that is able to take into account the community as a whole, as constituted by the urban area.

Yet, confronted with these issues, institutional cooperation is at best limited, and at worst nonexistent in many urban national parks in the Global South. This situation, which is behind recurring conflicts and challenges in the parks' management policies, cannot represent a politically viable long-term solution, especially if the modes of participation do not take into account the urban population as a whole, and do not resolve in a satisfactory manner all imbalances in park access, and more generally income as well as social inequalities. Even in the most extreme cases, where local authorities ideologically do not favour public participation and support centralised policies, participative practices are needed, for simple empirical reasons of efficiency and security.

This necessity leads to at least two difficulties. On the one hand, it might not be able to get all interested parties to agree. Maintaining a very conflictual park, like the one in Mumbai, requires the eviction of all squatter camps as well as holding out

[1]Norman Johnson (SANParks), *BiodiverCities 2015* Conference, Marseilles, August 17, 2015.

against property developers. As Wollenberg et al. (2001) argue, "pluralism" and "syncretism" do not always make it possible to accommodate all interested parties involved in the management of protected areas. There is not always the possibility of a win-win situation in the nature conservation-societal development pair: the compromise principle must be introduced during negotiations (Redford and Sanjayan 2003). On the other hand, this difficulty is reinforced by another dimension, due to the fact that the "national" park designation introduces nonlocal, national and international actors in the political process, and therefore complicates the political participation process by multiplying scales. The intervention of levels external to the city is mainly due to their technical or scientific skills, while urban mobilisation is based more widely on urban planning, including the legitimacy of residing there. In addition to the multiplicity of actors and interests at play peculiar to urban spaces, participation must therefore endeavour to cross vertical and horizontal scales.

1.1.3 Horizontality and Verticality: Beyond the Park Entry Gate

Let us say it from the outset, the expression "urban national park" is in itself an oxymoron or, put more simply, a challenge. Being of national interest, a park corresponds to legal frameworks and objectives defined by a State meeting various national considerations, and sometimes even global objectives, such as conserving certain endemic species or reducing climate change. Yet, a park is also "urban" in that it is locally situated and must therefore meet demands from the city, just as it must resist certain pressures. This, understandably, does not facilitate the task of a park's managers.

As such, a park must not be studied or managed in isolation. Horizontally, the urban area as a whole must be taken into consideration, or even the (bio-)region, due to ecological processes as well as traffic flows. Vertically, it is advisable to tackle the way a park is structured at local, national and international levels. In this light, it seems that a park is situated at the intersection of two dimensions, a horizontal dimension (which concerns the issue of its integration into the urban region), and a vertical dimension (which concerns the dialectics between the local and provincial, as well as national scales and beyond), in logics that are not structured only according to oppositions, but also to complementarities. In any case, the institutional, social, ecological and functional distinction between a park and a city, as well as the spatial segregation policies resulting from this distinction, turn out to be conceptually erroneous and practically unmanageable.

In Table 1.1, parks are classified through two entries. The X-axis defines high or low horizontal integration at the level of the urban area or region between park and

Table 1.1 National parks between globalisation and urban integration

	High city-park integration	Low city-park integration
High globalisation of the park	Rio, and to a lesser extent Cape Town	Nairobi
Low globalisation of the park		Mumbai

city. What we call integration corresponds to relations that are above all institutional, and that bring park and city managers to join forces or confer, and the way parks' and cities' objectives and interests can be similar or compatible at least. As we shall see in this book, in Rio de Janeiro, and to a lesser extent Cape Town, the authorities meet with a view to coordinating their efforts. This is not the case in Nairobi or Mumbai, where the two spheres tend to operate separately, ignoring each other.

On the *Y*-axis, we represent what we call the degree of globalisation, i.e. the importance of supra-local scales, and the global scale in particular, among the actors and objectives of national parks. TNP and TMNP are obviously highly globalised sites, with worldwide renown and a high turnover of tourists from overseas. NNP is not similarly globalised but, by virtue of its urban proximity, serves as an urban ambassador for Kenya's wilderness parks, though it is hardly ever a destination in itself. As for Mumbai, SGNP is not a destination for foreign visitors and, unlike in Rio and Cape Town, it is not a landmark for the city to project itself onto the global arena.

1.1.4 Temporalities: The Weight of History

Relationships to time are an important dimension to take into consideration, when interpreting the links between cities and their national parks. These temporalities vary according to the different actors involved in city/park interactions. In some cases, the city changes at a faster rate than the park does (e.g., with some recent exceptions the borders of the NNP have remained the same since the park's creation); in other cases, the opposite takes place (e.g., TMNP firmly intends to expand its influence over most of the Cape Peninsula). Confronted with the weight of historical heritage, it is important to distinguish long-term processes and analyse often determining breakdowns taking place in the short term.

Several steps regarding the way urban national parks have been formed can be pointed out: a first long-term temporality refers to their being recommended and designed by different types of actors, often following a top-down and sometimes bottom-up logic; a second temporality, characterised by short-term political decisions, concerns a park's implementation; a third temporality, stretching over a more or less long time period depending on conflict levels, concerns a park's gestation and appropriation by urban dwellers. A last temporality concerns a park's evolution: reassertion of the park in a renewed context of labelling and international promotion or, on the contrary, conflict between protection and urban development requirements—as exemplified by the establishment of over 500,000 squatter camp residents inside SGNP during the 1990s. For each one of these steps, it is important to fully understand the different local (metropolitan), national and international contexts influencing these temporalities.

1.2 Urban Nature + Natural City = *Naturbanity*

1.2.1 The Different Values of a Park

History informs us that it is rare for a national park to be created only for ecological purposes (Brockington et al. 2008; Byrne and Wolch 2009). The environmental value of a park takes on several meanings: it does not only concern its ecological value (i.e. the park's content), but also that which takes into account the urban "environment" (i.e. the park's container) (Dearborn and Kark 2010; Rosenzweig 2003). As such, the park sometimes appears to be protected for its urban cover as much as for itself. It is a mistake to reduce the ecological value of a protected area to the species it contains. The processes it helps sustain are just as important, if not more so.

The park serves as landscape to the city while offering many ecosystem services, such as recreation and tourism, water, micro-climate and an international landmark among others. A park can also do potential disservices or cause inconveniences for the urban space, such as a loss of land that could be urbanised and fire hazards etc. In Mumbai, where the park creates serious traffic jams by blocking the east-west traffic in the north of the city where it has to be bypassed, the city is beginning to dream up a tunnel going under the park. In Nairobi, some authorities wish to reduce the size of the park in order to introduce a ring road and railway line. In Rio, on the other hand, it has been a while since tunnels had to be dug up in order to facilitate traffic between the *morros*.

Urban national parks seem to have strong specificities compared to "normal"—i.e. rural—national parks. What makes the "urbanity" of an urban national park? Chapter 11 offers an answer to this question theoretically as well as empirically (an urban national park is for example a park where rangers can drive their children to school easily). Reciprocally, what makes the "parkness" or naturalness of a city endowed with an urban national park? Do these cities share distinctive characteristics? In Cape Town, and even more so in Stockholm for example (as briefly mentioned in Chap. 13), the park spreads its ecological values in the city; this is not at all the case in Mumbai. This alludes to a further value of these parks: their educational value, both locally and more broadly as ambassadors for nature conservation elsewhere.

1.2.2 A Park-Environmental Policy Dialectics that Does not Always Go Without Saying

The issues posed by urban national parks partly—and only partly—cover those of the "sustainable city". Where the former focus on protecting vegetal and animal species and ecosystems, in areas specifically dedicated to that end, the latter adopts "greening" strategies of urban policies in various domains. This distinction confirms a difference in terms of public policy sectors, where conservation as an institutional sector taking initiatives on nature, is historically distinguished from a vaster and less sectorial set of environmental policies (Dumoulin and Rodary 2005). For this reason, *naturbanity* questions explicitly the argumentation underlying the integration of conservation into wider urban sustainability policies. Issues in this regard are not speculative only: initiatives in favour of sustainability have these last years constituted a domain where cities emerge as political actors that, in some cases, can go beyond or compete with the role of the State, particularly in the globalisation context (Emelianoff 2007). Confronted with these transformations in urban policies, conservation for the time being seems isolated and sometimes exploited. The presence of a national park in a city does not necessarily indicate significant dynamism, as far as urban environmental policies are concerned. Sometimes it is even the opposite, and many cities seem to be snowed under with their environmental problems (e.g., waste management, makeshift housing, pollution, etc.) as is the case in Mumbai, Nairobi or even Marseilles. In some cases, urban authorities can even be tempted to offload problems onto the national park, by counting on it to benefit from a "green" image that will take the focus away from urban problems, i.e. using the national park as a "greenwashing" tool. Some of the cities under study, such as Cape Town or Rio, use existing corridors and ecological connectivity between the park, metropolitan and provincial nature reserves as well as certain open spaces, in an attempt to combine different types of environmental policies, by making the national park one of the components, among other things, of these policies; however, this does not necessarily influence the actual "greening" of the city. While the environment has been used by cities trying to assert themselves as global political actors, opposite governments or firms, can the presence of a national park be enough to conceal certain shortcomings as regards environmental policies?

1.2.3 Urban Nature

Urban nature is the first expression of what we call *naturbanity*. Current discourses on "nature in town" tend to consider that reciprocal relations between these two terms are obvious. By "nature in town", some understand a continuation of rural nature or even wilderness, and try to actually recreate spaces dedicated to this type of nature within cities (Dubost and Lizet 2003). Others understand the city as an ecosystem where biodiversity elements form an integral part of the urban ecosystem, and insist on the vigour of spontaneous vegetal and animal life in a world which is not only mineral

(Arnould et al. 2011). In the face of these two positions, it seems imperative to reassert that urban nature is not similar to rural nature. Not that there is always a reality peculiar to urban biodiversity, but connections of any kind caused by urbanity create configurations peculiar to urban spaces. This should dissuade any attempt to recreate "wilderness" in cities, via national parks in particular. This point is all the more significant since urban biodiversity is not only present in these parks, but is also structurally and functionally connected to other places, whether these are parks or corridors, and to isolated species or individuals. All of this has direct effects on the city (which scientists are trying to define and assess through the notion of urban ecosystem services) which, in turn, have effects on biodiversity. Without taking an interest directly in the specific forms and functioning of urban biodiversity (Farinha-Marques et al. 2011; Magle et al. 2012; Savard et al. 2000), we note that the urban context compels us to define specific modes of biodiversity conservation and management (Dearborn and Kark 2010). This also concerns national parks: pastures are from now on being grown inside the SGNP, unlike the previous policy that advocated afforesting it to a maximum, so that leopards can feed on more herbivores within the park instead of being tempted to go to town to hunt dogs or rodents.

1.2.4 Natural City

Through the idea of a "natural city", the second expression of *naturbanity*, we can theorise that an urban national park is an excellent laboratory to study the city around it. Such a park, which can almost be seen as a metonymy of the city, is a highly media-covered stage where actors—who are above all urban—interact: by looking at the park, one (almost) knows already what the city is like. It reveals social logics: integration or exclusion of social or ethnic practices in the park, and potential rivalries between urbanities and ruralities. As such, the political emphasis placed on accessing TMNP, which must be facilitated for the poor who are mainly black and coloured, reflects the weight of the apartheid legacy in the urban area of Cape Town, and beyond in the whole of South Africa; while the rather miserable lot of the Adivasis living in SGNP, expresses the lower status of the indigenous populations in Indian society.

When a national park owes part of its existence to relief (as is the case in Rio, Mumbai, and Cape Town), it structures part of the urban area due to its steep slopes. It can also represent a major element of the urban structuring and its evolution for other reasons altogether, particularly from the point of view of economics and identity. For this reason, our hypothesis is that a city housing a national park can enjoy or build for itself a "natural" identity which is stronger than any other cities of similar size. We will see in this publication that this hypothesis is validated in Cape Town and Rio de Janeiro, although not so much in Nairobi, and even less in Mumbai.

It is possible to measure the way urban dwellers appropriate the park's identity according to its more or less socially separated urban practices. A city can rightly advertise itself as "natural", when its residents develop a specific relationship with (protected) nature, making it possible to differentiate the city from other large cities devoid of a national park. However, there is a lack of reliable indicators in trying to assess this relationship with nature: it is interesting to measure the number of local visitors as opposed to external visitors (which is huge in Mumbai, but low in Cape Town), but many of the visitors confine themselves to picnic and recreation areas, not showing any real interest in the biodiversity and its ecosystems. In any case, it remains to be proven that visitors who respect protection measures in the park actually show concern with the environment in their daily way of life.

The fact remains that national parks are places of social activities. The presence of biodiversity in town compels authorities to develop specific political mechanisms that can take nature into account, while accepting the presence of many visitors as well as the diversity of actors involved. This is not without creating tensions and conflicts.

1.3 Conflicts and Injustice

1.3.1 Environmental Conflicts

The co-presence of a city and an urban national park can be a source of conflict. In the broadest meaning of the notion of environment—an environment characterised by the interface between park and city—conflict can result from the city extending beyond the park's borders (what we call an urban frontier, cf. Chap. 8), or on the contrary from the park extending beyond the city walls (what we call an ecological frontier, see Guyot 2011). The potential for conflict is high when it comes to land, as when parts of the national park are being "eaten away" by the city; or when urbanized spaces are being protected to ensure ecological continuity with the park, among others. But what we understand by "frontier" is not only tangible: it also includes policies, education and representations, among other things, which can also be conflictual.

These conflicts are characterised by and mobilise a wide diversity of scales, actors and processes. As such, this research is one more proof of the fertility of the "political ecology" approach (Robbins 2012). However, our various case studies show that the intensity of environmental conflicts is not directly proportional to the environmental value of the habitats concerned. In Nairobi, the conflict concerning the southern bypass expressway was perhaps less about the strip of land which the national park was going to lose, than about the *principle* of delisting a gazetted area without the Parliament voting on it, and about the risk of creating a precedent. Consequently, it is advisable to determine the factors influencing conflict intensity, whether these concern the violence of the urban sprawl, the magnitude of the social gap, the degree of collective heritageisation of the protected biodiversity, or the park's history, etc.

1.3.2 The Issue of Access to Parks

When access to urban national parks is too limited, it has an impact on their social acceptance. The dilemma of many a park manager is as follows: they can privilege closing a park in the name of protection, in order to limit cases of destruction and disruption linked to urban influences; but closing a park will mean running the risk of paradoxically undermining it if, where it is not sufficiently appropriated by the local population, it is not considered as public goods and subsequently might be degraded. They can, on the contrary, privilege opening a park, either in the name of appropriation and social acceptance or because of the income expected from visitors. But opening a park can also undermine its very existence, if it turns out to be overvisited, as is the case with the beaches and dunes of Noordhoek, in Cape Town, where the balance is threatened by the excessive presence of strollers and pets, such as dogs and horses.

Issues linked to park access are not limited to their direct peripheries, but involve the entire urban area. In Cape Town for example, the challenge consists in ensuring that the park is not being used only by the white populations living around it, but also by populations from distant—mainly black and coloured—suburbs. In Mumbai, the park is situated too far up in the north of the urban area for the residents of the south to appropriate it—residents who sometimes have to spend over two hours to travelling to get there.

Moreover, segregation strategies of a social or ethnic nature etc., can be deployed around national parks, their location and situation. Maasai herders, for example, are banned from crossing the river giving access to NNP, even in the case of drought; yet urban white collars, who benefit from a yearly pass paid by their firm, can drive through the park every day to avoid traffic jams in the capital city. The park management justifies these double standards with the fact that visitors, who are channelled along trails or on picnic areas, are less harmful to conservation than wandering cattle competing with wild herbivores and potentially transmitting diseases. This argument is challenged by those who advocate rejuvenating park grass by the cattle and fires of the Maasai (in the name of ecological efficiency), or authorised access for the Maasai at least during the drought periods (in the name of social justice).

1.3.3 Justice and Inequalities

The presence of nature in cities often raises the problematic of social inequalities. Studying this interaction makes all the more sense in the case of metropolises which are characterised by considerable socio-spatial segregation phenomena. In Rio, the national park includes Tijuca Mountain, which makes up a natural barrier between the popular and industrial north of the city, and the wealthy and touristic south of the city. Furthermore, banning the specific links maintained by "indigenous" communities with nature can lead to processes of economic and political marginalisation. Indeed, conserving biodiversity and satisfying the needs of the destitute are inseparable, and call for the implementation of specific public policies. It is necessary to point to the different and sometimes divergent representations of what an urban national park ought to be, in order to meet the paired ecological and social issues.

Social justice can also be questioned through the prism of differences in temporality, by examining the opposition between, on the one hand, the heaviness of a deeply unequal urban space and, on the other, the rapid emergence of national parks as international touristic icons promoted by wealthy and influential minorities. The idea is to look for factors and dynamics that can make the park a driving force for the positive development of the city, without however neglecting the influence exercised on the future of the park by "the pace of the city" and its legacies.

Finally, the notion of "environmental justice" can be used in trying to shed light on many processes (Moreau and Gardin 2010). The creation of parks has often generated spatial *injustice*, by evicting local populations, or at least depriving them of access to most of the park's resources. In all cases, these parks can also be questioned in terms of spatial *justice*. If the squatter camps of SGNP must be destroyed, does that mean that the tribal hamlets of the Adivasis must also be demolished, where these represent a population probably fifteen times smaller than the squatters, and are especially less harmful and even very useful for conservation (Zérah and Landy 2013)? To a certain extent, do not the greening of urban discourses, and the advancement of national parks as compensating elements of environmental justice for populations as a whole, take part in maintaining far more blatant inequalities, whether in water access or failing waste management (Blanchon et al. 2009)? Nonetheless, parks can also represent *tools* of justice (Landy et al. 2017): this is the case for TMNP with its slogan "A park for all, forever", which shows the park's desire to build a more equal South Africa, favouring access to a park that was once situated in a white area, and developing ecological awareness leading to the promotion of African knowledge (cf. Chaps. 4 and 5).

1.4 Book Structure

This publication is structured as follows.

In this introduction, that is followed by four synthetic maps the neologism "naturbanity" has been discussed as expressing the necessity for cities endowed with a national park to integrate it into their functioning. Conversely, such parks must take into account their location in an urban environment, both as a source of heavy pressures on nature and as a nexus of incentives to support its conservation. What we call the principle of non-exclusivity (neither the city nor the park has a right nor even the possibility to negate the other's presence) summarises the main argument of our book.

Four years of field research have also shown how necessary it is to disaggregate the notions of "park" and "city". Both park and city management are made difficult by the diversity of actors operating at different levels (from the local to the global), each bringing often divergent representations of what "nature" is, meeting often contradictory interests endowed with unequal powers. Furthermore, an urban national park has to meet national objectives, and even global goals of biodiversity conservation, whereas it is also locally situated and must satisfy some demands of the city. When a metropolis competes internationally to obtain the status of "global city", a national park can be an efficient logo, a symbolic branding image attracting tourists, capital and international conferences (e.g., Cape Town and Rio de Janeiro). But in other cities, such as Mumbai and Nairobi, the parks can seem largely neglected. In the former case, "emerging parks" attached to "emerging cities" appear as a driving force for "emerging countries". In the latter, parks remain marked by the old "fortress" approach to conservation, with very little integration taking place with the metropolitan area. Our book explores the logic of this pattern.

The book is divided in three sections. The first section, "Actors. Whose parks?" begins with Chap. 2 describing the historical and spatial settings of the four case studies. The date of creation of the parks, the fact that some of them were municipal parks before becoming national parks, the size and history of the cities, the size of the parks and their location more or less close to the urban edge, the national social and political contexts, etc., are among the factors contributing to the various trajectories of the city-park interrelationships.

The following two chapters deals with the respective relationships of "the rich" and "the poor" with the national parks. According to Chap. 3, the interrelationships between middle-upper classes and the existence of an urban national park may seem obvious in Cape Town and Nairobi, where colonial systems are key factors explaining the location of rich and white inhabitants adjacent to the protected areas. They are much less obvious in Mumbai and Rio, as history bears out. We argue that the influence of "the rich" on the creation of the national park, and later on its management, is often crucial. Yet the presence of the park within the urban agglomeration cannot alone explain territorial strategies employed by "the rich".

The last point is further elaborated and complemented the Chap. 4. "The poor", so numerous in the "emerging cities" under study, are often described by "non-poor" local stakeholders as having mainly short-term perspectives, as being unaware of many environmental issues and as struggling for their livelihood. However, we argue that their practices as well as their representations of nature are often in line with conservation policies. Our case studies show that, in any case, the poor are no more injurious to the national parks than are the rich.

This issue is elaborated in Chap. 5 through the specific experience of indigenous peoples. National parks are areas placed under specific rules, generally prohibiting any human being to live within its borders: their categorisation in terms of spatial boundaries as well as management rules can alternatively label the local dwellers with the tag of "encroachers" or see them as legal inhabitants. Similarly, "indigenous peoples", when politically recognised, are the result of a political decision that can provide or take away some rights to the concerned groups. This chapter considers linkages between these two categories of space and people, to wit, urban national parks and indigenous peoples. Both can be classified in terms of domination—domination of the park by the city, and domination of indigenous peoples by conservation institutions. The second part of our chapter uses the analytical concept of eco-ethnicity to assess the possible recognition and instrumentalisation of the claimed identities of indigenous peoples, and concludes by considering ethnicity as a marker of environmental governmentality—what is called by Agrawal (2006), following Foucault, "environmentality".

The second section of the book is entitled "Beyond local challenges: the city-park relationships". It starts with the Chap. 6, which examines the impact of the national park on residential prices in respective cities. This influence is assessed in combination with other factors which might play an important role in determination of housing prices. The chapter uses secondary data that have been mapped appropriately to show locations of various social and environmental facilities as well as the residential prices. A regression method has been used to assess the individual factor influence on the residential prices in Mumbai and Cape Town. The data used for Rio and Nairobi are more qualitative in nature. One of the results is that whereas in Cape Town the proximity of the park is correlated to higher estate prices, in Rio, and above all in Mumbai, the poorer sections of society are concentrated around the park, the reason being the cheap availability of land or illegal occupation of government forest areas.

Chapter 7 brings out another possible factor determining real estate prices, namely the presence or absence of wild carnivores. Why in Mumbai were many people "eaten" by leopards whereas, in Nairobi, leopards, lions, hyenas or hippos have never killed anyone in living memory? In Mumbai even more than in Nairobi, the carnivore represents a double conceptual problem. It shakes the nature-culture dichotomy that leads to the delineation of protected areas and the city-park disconnect. Following Latour (1991) the nature/culture purification is endangered by the Mumbai leopards which refuse it and create hybridisation. For most people and park managers, leopards outside protected areas are "stray" animals that must be "rescued", i.e. trapped and deported to protected areas. We argue that wildlife management and more generally the way of living with wildlife must not be seen as neutral techniques or practices but as elements of an institutional framework and, beyond that, of a protection paradigm and an ontology of nature.

Chapter 8 goes further by theorising and contextualising—mostly from the Cape Town case study—the notion of frontier, defined as both a boundary and an area of contact. We look here at a specific frontier, the one dividing a city from a protected area. We ask if ecological frontiers are active within a city where new spaces of eco-conquest appear to be quite scarce. If such a conquest seems sometimes physically difficult due to land tenure and users constraints, we argue that the ecological frontier can be adapted by urban policies and city dwellers, directly at the contact of the park or further away. How is the urbanisation process, in a socially and often ethnically divided city, using and impacting the park boundary? Is the frontline area between the park and the city a space of contest or a space of confidence and interactions—what we will call a space of hybridisation?

Chapter 9 places the urban national parks on the "metropolitan stages": which systems of actors can exist in incomplete models of governance? As shown by the Nairobi case study, decentralisation generates new authorities at the local level that might perceive the national park as part of their competence; urban stakeholders and the "civil society" intervene more often in or about the management of the park. Yet there is little coordination between these actors; similarly, the parks are rarely granted an important place in the management plan of megacities.

The last section of the book engages with the "Circulation of practices, discourses and models". Chapter 10 aims to set the four national parks in a global typology classifying the governance systems of protected areas according to the place they give to parties in decision-making processes related to biodiversity conservation, but also to the integration of urban issues and social and environmental justice. Whereas SGNP—and to a lesser extent NNP—may seem to refer to the traditional paradigm of "reservation ecology", TMNP is officially considered as a tool for policies of political reparation, compensation and nation rebuilding. Yet TNP seems to be the only park with a—partly—joint management guaranteed by the federal Constitution.

Chapter 11 proposes a theoretical grid for evaluating the "urbanity" of national parks within the national systems of protection, and focuses mostly on NNP. Which criteria characterise an urban national park? As such, any national park in the world is "urban", since they were created by urbanites and are mostly visited by them. But there do exist gradients of urbanity. Urbanity implies specific types of management (e.g., in Nairobi poachers can easily escape and vanish into the city), specific functions of the park, and also specific narratives and images: in Cape Town the park is iconic and represents the logo of the city, whereas in Mumbai SGNP is to a large extent ignored by the city.

Is not a specific environmental education another feature of urban national parks? Chapter 12 considers that environmental education may be an education *for* the environment, so that individuals and groups may better protect it. But it may also been considered more ambitiously as an education *by* the environment, with "environment" used as a tool for personal development, and above all for socialisation and (re)construction of individual, group, and national identities. Lastly, a more politically incorrect view is to see environmental education as a way to impose exclusive paradigms that have nefarious impacts on the life of marginal groups whose conception of environment differs from the dominant narratives.

The conclusion Chap. 13 summarises the book by highlighting the commonalities between the four sites under study, and their differences. Trying to understand these differences turns out to be fruitful since new policies and regulations could emerge from our conclusions—in particular as far as the dialectic relationships between flows and fences are concerned. A final typology is proposed, opposing the "fortress parks" (yet with more and more breaches, willingly or not) and the "multidimensional parks".

As a kind of epilogue, Chap. 14 analyses the uneasy collaboration between researchers and practitioners. From the very start, our UNPEC research project has been preoccupied with the so-called science-policy interface. This is to say, as social scientists, we intended to closely collaborate with the practitioners who manage and govern the parks and cities in which our research was focused. To be sure, our work would not have been possible without the cooperation of these colleagues, but in practice, their role in the project remained primarily as informants and facilitators rather than as co-creators or co-designers of the research. Similarly, while some actively participated in our periodic "BiodiverCities" workshops, the research we've undertaken remained—at best—peripheral to their daily work. This chapter draws lessons from our experience, recognising with hindsight the naivety of our intentions, the obstacles and opportunities we encountered, and takes stock of the consequences of our approach—both positive and negative, intended and unintended.

Cartographic Appendix

The National Parks of Nairobi and Mumbai at a (Complex) Glance[2]

These two maps are intended to show as synthetically as possible the complexity of the processes and factors of evolution in the inter-relationships between the national parks of Nairobi and Mumbai and their respective cities.

A few words of explanation:

The sources: land cover was derived from Landsat images, other elements from various sources (e.g., government, NGOs, etc.).

The scale: we have mapped areas largely beyond the national parks, since they are dependent on other spaces. Some spaces are natural and their role in ecological connectivity is essential (Kitengela's wildlife migration zone for Nairobi and Tungareshwar Wildlife Sanctuary for Mumbai); other spaces are anthropised and interactions between natural and urban spaces are dense in terms of flows of fauna and flora as well as flows of people (visitors) and pollution, etc.

The key:

To enhance comparative map reading between the Nairobi and Mumbai projects, a colour code has been chosen: green for "natural" items, red for tourism, black for the city, etc. The first half of both legends deals with land use and administrative boundaries of the protected areas and cities. Maps appear relatively static but this should not be misleading: given strong urban dynamics, land use is constantly evolving, the boundaries of which might thus be challenged. In any case, they remain under constant pressure.

The second half of the legend is dedicated to conservation policies and tourism. Interactions between the two are analysed as well as urban threats and various solutions already locally implemented.

As rich in content as a text, the legend is structured around descriptive parts and subparts to offer substantial information to the reader. It was collectively agreed that the result of our research was best conveyed using semi-narrative legends.

[2]This appendix section has been coauthored by Emilie Edelblutte, Melody Rosdahl, Julien Dellier, and Frédéric Landy.

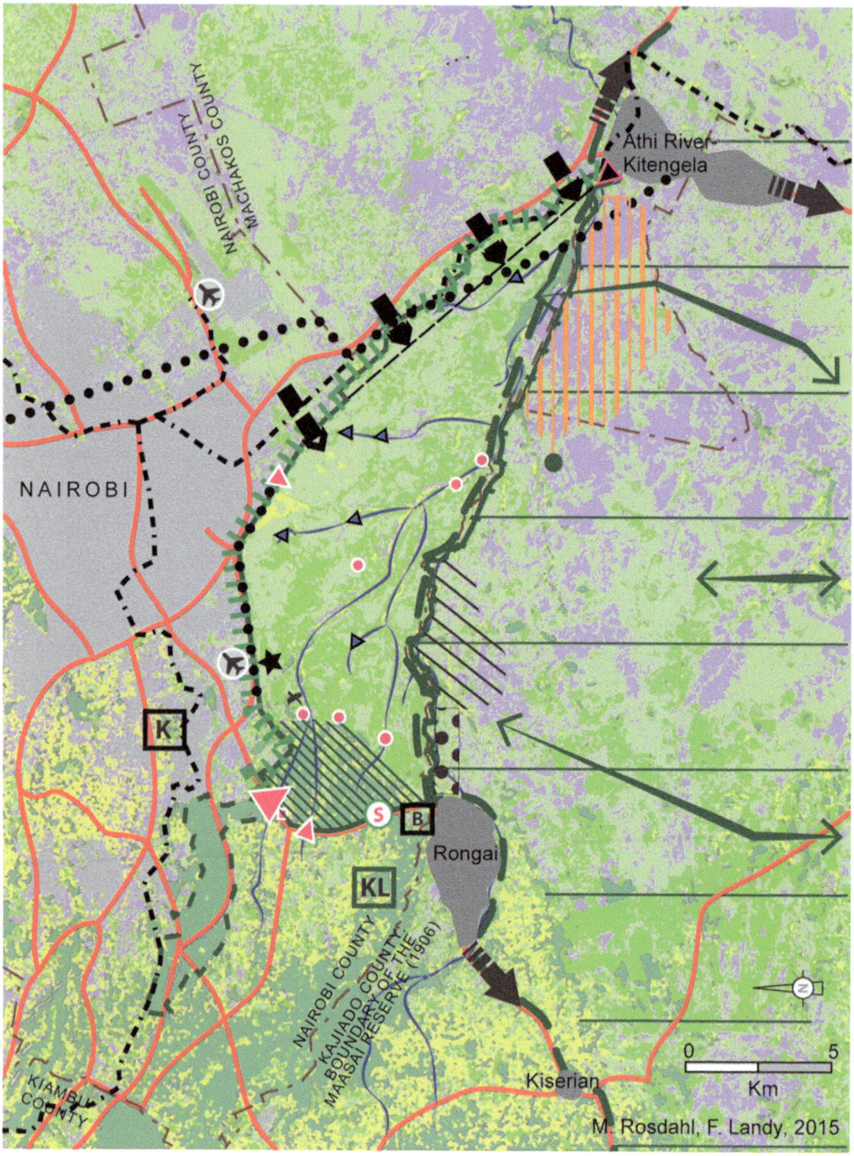

Map 1.1 Nairobi National Park, Nairobi
Source Authors

Key to Map 1.1:

I. A metropolitan area
A. Various land uses
Built-up area
Main towns
Cropland or woodland
Forest
Grassland (savanna or field)
Shrubland
Bare soil
Permanent river
II. Can animals respect official boundaries?
B. Official urban boundaries
County boundary
MACHAKOS County boundary
C. Major protected areas
Nairobi National Park (NNP)
Ngong Road Forest Sanctuary
"Conservancy" (planned)
D. Migratory wildlife
Main animal migration routes
"Dispersal area"
Open boundary of the NNP
III. Tourism and conservation
E. Welcoming tourists in the park...
Main gate of NNP
Other open gates
Safari walk
Picnic spots
F. but conservation first
x
Ivory burning site: poaching control
Forested area in the park
Dams: fauna attraction
S
Sheldriks orphanage
IV. Urban pressure on the park: to contain, to yield, or to negotiate
G. To contain
Major industrial pollution
Airport
Railway
Gate closed because of pollution
Main road
Green Line Project
H. To yield
Main axes of southern urban sprawl
Park area severed by the southern bypass
High voltage electric line
Location of most fenced plots in former ranches
Pipeline
I. To negotiate
Proposed extension of the park in compensation of the bypass
Lease programme area
J. The rich and the poor, the urban dwellers and the pastoralists
KL
Karen-Langata: main affluent area (green connectivity)
Attacks on livestock: 78% of the 72 predation cases by lions (Oct 2011-June 2012)
K
B
Kibera, Bangladesh: main adjacent slum

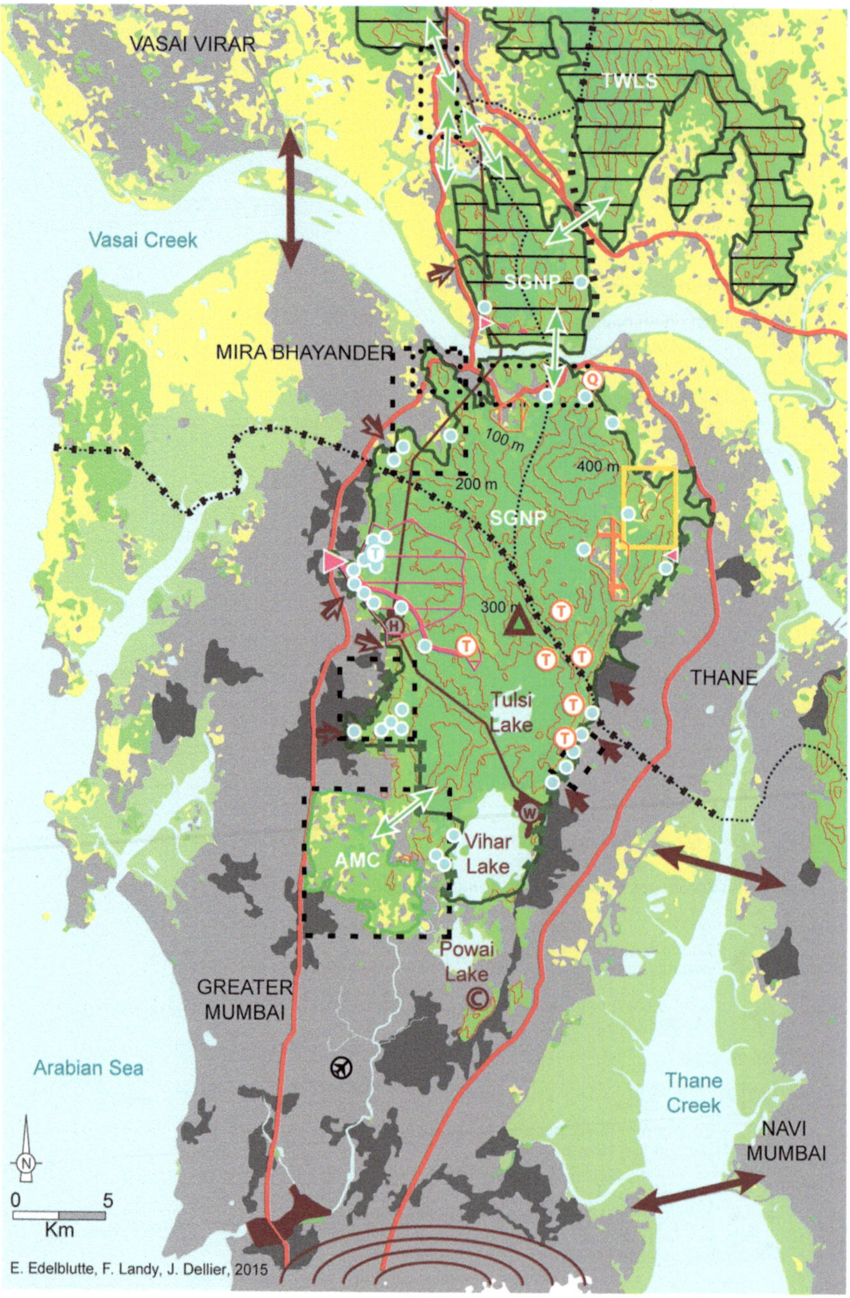

Map 1.2 Sanjay Gandhi National Park, Mumbai
Source Authors

Key to Map 1.2:

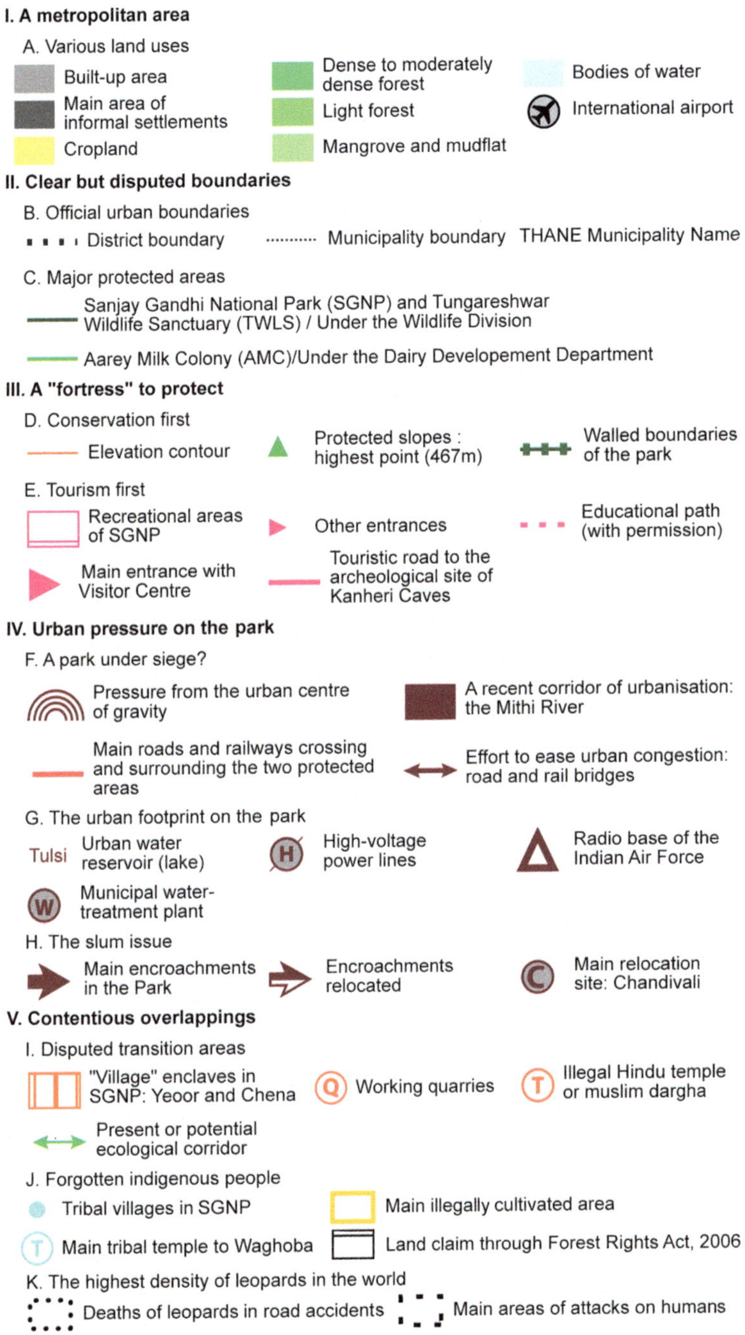
I. A metropolitan area
A. Various land uses
Built-up area
Main area of informal settlements
Cropland
Dense to moderately dense forest
Light forest
Mangrove and mudflat
Bodies of water
International airport
II. Clear but disputed boundaries
B. Official urban boundaries
District boundary
Municipality boundary
THANE Municipality Name
C. Major protected areas
Sanjay Gandhi National Park (SGNP) and Tungareshwar Wildlife Sanctuary (TWLS) / Under the Wildlife Division
Aarey Milk Colony (AMC)/Under the Dairy Developement Department
III. A "fortress" to protect
D. Conservation first
Elevation contour
Protected slopes : highest point (467m)
Walled boundaries of the park
E. Tourism first
Recreational areas of SGNP
Other entrances
Educational path (with permission)
Main entrance with Visitor Centre
Touristic road to the archeological site of Kanheri Caves
IV. Urban pressure on the park
F. A park under siege?
Pressure from the urban centre of gravity
A recent corridor of urbanisation: the Mithi River
Main roads and railways crossing and surrounding the two protected areas
Effort to ease urban congestion: road and rail bridges
G. The urban footprint on the park
Tulsi
Urban water reservoir (lake)
H
High-voltage power lines
Radio base of the Indian Air Force
W
Municipal water-treatment plant
H. The slum issue
Main encroachments in the Park
Encroachments relocated
C
Main relocation site: Chandivali
V. Contentious overlappings
I. Disputed transition areas
"Village" enclaves in SGNP: Yeoor and Chena
Q
Working quarries
T
Illegal Hindu temple or muslim dargha
Present or potential ecological corridor
J. Forgotten indigenous people
Tribal villages in SGNP
Main illegally cultivated area
T
Main tribal temple to Waghoba
Land claim through Forest Rights Act, 2006
K. The highest density of leopards in the world
Deaths of leopards in road accidents
Main areas of attacks on humans

Tijuca National Park and Table Mountain National Park in Their Urban Environment[3]

Similar principles were chosen for mapping Tijuca and Table Mountain National Parks: including the whole urban agglomerations to better understand the interrelationships between the parks and their environment, and using a similar colour code in the key. The legends are slightly less rich than in the former two maps, but are easier to read. In the case of Tijuca, for instance, the "narrative aspect" of the legend is weak, but to the advantage of simplicity.

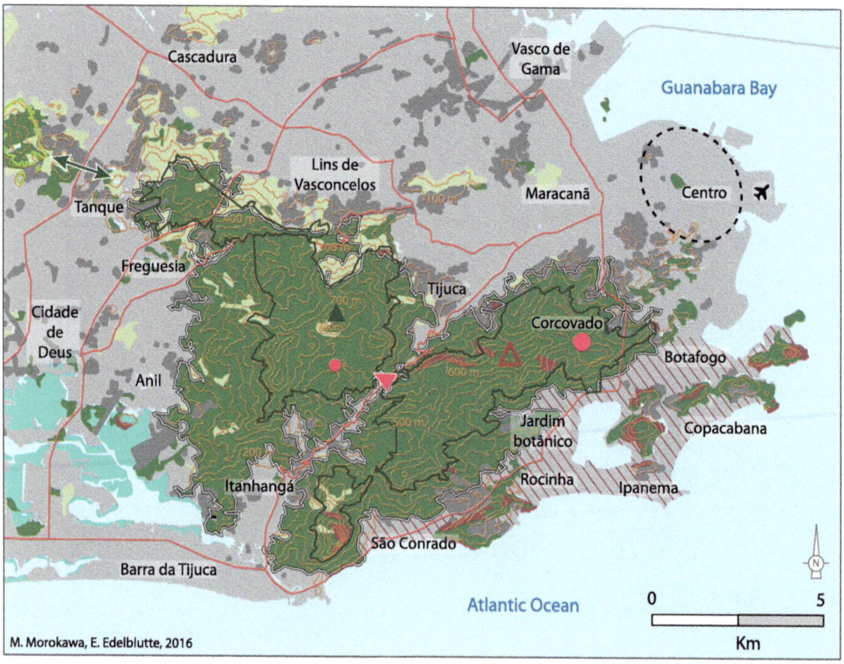

Map 1.3 Tijuca National Park, Rio de Janeiro
Source Authors

[3]This appendix section has been coauthored by E. Edelblutte, M. Morokawa, J. Dellier, S. Guyot, E. Rodary.

Key to Map 1.3:

A. Various land uses

- Built-up
- Dense forest
- Grassland
- Rocky outcrop
- Mangrove and mudflat
- Bodies of water

B. Major protected areas

- Boundaries of Tijuca National Park
- Boundaries of Pedra Branca Park
- Elevation contour
- Highest point (1021m)
- Potential ecological corridor

C. Tourism and urban infrastructures

- Main entrance with Visitor centre
- Main tourist attractions
- Radio base station
- Main roads crossing and surrounding the park
- Airport

D. Urban threat or urban asset?

- Historical centre, business district
- Main informal settlements
- Rich neighbourhoods
- Boundaries of buffer zone: A city-forest interface?

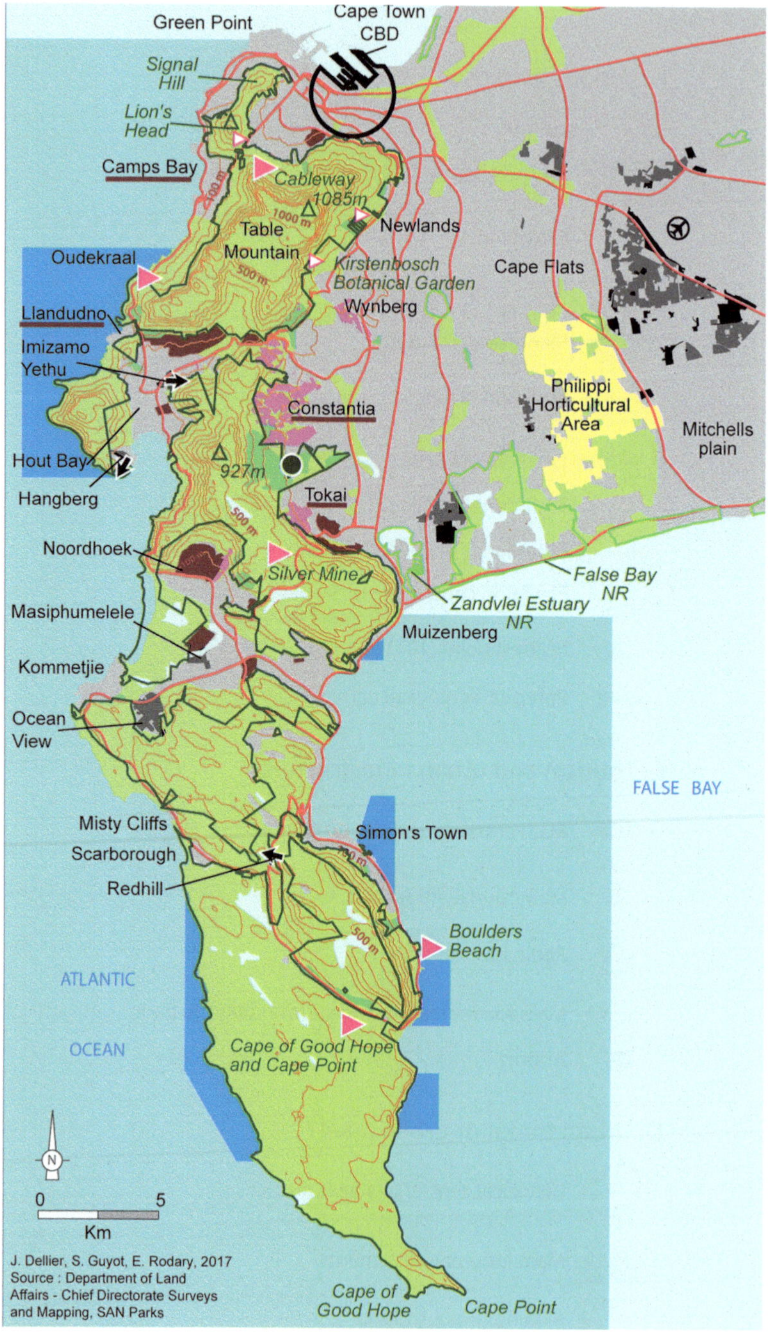

Map 1.4 Table Mountain National Park, Cape Town
Source Authors

Key to Map 1.4:

I. A metropolitan area

A. Various land uses

- Built-up area
- Cropland
- Vineyard
- Forest
- Grassland/Fynbos
- Bodies of water

B. Local and international connections

- International airport
- Main roads crossing and surrounding the park

II. An archipelago park

C. Major protected areas

- Boundaries of Table Mountain National Park (TMNP)
- Nature reserve managed by the City of Cape Town
- SanParks Headquarters
- TMNP Marine Protected Area
- TMNP Marine Protected Area (Restricted Zone)

D. Altitude

- Elevation contour
- Highest points

E. Tourism

- Paying access, main attractions
- Free entrances

III. Who is living next door?

F. Main poor residential areas

- Low-cost housing with backyard shacks
- Main areas of informal settlements

G. Main wealthy residential areas

- Tokai Upmarket residential area
- Gated communities and "eco-estates"

H. Encroachment issue

- Slum encroachment

References

Arnould, P., et al. (2011). La nature en ville: L'improbable biodiversité. *Géographie, Économie, Société, 1*(13), 45–68.

Balmford, A., & Cowling, R. M. (2006). Fusion or failure? The future of conservation biology. *Conservation Biology, 20,* 692–695.

Balvanera, P., Daily, G. C., Ehrlich, P. R., Ricketts, T. H., Bailey, S.-A., Kark, S., et al. (2001). Conserving biodiversity and ecosystem services. *Science, 291,* 2047–2047.

Berkes, F. (2007). Community-based conservation in a globalized world. *Proceedings of the National Academy of Sciences, 104,* 15188–15193.

Blanchon, D., Moreau, S., & Veyret, Y. (2009). Comprendre et construire la justice environnementale. *Annales de Géographie, 665–666,* 35–60.

Boisvert, V., Méral, P., & Froger, G. (2013). Market-based instruments for ecosystem services: institutional innovation or renovation? *Socity & Natural Resources, 26,* 1122–1136.

Brockington, D., Duffy, R., & Igoe, J. (2008). *Nature unbound: Conservation, capitalism and the future of protected areas*. London: Earthscan.

Brosius, J. P., Tsing, A. L., & Zerner, C. (2005). *Communities and conservation. Histories and politics of community-based natural resource management*. Walnut Creek: AltaMira Press.

Byrne, J., & Wolch, J. (2009). Nature, race, and parks: Past research and future directions for geographic research. *Progress in Human Geography, 33,* 743–765.

Dearborn, D., & Kark, S. (2010). Motivations for conserving urban biodiversity. *Conservation Biology, 24*(2), 432–440.

Dellier, J. (2010). Les écueils de la Wild Coast, Afrique du Sud—la co-gestion des réserves naturelles dans un contexte post-apartheid. EDYTEM n° 10 *Espaces protégés, acceptation sociale et conflits environnementaux*, Chambéry, pp. 197–208.

Dressler, W., Büscher, B., Schoon, M., Brockington, D. A. N., Hayes, T., Kull, C. A., et al. (2010). From hope to crisis and back again? A critical history of the global CBNRM narrative. *Environmental Conservation, 37,* 5–15.

Dubost, F., & Lizet, B. (2003). La nature dans la cité, de l'hygiénisme au développement durable. *Communications, 74,* 5–18.

Dumoulin, D., & Rodary, E. (2005). Les ONG, au centre du secteur mondial de la conservation de la biodiversité. In C. Aubertin (Ed.), *Représenter la nature?* (Vol. Editions de l'IRD, pp. 59–98). Paris: ONG et biodiversité.

Emelianoff, C. (2007). La ville durable: L'hypothèse d'un tournant urbanistique en Europe. *L'Information Géographique, 71,* 48–65.

Farinha-Marques, P., Lameiras, J. M., Fernandes, C., Silva, S., & Guilherme, F. (2011). Urban biodiversity: A review of current concepts and contributions to multidisciplinary approaches. *Innovation, 24,* 247–271.

Ghimire, K. B., & Pimbert, M. P. (1997). *Social change and conservation*. London: Earthscan; UNRISD.

Guyot, S. (2011). The eco-frontier paradigm: rethinking the links between space, nature and politics. *Geopolitics, 16*(3), 675–706.

Kepe, T. (2007). Land claims and comanagement of protected areas in South Africa: Exploring the challenges. *Environmental Management, 41,* 311–321.

Landy, F., Belaidi, N., & Gaudry, K. H. (2017). Urban protected Areas: Forces of justice or injustice for Indigenous populations? The cases of Xochimilco and the national parks of Mumbai and Cape Town. *Justice Spatiale|Spatial Justice*, n° 11. http://www.jssj.org.

Laurans, Y., Rankovic, A., Bille, R., Pirard, R., & Mermet, L. (2013). Use of ecosystem services economic valuation for decision making: Questioning a literature blindspot. *Journal of Environmental Management, 119,* 208–219.

Magle, S. B., Hunt, V. M., Vernon, M., & Crooks, K. R. (2012). Urban wildlife research: Past, present, and future. *Biological Conservation, 155,* 23–32.

Mathevet, R. (2012). *La solidarité écologique*. Actes Sud, Arles: Ce lien qui nous oblige.

Moreau, S., & Gardin, J. (2010). Obviously …, *Justice Spatiale/Spatial Justice*, 2. Online: http://www.jssj.org.

Redford, K., & Sanjayan, M. A. (2003). Retiring Cassandra. *Conservation Biology, 17*(6), 1473–1474.

Robbins, P. (2012). *Political ecology: A critical introduction*. Wiley-Blackwell.

Rosenzweig, M. L. (2003). *Win-win ecology: How Earth's species can survive in the midst of human enterprise*. New York: Oxford University Press.

Savard, J.-P. L., Clergeau, P., & Mennechez, G. (2000). Biodiversity concepts and urban ecosystems. *Landscape and Urban Planning, 48,* 131–142.

Shackleton, C. M., Willis, T. J., Brown, K., & Polunin, N. V. C. (2010). Community-based natural resource management: designing the next generation (Part 1). *Environmental Conservation, 37,* 1–106.

Shwartz, A., Turbe, A., Julliard, R., Simon, L., & Prévot, A.-C. (2014). Outstanding challenges for urban conservation research and action. *Global Environmental Change, 28,* 39–49.

Wollenberg, E., Anderson, J., & Edmunds, D. (2001). Pluralism and the less powerful: Accommodating multiple interests in local forest management. *International Journal of Agricultural Resources, Governance and Ecology, 1*(3/4), 199–22.

Part I
Actors: Whose Parks?

Chapter 2
The History of the Four Parks: Favouring or Protecting from Urban Growth, Different Successive Conservation Policies

Estienne Rodary, Louise Bruno-Lézy, Frédéric Landy, Mayara Morokawa and Janie Swanepoel

2.1 Introduction

The history of urban biodiversity conservation and national parks in cities is still in its infant stage. The historical work presented in this chapter on the four sites of the UNPEC project, shows nevertheless that the lack of historiography does not mean a lack of actual history. All four parks studied here reflect histories that can be traced back to the colonial times. These trajectories are furthermore not isolated in their geographical places but have been largely shaped by their inclusion and connections at the colonial and global levels. As an ideology, conservation matured in the eighteenth century encounter between Europe and the tropics. From the outset, conservation debates in the colonial context were not a matter of wholesale importation from the United States or Europe, but a confluence of colonial expansion (and consequently environmental degradation), scientific philosophies,

E. Rodary (✉)
French Research Institute for Development (IRD), Nouméa, New Caledonia
e-mail: Estienne.Rodary@ird.fr

L. Bruno-Lézy
Environmental Projects & Partnerships, Essonne, France

F. Landy
University of Paris Nanterre, Nanterre, France

F. Landy
French Institute of Pondicherry, Pondicherry, India

M. Morokawa
University of Lyon, Lyon, France

J. Swanepoel
University of the Free State, Bloemfontein, South Africa

J. Swanepoel
University of Cologne, Cologne, Germany

F. Landy (ed.), *From Urban National Parks to Natured Cities in the Global South*,
https://doi.org/10.1007/978-981-10-8462-1_2

Table 2.1 Some features of the four sites

Name of the national park	Sanjay Gandhi National Park, Mumbai	Nairobi National Park, Nairobi	Table Mountain National Park, Cape Town	Tijuca National Park, Rio de Janeiro
Date of creation	1950	1946	1998	1961
Original purpose of first protection measures	Water supply	Wildlife	Water supply, tree plantation, mountaineering	Water supply
Present area	104 km^2	117 km^2	250 + 1000 km^2 marine	39 km^2
Highest altitude/height difference	466 m/436 m	1760 m/237 m	1085 m/1085 m	1021 m
Managing institution	Maharashtra Forest Department	Kenya Wildlife Service	SAN Parks	ICMBio (Instituto Chico Mendes para Biodiversidade, ministry of Environment)
Size of urban agglomeration (million)	20	4	4	12
Global ranking of the city (GaWC 2016)[a] and main characteristic	Alpha. Economic capital city of India	Beta− (2012). Economic and political capital of Kenya	Beta+. Second South African agglomeration	Beta−. Second Brazilian agglomeration, former capital city
Main protected ecosystem	Moist deciduous forest	Savannah. Focus on wildlife	Fynbos (Cape floral kingdom)	Tropical Atlantic forest
Famous landmark	Kanheri caves, tiger safari	Open savannah with grazing animals	Table Mountain (landscape)	Corcovado Christ
Emblematic species	Leopard, spotted deer	Rhino, wildebeest, lion…	Baboon, African penguin, *Proteaceae*	*Ara ararauna* (Parrot)
Access monitoring/number of visitors	Encouraged only in recreation area. 1 M paying visitors/ year + numerous local walkers. Very few foreign tourists	Selectively encouraged (high entry fee, esp. for foreigners). 100,000 pers/year, more than 40% foreign	Encouraged. 4 M pers./year, many foreign tourists (ca.60–90% at the cable)	Encouraged. 2.7 M pers. (2013) incl. more 1 million Corcovado. Many foreign tourists (ca.30%)

[a]http://www.lboro.ac.uk/gawc/world2016t.html

the colonial experience, soils, weather, animal patterns and game hunting practices (Grove 1995). While one can surmise that the history of biodiversity conservation in urban contexts is likely to be very specific, compared to the history of conservation policies in rural or wild areas, the four park and city histories presented here also show a complex interaction between local trajectories and long-distance links.

Across different colonial empires, several continents and divergent urban settings, similarities emanated from the comparative histories of the parks. First, the history of the actual national park is but the last step of a long chronicle of public policies and institutional transformations that took place in the last centuries. The protection of the places and habitats that eventually became national parks has been endorsed and advocated by different groups and institutions, in which the national level is only one among many layers of actors. Besides, the nationalisation of the parks has systematically raised concern among stakeholders and spawned political issues at the intersection of environmental and urban management.

Second, the diversity of institutional arrangements reflects the long history of different practices, focused around diverse dimensions of environmental protection that came before the actual foundation of a national park. It appears from the findings that in big cities the choice of privileging a national park as the institutional form is closely correlated with the focus on biodiversity conservation. Prior to this, the consecutive structures put into place to protect nature had in their mandate different environmental focuses. Hence, with the gradual and increasing focus on biodiversity conservation policies that would eventually bring the nationalisation of urban protected areas came a similar shrinking of institutions in charge of nature in the city.

Third, among the conservation issues that came before biodiversity, two clearly emerge as recurrent issues across the three continents: water governance and forest protection. Water supply for the neighbouring city is a crucial historical reason for the protection of Rio, Mumbai and Cape Town parks. At that time, protection was to support urban growth rather than to guard against it. The focus on water and forest elicited several trends in the history of urban protected areas. Because the two topics were generally seen as interconnected, such as erosion due to deforestation affecting water supply and local climate, the first local policies were—in intention if not in application—systemic in their orientation. This echoes Richard Grove's claim that the first environment concern explicitly took a holistic approach and that only after conservation became a sectorial activity did the environment lose its integrative dimension (Grove 1995). Another trend relates to the long-term use of the places that would later become national parks. Far from being wilderness areas, nature in urban areas have always been profoundly transformed and affected by human use and exploitation, with the result that the ecosystems that are now protected are often far from being pristine forests.

Fourthly, the place and role of height in the conservation of urban environment is also of significance. As a limit to urban development and sprawl, hills and mountains were a strong component of early conservation policies, and have explicitly been incorporated into strategies and narratives, for instance in Cape Town where the Mountain Club claimed to be custodians of the park. The "politics of slopes" needs further investigation, but it seems that it played a strong role in most histories of urban protected areas, in a sense comparable to the function remoteness had in rural conservation policies. Here Nairobi stands in a different historical pattern, and not only because it is located in the lowlands: big wildlife concentrated the early environmental policies in a context where hunting was a highly symbolic issue for the imperial British elite (MacKenzie 1988).

Fifth, all urban parks experience new development and extension in the form of environmental networks and ecological connectivity that try to both link the park with the concrete city, and combine the park with other ecological units at the built-up area and regional levels. Here Kenya stands alone again (even though Mumbai's park might soon be similarly threatened), with the current dynamics of urban sprawl jeopardising the ecological connections that have been so instrumental in giving the park its exceptional density of wildlife. This latest trend and the diverse forms taken on by previous urban protected areas, seem to suggest that the urban national park, as we know it today, might be only one moment in the history of urban conservation policies that are very much into a process of becoming (Table 2.1).

2.2 Mumbai: 20 Million People and One Forest

While Sanjay Gandhi National Park (SGNP, Fig. 2.1) received its official name in 1981, the history of the policies that progressively made up this park goes back to the eighteenth century. It was built through a series of land acquisitions and area designations, which today form a patchwork of protected lands with varied statuses (Zérah and Landy 2013).

2.2.1 Waters

Historically, the central area of the park benefits from a specific status due to the presence of the most important landscape in Mumbai that reaches a height of 466 m above the sea lagoon (Paranjpye 1997). This is where, from the first century AD, Buddhist monks created a monastery and dug the spectacular Kanheri Caves in the volcanic basalt, which today are the responsibility of the Archaeological Survey of India. Water tanks and canals testify to the monks' knowledge of water management. During colonisation, with a view to meeting water supply issues in Bombay, which at the time was an expanding city, the British government created three storage reservoirs: Vihar (1860), Tulsi (1897), and Powai (1891). The first two were located in the area of the actual SGNP and the third outside, to the south. The lakes of Vihar and Tulsi, and the forest areas of their catchment areas were acquired by the Bombay Municipal Corporation (BMC), in 1860 and 1883 respectively. They became protected by the jurisdiction of the local government and were subjected to a scientific management plan as early as 1901, intended to perpetuate their water supply capacity for the city.

2.2.2 Forests

The archives pertaining to the current SGNP area mention, as early as the seventeenth century, the forests of Yeur and Nagla, which at the time were lands of the Maratha Empire. In 1808, the colonial government took over these forests (mainly moist deciduous forests) and with the creation of the Imperial Forest Department in 1864, they were monitored, with an administration responsible for a management plan. Nevertheless, at the time, the two forests were easy to access and regularly exploited, which contributed to their impoverishment. Moreover, forest management and administration were still not very structured.

The Indian Forest Act of 1865 was the first attempt at creating a piece of legislation for forests in this part of the British Empire. It was succeeded by the Indian Forest Act of 1878 and the National Forest Policy of 1894, which shows the importance given to ligneous resources in India. Behind a discourse insisting on satisfying the needs of local populations, it was mainly considerations based on revenue that guided the establishment of the law, by seeking a State monopoly on forest resources. The Indian Forest Act of 1927 came to consolidate this legislative structure by tackling the commercial stakes of forest production as well as taxes on timber and other forest products.

The oldest protected section of SGNP was initially notified as protected forest by the Act of 1927. In 1939, the parks were transferred to the Forestry Department of the Province of Bombay. However, the sections of the future SGNP that were not at that time protected belonged to absentee landlords and prior to 1950 were without much concern exploited to commercial ends. They were therefore particularly poor ecologically. Admittedly, while forests were previously exploited with management practices that included thinning and clear cutting, the legislation established more ecological practices. Yet the sale of deadwood and firewood was still allowed in the Management Plan of 1974, and for this reason the majority of the park's forests consist mainly of young trees, except for certain small plots in the northern section of the park that are difficult to access (Khati 2003).

A historical view of the growth of the park area shows that not only has urban sprawl brought Mumbai closer to the park, it has also resulted in the park coming closer to Mumbai. This partly explains the penetration of many informal settlements within the park's boundaries (more than 500,000 people in 1995 and probably still 100,000 remaining in 2016). The permanent extension of the protected area has encroached on rural areas adjoining the park. The core of the park always had a specific status, not so much because of its altitude as due to its steep slopes. But SGNP, with its current borders, is the result of a historical process whereby various tracts of forest and agricultural land gradually merged (Zérah 2007). Many components of the present protected area were private property until the recent past; Chena revenue village was exploited by absentee landlords due to a lack of felling regulations until 1956 (Management Plan 1974). In 1947, the municipal government took over an additional 47 km^2 and another 20 km^2 were added and transferred to the Forest Department in 1969. The National Park itself was created in

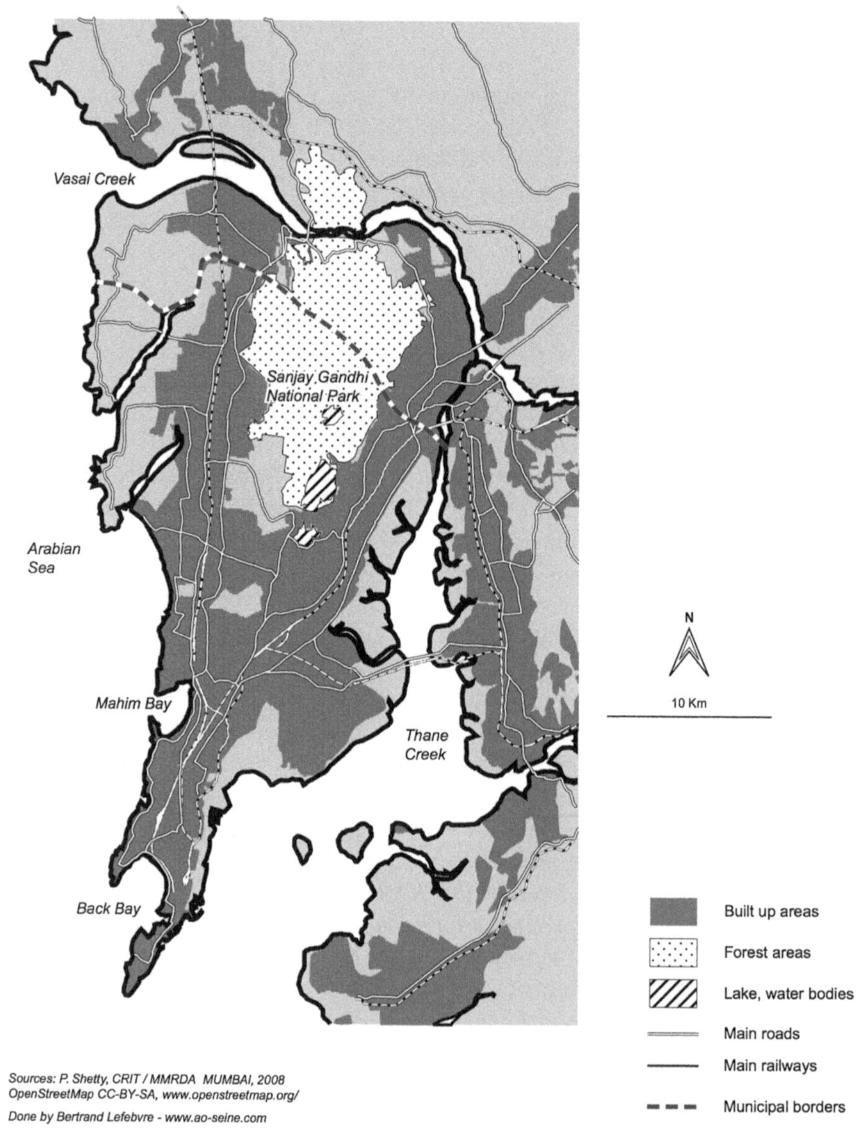

Fig. 2.1 Sanjay Gandhi National Park in the urban agglomeration of Mumbai

1950, through the Bombay National Park Act. What was then the Krishnagiri National Park covered around 20 km^2, of which 5.75 km^2 was a recreational area. In 1969, the park was extended by incorporating private, government and municipal lands. This was an important step for the Maharashtra Government, which took the opportunity to rename this 49 km^2 addition as Borivali National Park. The aim was then to protect the remaining parts of a forest landscape next to a large urban area, as well as to offer the urban dwellers a place for recreation and leisure. Land acquisition continued with the inclusion of 10 km^2 of a village in Thane district in 1974 and 15 km^2 of private forest land in 1975. In short, the park "came into existence by piecing together" patches of land (Paranjpye 1997, 8) in a process that went on for decades. The area of the park multiplied five times between 1967 and 1982, with different legal statuses over time, and a concern focused on forest conservation that lasted until the park took a more integrative approach to biodiversity conservation.

2.2.3 *Biodiversity*

With the unification and promulgation of the National and State Park Act (1971) and the Wildlife Protection Act (1972), the park acquired an additional legal basis and a higher status where, from that point on, it was determined at the national level. In a way, the decision to rename the park Sanjay Gandhi National Park in 1981, and pay homage to someone who was not local, namely the recently deceased son of Prime Minister Indira Gandhi, was a testimony to this.

Policies for the conservation of fauna followed the same path as those concerning forests. It was only really in the middle of the twentieth century that the two sectors actually joined forces. Legislation on conservation had been introduced in India under British dominion, with the main objective of regulating hunting. The premature Act for Preservation of Wild Birds of 1887 was, however, only applicable in areas directly administered by the British, and showed several shortcomings. It was replaced by the Wild Birds and Animals Protection Act of 1912. In 1953, the Bombay Wild Animals and Wild Birds Protection Act of 1951 was passed in the State of Bombay, and renewed by amendments in 1960. This law applied to the State's forest and non-forest areas.

Present-day India, unlike Kenya, for example, does not have forestry services that are distinct from wildlife protection services. The Wildlife Protection Act of 1972 changed the park's management in Bombay, with protection of the fauna becoming as important as protection of the forests. It was within this framework that the first management plan of Borivali National Park was established in 1974. This management plan was intended to be scientific, and was adapted from a report by W. Gratton, a consultant from the National Park Services of the United States. The report emphasised the preservation of ecosystems, unlike most Indian National Parks that were essentially oriented towards managing wildlife. As such, the park's main objective in the 1970s was to preserve the natural environment (and the vegetation in particular), in order to offer outdoor leisure activities to the citizens of Mumbai, and an alternative to urban routine. For this reason, the park was considered as a unique national asset that ought to be protected and maintained as a national heritage site for current and future generations (Management Plan 1974).

These new orientations had direct impacts on the management methods of the park's forest, and the Forestry Department had to change its policies accordingly.

The management of these forests, from then on undertaken by the State, was to take into consideration their ecological condition, along with forestry factors, in evaluating financial revenues. Access restrictions were imposed upon the Reserved Forest which makes up the majority of the park. In the entire national park, any exploitation was forbidden and an emphasis was placed on preserving natural landscapes and raising public awareness of natural heritage.

The first management plan, established for a ten-year period, was only renewed in 2001. The third management plan, which began in 2011, falls within the framework of the National Forest Policy of 1988 and the Reference Guide for National Parks and Sanctuaries in Maharashtra (2005). The plan defined the following objectives for the park: (i) conserving key species; (ii) protecting the catchment basins of Tulsi and Vihar lakes; (iii) conserving the mangrove ecosystems; (iv) eliminating all encroachments and consolidating the external limits of the park; (v) promoting environmental education for visitors; and (vi) managing nature tourism (Management Plan 2011).

The diversity of objectives is reflected in the multiplicity of the institutions responsible for different sections of the park. This institutional heterogeneity is also linked to the history of the park's expansion. First, of the 104 km^2 that are usually said to be the total area of SGNP, only 84% is a notified National Park. The remaining 16 km^2, mostly situated in peripheral locations, are practically managed as a national park but are not officially notified as such. Second, whereas the bulk of the park comprises "reserved forest", 20% is "unclassified forest" (mainly in the recreation zone), not to speak of a small patch of "protected forest" in the north-west (Pande 2005). Third, enclaves of revenue land exist in the east (Yeoor village) and the north (Chena village). A good part of the population growth and urbanisation is taking place there illegally, but the status of revenue land clearly prevents protection rules from being drastic, a situation that clearly helps the encroachers to claim their rights and win their case, should legal action be started against them. Finally, land belonging to the various public institutions (such as municipal lakes, Air Force base, etc.) are other factors that add to the complexity of the land tenure system of the park.

Several authorities are responsible for managing the park: in addition to the Forest Department and the Bombay Municipal Corporation, the Archaeological Survey of India is also involved in the perimeter of the Kanheri Caves, a tourist-attracting and archaeological enclave in the heart of the national park. With this fragmentation of the structures in charge, it is difficult to establish conservation initiatives and coordinate their management with organisations outside the actual park. And this only concerns the institutions responsible for the territory of the national park! The situation is even more complex when we take into account the edges of the protected area, for which no coordination has been planned. The absence of dialogue, until recently at least, with the urban or peri-urban authorities (Aarey Milk Colony), due to the "national" character of the protected space, is one of the major problems of the park—this problem is also found to a lesser degree in the three other parks under study.

However, a fledgling ecological networking policy has been initiated. The State Wildlife Advisory Board decided, in 2001, to create a new protected area to the north of the park, following the recommendations for the extension of SGNP made by the Wildlife Institute of India in 1988. For this purpose, the Tungareshwar Wildlife Sanctuary was created in 2003, and today covers 96 km^2 to the north of the urban area of Mumbai. Despite the presence of a river, a railway and a major road, it turns out that large mammals are transiting between SGNP and this sanctuary, which constitutes a crucial connection for a park which, in other respects, is very isolated, in an urban area where population densities are said to be the highest in the world.

2.3 Rio de Janeiro: From Coffee Plantations to Forest Protection Under Christ the Redeemer

Before the arrival of the Portuguese on the Brazilian coast, the land where Rio de Janeiro is today located was occupied by Tupinambás, Tamoios and Goitacás indigenous groups who generally had a very low impact on the landscape and who occupied mainly the flat coastal area (Guerra and Coelho 2012). Founded in 1565 with an initial military objective, the city of Rio de Janeiro was located between the ocean and the Bay of Guanabara, and between the Tijuca and Pedra Branca mountains (Fig. 2.2). This geographical configuration led to a long and intimate interaction with nature, resulting not only in exceptional natural beauty but also in constraints on the way the city developed.

The history of Tijuca National Park is directly linked to the growth of the city of Rio de Janeiro, and to cycles of agricultural development. From the sixteenth to the nineteenth centuries, the tropical forest was progressively replaced by urban expansion and the progression of cultivations and pastures. The exploitation and clearing of the forest began when Europeans arrived in the region and started cutting wood species selectively, particularly those intended for charcoal, building houses and ships. Some species were exploited more intensively, as was the case for the *tapinhoã* (*Mazilaurus navallium*), *peroba-rosa* (*Aspidosperma peroba*), *pau-brasil* (*Caesalpinia échinent*) and *cedro-rosa* (*Cedrela glaziovii*), these being considered almost extinct today.

It is likely that prior to 1750 the higher areas of the park remained intact, with only a few people residing there, such as lumberjacks, coalmen and fugitive slaves (Drummond 1997a, b). The first occupation register of Tijuca dates from 1569, with the establishment of sugar refineries and sugar cane plantations managed by priests from the Company of Jesus. The slopes of Rio were progressively occupied by these plantations that were spread via the *serras fluminense* and *mineiras*, and via the Paraíba Valley. As early as 1810, the development of coffee led to intense deforestation in the entire region. The current area of Tijuca National Park in particular, has been almost entirely deforested (Corrêa et al. 2001).

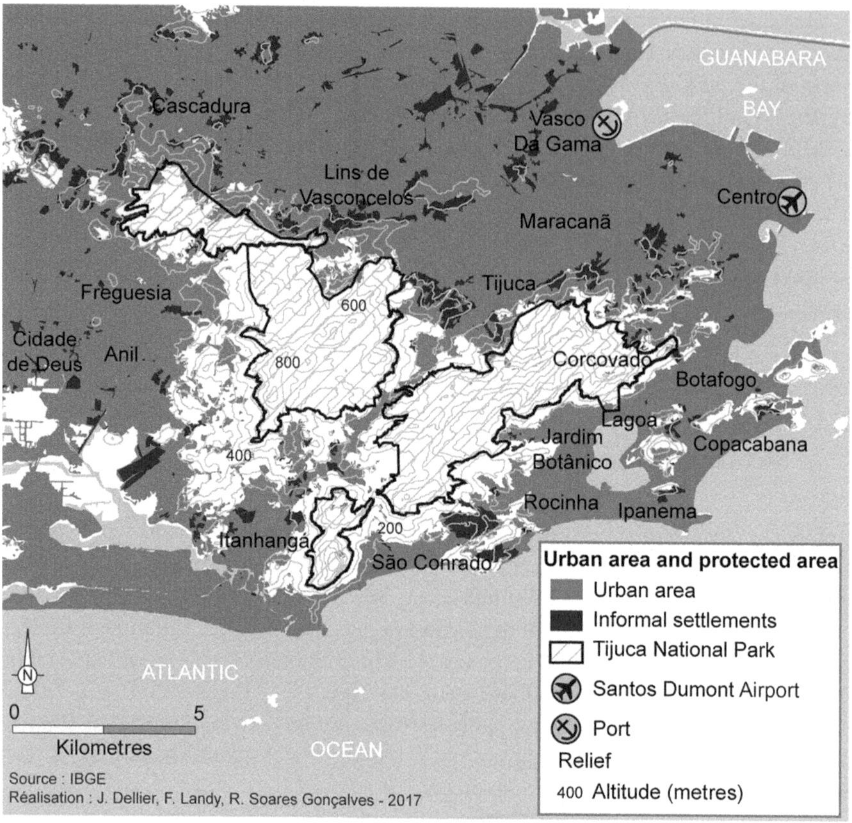

Fig. 2.2 Tijuca National Park in the agglomeration of Rio de Janeiro

2.3.1 Water

Historically, Tijuca Mountain was fundamental to the development of the city, thanks to its water which was first harnessed from the Carioca River that flows from the Corcovado forest. Later, in 1720, it was canalised towards public fountains, and in 1750 the aqueduct of Carioca was built (today known as the aqueduct of the Arcos). Other streams having their source on Tijuca Mountain, such as the Maracanã and Paineiras, were also used. Supplying crises, attributed to the slopes being occupied and to droughts, led to decrees in 1817 and 1818 suspending all woodcutting activities around sources supplying the aqueduct. Nonetheless, these preventive and punitive measures were difficult to enforce. The droughts of 1824, 1829 and 1833 provoked, in 1830 and 1834, the enactment of new laws aimed at forbidding tree cutting.

The repetition of problems linked to the supply of water was associated with the deforestation of the Tijuca forest. In this context, the only solution envisaged was to promote reforestation upstream. In 1844, the expropriation of a few properties was envisaged to ensure the conservation of the forests of Paineiras and Tijuca, situated in the upstream sections of the Carioca and Maracanã rivers. In 1850, the government promulgated a law on land expropriation in water source areas, particularly those in Serra de Tijuca. Again in 1854 and 1856, the State proceeded with the acquisition of a few properties on the River Maracanã (which at the time was being canalised), and organised trees to be planted with a view to rehabilitating degraded lands.

Just as in Mumbai, the water supply crisis in the capital city, caused by urban growth and the expansion of coffee cultivation, was the main factor that initiated the reforestation process of the area.

2.3.2 Forests

However, these reforestation policies were also motivated by other factors, such as the beginning of forestry as a science, the arrival of eucalyptus and the recreational value given to forests.

In 1808, the arrival of the court of Portugal and the establishment of the king around town (*Quinta de Boa Vista*) provoked important changes in the way the colonial city and society functioned. The newly arrived aristocracy brought a new form of living close to nature, of romantic inspiration, and developed coffee cultivation on the slopes of the mountain. A group of French aristocrats, exiles or deserters from the court of Napoleon, settled in the forest of Tijuca, while English nobles preferred the Corcovado area. However, the mountains were also the refuge of many fugitive slaves who lived there in villages (*quilombos*). Their presence represented a threat to landowners on the mountain and to the urban populations of the adjoining valleys, where they made regular incursions. Two successive epidemics of yellow fever and cholera (1850–1853) provoked the migration of the well-off classes from the infested urban plains to the healthy heights of the forest.

As such, the policies for the protection of water resources and the preservation of the forest benefitted mainly the aristocracy that owned the land on the mountain and which was close to the emperor. The list of families expropriated in 1856, contained only Luso-Brazilian names, while the names of aristocrats of French or English origin were absent from it altogether. In parallel, the cultivation of coffee in Rio was on the decline from the 1860s as the plantations moved inland. The large properties that became less profitable were subsequently divided into *chácaras* of around 10,000 m^2, and sold off to the capital's rising bourgeoisie.

In 1861, the idea of establishing a forestry service materialised, as in other European colonies, and in the British Empire in particular. In this context, an order created the National Tijuca Forest and the National Paineiras Forest. The order defined reforestation and conservation rules, and established the regular plantation of trees and prohibition of woodcutting. Following appointments by the Emperor Dom Pedro II, the Paineiras sector was placed under the administration of Tomás

Nogueira da Gama, and the Tijuca sector under that of Manuel Gomes Archer. Today these two sectors constitute the basis of the current Tijuca Forest National Park (TFNP).

Among these new regulations, the measure with the greatest impact was the reforestation of the slopes of the Tijuca sector, under the administration of Manuel Archer. While Frederick Olmsted initiated urban landscaping with the creation of Central Park in New York, and while Napoleon III developed the Bois de Boulogne in Paris, Archer began his work in 1862. With no prior specific training, and supported by six slaves, he went on to deploy sylviculture in the country. His reforestation work was characterised by the diversity of species, the majority of which came from the local Atlantic forest, accompanied by a few exotic species. The latter gained in importance in the forest under the influence of the botanical garden of Rio. Ornamental species, such as *Dracaena sp.*, species with a high forestry value, such as the eucalyptus, and fruit trees, such as the jackfruit tree from India were, for example, introduced in reforestation areas. They reflect the amazing history of plant transfers and acclimatization during the colonial era.

At the end of his undertaking, Archer demanded that adequate forestry legislation be enforced and that a forestry education institute be established. In doing this, he was the first Brazilian to develop this field of applied sciences which we would now call ecological rehabilitation. Archer also highlighted the importance of the joint promotion of reforestation and natural regeneration, which is conceivable when preservation follows reforestation. He left his post in 1874. TFNP as we know it today is for the most part the legacy of Archer and his slaves, which also testifies to the fact that ecological rehabilitation is labour intensive.

The Baron d'Escragnolle took on the administration of the forest after Archer, until 1888. He took care in embellishing the forest and transforming it into a pleasant and welcoming park. In parallel to pursuing reforestation, under d'Escragnolle's management, the rehabilitated domain was equipped with bridges, belvederes and artificial lakes. Between 1861 and 1888, around 90,000 trees were planted in Tijuca Forest. The end of that period was characterised by the abolition of slavery in Brazil (1888) and, one year later, by the transition from a monarchy to a republic. At a more sectorial level, an 1874 report drawn up by Paula Freitas and Buarque de Macedo, advocated the construction of new water reservoirs in other sectors beyond Tijuca Mountain. The urban developments of the *Alto da Boa Vista* sector had begun at the same time as the first expropriations, in 1856, with the construction of a road suitable for vehicles, linking this sector to the botanical garden. A transport concession was attributed as early as 1859 and, in 1897, the sector was served by a tramway service. All these conditions combined to make the mountains a privileged sector of the city.

However, Tijuca Forest was subsequently abandoned for almost half a century under the impact of the country's political transition and the new ideology that tended to break away from any relics of the monarchy. Still, in 1931, the statue of Christ the Redeemer was completed, an enclave belonging to the Church in the middle of the park. On the other hand, at the beginning of the twentieth century, the hygienist urban reform orchestrated by Mayor Pereira Passos, led to the destruction of the slums and other popular housing in the city centre. The evicted population increased the size of the first *favelas* taking shape on the hilltops. For this population excluded from the city, the slopes of the forest-covered mountains represented an empty space, outside the real estate market. The unoccupied public space was, during the twentieth century, occupied? by the most disadvantaged populations.

Only in 1943 did Tijuca Forest become the centre of attention again. It benefitted from new rehabilitation actions until 1945, under the supervision of Raymundo Ottoni de Maya Castro, who was in charge of regenerating the forest. The forest borders were redefined and with the construction of new access roads, restaurants, picnic areas, children's corners and sites of touristic interest, the forest was transformed into a park frequently visited by the Carioca population and tourists alike.

In 1961, the forests of Union, Tijuca, Paineiras, Corcovado, Gávea Pequena, Trapicheiro, Andarai, Três Rios and Covanca were grouped together to form the National Park of Rio de Janeiro, not only for their natural beauty, but also for the protection of water sources. In 1967, the park was widened (to 3300 ha) and received its current denomination. While some areas were considered beyond rehabilitation and were excluded as a result (e.g., the forest of Covanca and part of Andaraí), other, new areas, were added (e.g., Pedra Bonita, Pedra da Gávea and parts of the Morro de Dona Marta and the Corcovado).

2.3.3 Biodiversity

The 1970s saw the development of a fauna restoration policy. The project consisted in reinforcing reptiles, birds and mammals, as well as the vegetal species found in the diet of these animals, by reintroducing species that had disappeared and in controlling predator populations as well as hunters. Although these measures led to imbalances in the populations of certain species, these two restoration projects were the first of their kind in this park, notably because they involved a large urban park not based on the preservation of the indigenous forest of this biome (Mata Atlântica forest) as found in the hinterland, but on degraded secondary tropical forest with a high frequency of exotic species (Drummond 1997a, b).

Up until 2000, the management of protected areas in Brazil was governed by a decree regulating Brazilian national parks. Following the management methods proposed by the International Union for the Conservation of Nature (IUCN), the Directorate for Natural Resources of the Brazilian Institute for Forestry Development (IBDF) and the Brazilian Foundation for the Conservation of Nature (FBCN) elaborated, in 1981, the first management plan of Tijuca Park. With the national system of conservation units established in 2000, the requirements for the elaboration of a management plan for each conservation unit came into effect at the federal level. By law, the management plan is a technical document, through which zoning and the standards that must govern the use of space and the management of natural resources are established. This includes implementing the required infrastructures. The first plan for Tijuca was reviewed between 2004 (with, in particular, the integration of Lage Park and the Covanca/Pretos Forros sector, namely 3973 ha) and 2008. The guidelines of the current plan are governed by specific objectives, obtained from an environmental diagnostic analysing the park and its region at the same time, or in other words, the area surrounding it, which can have an external influence on the milieu. They also take into account information on special projects for the region, such as the 2016 Olympic Games.

Managing TFNP was, between 1999 and 2005, carried out by a partnership with the municipality of Rio, in an innovative procedure of collaboration between State services and local authorities. This collaboration was temporarily interrupted for political reasons, and finally resumed in May 2009, under the responsibility of the Chico Mendes Institute for the Conservation of Biodiversity (created in 2007), which was from then on responsible for managing all the federal units of conservation. In 2014, a federal decree led the park to expand its borders to what they are today.

2.3.4 Networking

Finally, in 1993, the park was classified as a Biosphere Reserve by UNESCO, as part of the vast Mata Atlântica reserve, which covers around 300,000 km^2 spread over 15 Brazilian States. In 2012, the carioca landscape, of which TFNP represents a typical aspect, was classified as a World Heritage site by UNESCO, thereby defining an important ecological network on the scale of Rio's urban area.

2.4 Nairobi: Nationalisation from the Start

Nairobi National Park (NNP) has a markedly different history to that of the other parks presented here: the progressive attention shift from water and forest towards biodiversity did not take place. In Kenya, the focus was on the large fauna, as a symbolic element of British dominion and then as a national economic issue, in a context marked by high segregation between communities and which involved protected areas as buffer zones.

As early as 1900, the British colonial power wanted to create a Southern Game Reserve, south of Nairobi, up to the Tanganykan border, so as to forbid hunting in the area. The northern tip of this area, corresponding to the current national park and called Nairobi Commonage, was supposedly a part of it (Boy 2012) but nothing came of it in the end. The Commonage began to house Somali ex-servicemen, who were settled there for retirement. During WWI, the Commonage became a base camp and training ground, where wildlife was plundered to feed the troops. After the war, cohabitation between herds which had recovered from epizootic diseases, their Maasai owners, the new settlers and wildlife became increasingly difficult. The supporters of the creation of a national park were heard, and a Game Policy

Committee was convened to try to control the area more efficiently. However, WWII broke out, putting a stop to the project, and the Nairobi Commonage became a military base once more, suffering from the creation of barracks, ammunition stores, shooting ranges and quarries. It was only from December 1946 onwards that, after the costly construction of roads, bridges, dams and salt licks, the Nairobi Royal National Park was officially created (Boy 2012). At the time, this was the first ever national park in Eastern Africa (Fig. 2.3).

This was an eminently colonial creation, resulting from the diffusion of the national park model, as a world natural heritage conservation system. Two ideas directed this creation: first of all, that the best wardens of African nature were the heirs of the great hunters of the first colonisation; and second, that the best spatial system was the national park, exclusive and with prohibited access, organised around game viewing tourism rather than hunting tourism, and different from the hunting reserves that had spread in British Africa during previous decades. Because these policies were supported by settlers, they excluded de facto Africans from nature conservation. The park of Nairobi is above all colonial and national. It has taken on a geo-iconic value, reflecting the will of the whites to preserve an African Eden, even before the creation of major parks such as that of Tsavo, which was able to accommodate greater numbers of large fauna. For this reason, NNP is a symbol, and the fact that the ceremony of Independence took place specifically in Uhuru Garden, facing the savannah of the park, is not insignificant. The objectives and management methods of the park reflect this genesis, particularly in the very exclusive and orthodox way in which the Kenya Wildlife Service (KWS)—the heir to the colonial conservation services, very close in this respect to the Forest Department which is managing the park of Mumbai—is envisaging its territory and mission.

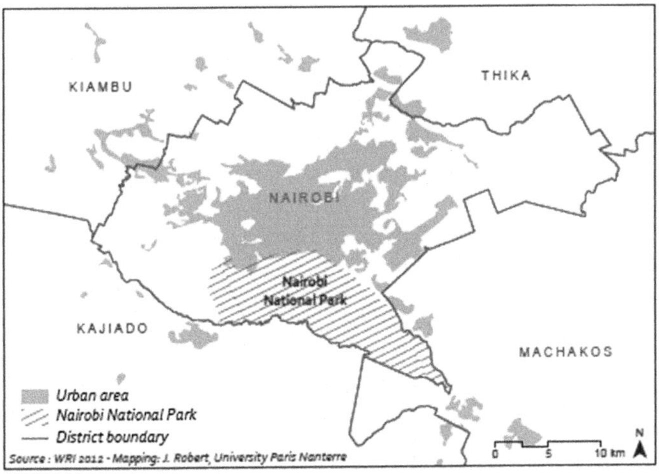

Fig. 2.3 Nairobi National Park in the agglomeration of Nairobi

2.4.1 Water as a Factor of Location

The name Nairobi comes from a Maasai expression which means "cool water". While the town was indeed created in 1899 by the British as a major railway stop for the construction of Uganda Railways, almost halfway between Kampala and Mombasa, the choice of this settlement also corresponded to an environment that was relatively appropriate, and where water had its importance, as was the case for the other parks under study. The town became the capital of the British East Africa Protectorate in 1907. While it was growing economically and administratively, it also attracted many European and American tourists, as a result of having one of the highest concentrations of large game in Africa. Dams created in the park reinforced its hydrological resources, and the Mbagathi River, which makes up the south border of the protected area, was regularly crossed by fauna that came to take advantage of the water resources in the park, especially during the dry season. The park constitutes the northern end of a vast area of annual migrations of herbivores: the Kapiti Plains which stretch to the south up to the outskirts of the Chyulu ranges and Amboseli, and even to the Serengeti in Tanzania (Gichohi 2000), a vast savannah land with abundant species of wild herbivores which attract large carnivores, especially lions. The corridor of Kitengela-Kaputei links the park to these plains. South of the park, the absence of fencing contrasts with the fearsome electric fences of its other borders: this southern opening makes it possible to maintain the migrations of herbivores and the integrity of the park's ecosystem in relation to the Kapiti Plains, a link which is threatened today, as we will see further on.

2.4.2 Before Biodiversity: Wildlife and Its Wardens

As such, the history of the park of Nairobi was organised partly around the presence of water, but it was the specific focus on the large fauna that was going to justify its existence and organise the different policies established prior to today. While the park is currently separated from the urban area by a mere boulevard, when it was created the park was located around seven kilometres south of the city, at the southern tip of the white area: it was, in a way, used as a "wild" buffer zone between the city and the area left to Maasai pastoralists by the British. During colonisation, the Maasai Extra-Provincial District (which covered the current districts of Narok and Kajiado or thereabouts) had a separate status: it corresponded approximately to the Southern Reserve, where the Maasai had been confined during the treaty of 1906 (Fig. 2.4). A special permit was needed to enter the Reserve, which was to protect the Maasai from the rival Kikuyu ethnic group: between 1906 and the end of the 1960s, the Maasai had exclusive rights to occupy the present Kajiado and Narok districts (Rutten 1995). Indeed the Maasai occupied a special position in British ideology. Unlike the Kikuyu farmers, they were considered as being more coarse and primitive, and their diet consisting mainly of meat reinforced

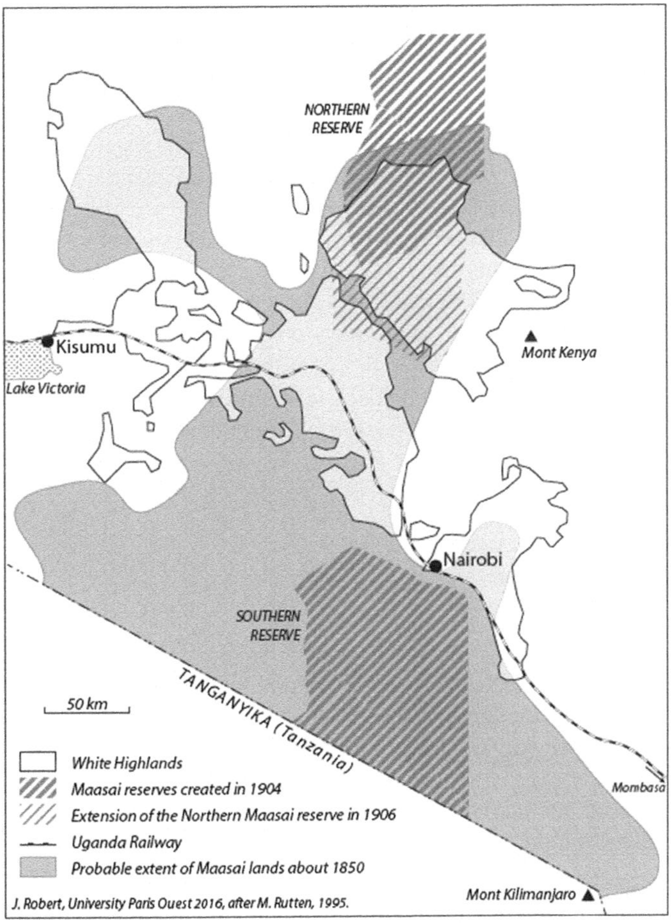

Fig. 2.4 The Maasai reserves

this representation. This might have played in their favour subsequently, to create an environmental image, an "eco-ethnicity" (see Chap. 10) that could be developed to seek the favour of ecological organisations. As early as the colonial era, this was also relatively useful. Indeed, creating a Maasai Reserve was not only the mark of indirect rule that was typical of British colonisation, but was also a means of protecting the fauna by proxy, by allowing the tribes that were closest to nature manage nature—it being understood that the more "civilised" groups were less qualified to that end: "Natural man was bound together with wildlife in nature" (Knowles and Collett 1989, 472). However, at this stage, colonial policy being ambiguous, we need to qualify the situation further. The primitive state of the Maasai could also be considered as potentially destructive as far as nature was concerned, and as going against the protection of wildlife as game for white hunters. Therefore the objective was to try to transform these pastoralists into ranchers, before and after colonisation; it was necessary in any case to create national parks, since modern ranchers did not get on well with the wildlife.

On independence in 1963, conservation policies took a turn in Kenya as well as in the rest of Eastern Africa (Matheka 2005). On the one hand, the elites and the colonial authorities apprehended the consequences of a change of power on conservation policies and envisaged the need to involve local elites more in these policies, so as to ensure that Independence did not mean the end of protected areas (Huxley 1961). On the other hand, the local populations, and the Maasai in Kenya in particular, feared that Independence meant that their communal lands would no longer be secure. As a result, they often displayed increasingly conservationist discourses and practices, so as to legitimise further their rights to the land, thanks to the support of international ecological organisations (Matheka 2005). However, these dynamics, for taking into account local populations, did not directly affect the park of Nairobi.

2.4.3 Disconnection

Unlike the other parks under study, the recent dynamics around the park of Nairobi do not favour its connections with the natural environments around it. On the contrary, the park is increasingly being subjected to pressures, urban pressure in particular, which tend to isolate it ecologically despite the measures taken. The urbanisation frontier (see Chap. 8) is more or less channelled on the northern limits of the park, thanks to a wall doubled with an electric fence, but even so, projects have encroached on the park (e.g., a ring road, high-voltage lines, etc.). As for the southern border, which has remained open for animal migration, different closing processes are under way. The transformations of the lifestyle of the Maasai (Homewood et al. 2009), the division and privatisation of plots, and the development of cultivated fields, houses or quarries, are increasingly reducing the migrations of large herbivores (Ogutu et al. 2011), which has led conservationists to adopt the leitmotiv “NNP, a park under siege”. One of the working scenarios of the KWS consists in imagining NNP as totally enclosed, with annual herbivore migrations having stopped, so that a totally anthropised renewal of the genetic heritage takes place by translocating animals bought from the private reserves of the Laikipia or the Rift Valley.

This threat to the natural functioning of the park is emphasised by the fact that its southern border also matches the municipal and provincial borders. Furthermore, the migration area of the herbivores, on which the normal functioning of the park depends, is under the administrative and political responsibility of different authorities, with often diverging views on nature conservation. While the ecological system as a whole functions in a coherent manner, the area covered by it is fractured into different administrative territories. This disjunction between ecological and administrative functioning is a source of contradiction and even conflicts. The national park is cut off administratively from its ecological extension in the south, and is also cut off from the urban district of Nairobi in the north, because of its "national" status and imposing material border. Moreover, the relatively strict municipal bylaws on construction, which are enforced in the territory of Nairobi City County, do not apply outside. As such, candidates for property ownership little prepared to observe these fairly strict town planning standards, are flocking there to take advantage of the more relaxed building requirements. The resident sectors of the municipal perimeter, at the end of the roads of Ngong and Athi River which go around the park, barely missing it in the north-west and the south-east, are good examples of rapid urbanization. The villages of Ongata Rongai, Athi River and Kitengela have become dormitory towns around Nairobi, and their constant expansion is eating away at the migration corridors of herbivores. Finally, the integrity and functioning of the park might be fatally damaged by the construction of the Mombassa–Nairobi Standard Gauge Railway, which is crossing right through it (see Chap. 12).

2.5 Cape Town: From White Mountaineering to National Reconstruction

From what we know, in the pre-colonial period of the areas surrounding the Cape Peninsula (Fig. 2.5), the indigenous pastoralists utilised the environment for extensive seasonal grazing routes and for harvesting food from the ocean. These livelihoods were seriously disrupted with the settlement of Europeans (Beinart 1989). With the arrival of the first colonialists, the place that would become Cape Town was perceived as a "wilderness", this word conferring different qualities at the time. For these early settlers, the African environment lay beyond the borders of civility and was feared (Coetzee 2007). Cape Town in the seventeenth century was limited to a vegetable garden surrounded by an administrative centre, a few estates and hamlets. Creating a garden within this feared wilderness was about domesticating the environment. Yet the planting of pines, firs and oak trees was also about compensating for the large-scale removal of local indigenous trees. The establishment of the Dutch East India Company (VOC) was built on the supply of wood harvested by slaves and, by 1772, it was reported that there was virtually no wood left in the vicinity of the settlement (Hey 1978).

As the boundaries of the settlement expanded, it came increasingly into conflict with local human and nonhuman livelihood patterns. The threat posed by predators and the clashes with the "marauding Hottentots" who defended their grazing land and animals, resulted in what was probably one of the first borders in the Cape. Van Riebeeck's "circumference of the Cape" was a sturdy wild almond hedge built around the settlement. It could translate into a social border that allowed settlers to imagine that it was possible to define anything outside the border as heathen and wild. Outside the hedge, there was a general disinterest in exploring the Cape

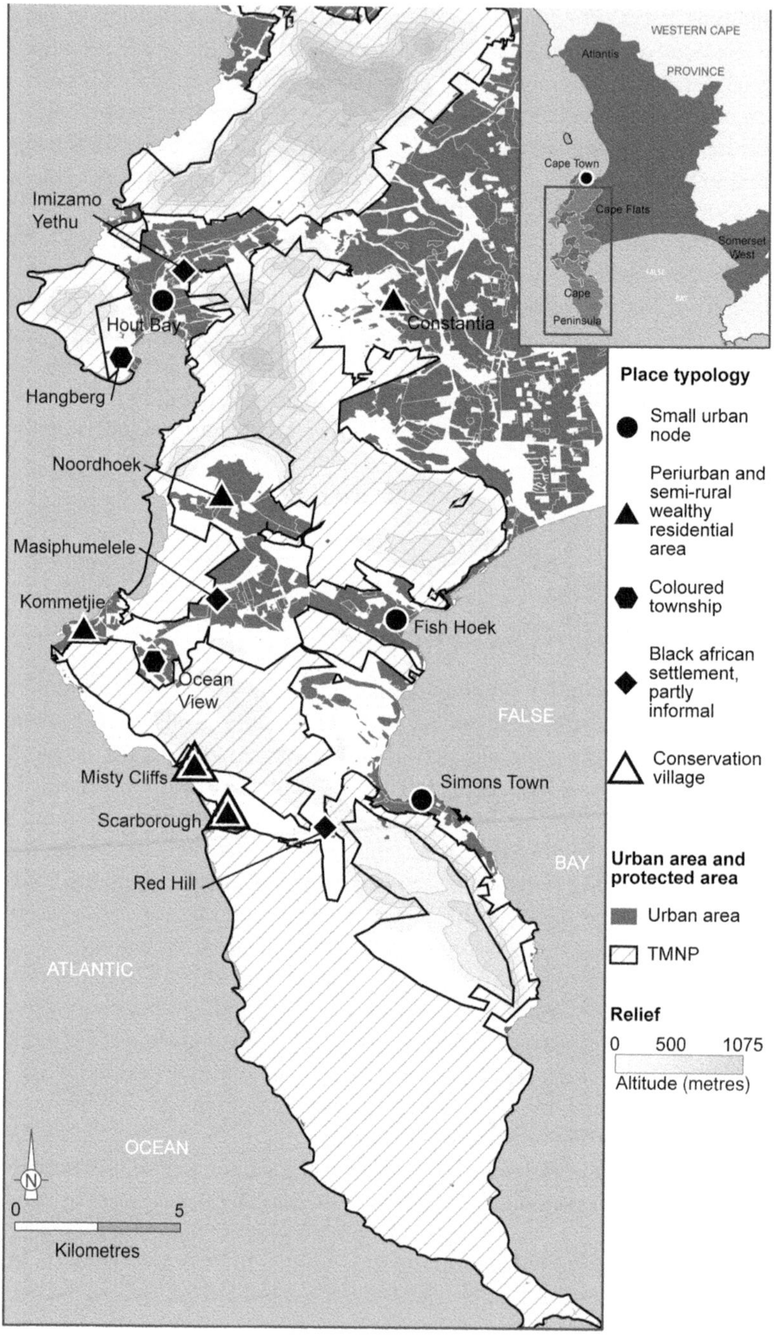

Fig. 2.5 Table Mountain National Park in the Cape Town agglomeration

Peninsula during the VOC period, save for harvesting wood (Lückhoff 1951). The growth of the settlement was also relatively slow in comparison to the period of the British colony. Urban conservation was limited to minimising water pollution and early laws pertaining to the forests, such as the starting of fires (which were largely unenforced). A sense of environmental concern or protection was limited to a few individuals (Grove 1989).

By the late nineteenth century, in contrast, the modernisation process had largely tamed the Cape Peninsula wilderness. These processes gained ground during the mineral revolution of the 1880s. The revolution energised Cape Town's emerging capitalist economy, extending and expanding its dependence on the resources of the Peninsula. The introduction of alien trees and the destruction of indigenous trees were but one of the many processes involved in the urbanisation of nature within and around Cape Town. The city's demand for water was entirely satisfied from the streams running from the Cape Peninsula, and protecting this valuable resource was a high priority. In 1840, the city's colonial administration was made responsible for providing water to the city. In order to secure this provision, the government built four additional reservoirs on the back of Table Mountain, as well as a tunnel for funnelling the water to the city (Juuti et al. 2007). Water provisioning was one of the city's earliest dependencies on the Peninsula and was strictly controlled by the government because of its value.

Yet it was the commencement of modern forestry that really altered the face of the Cape Peninsula. It was a development in which timber production shifted from being a relatively small-scale activity to become a large-scale process, intimately intertwined with the exhilaration of global capitalism (Anderson and O'Farrel 2012). During the first years of European occupation, most of the indigenous trees were exhausted. Later, the continuous demand for wood led to the introduction of a tremendous amount of alien trees and, subsequently, to the establishment of modern forestry plantations on the lower slopes of the Cape Peninsula (Lückhoff 1951). The Department of Forestry was established in 1880 and was the first of its kind in the country (Brown 2003). Stone pines were especially suited to the Cape climate and remain a familiar (and for some a sentimental) feature of the Peninsula landscape. The introduction of alien trees also coincided with early urban conservation practices, with the planting of alien trees on eroded areas through the establishment of a nursery close to the area known today as Kirstenbosch (Moll 1989). This task of reforestation was influenced by soil conservationists who believed that the colony's deforestation resulted in severe droughts (Grove 1989).

2.5.1 Mountaineering in the Colonial City

In his reading of the earliest conservationist debates in the Cape Colony, Grove (1989) illustrates how the first conservation ideology was developed into policy because it suited the government of the time, and was granted enough support from the white settler community. These conservation policies were concerned with land, forest and water degradation and were, at times, in conflict with settler enterprises in agriculture that were resource intensive. The Forest and Herbage Preservation Act was promulgated in the mid-nineteenth century. While this act was considered to be the "most advanced of its time" because of its holistic nature, its effectiveness was questionable (Beinart and Coates 1995).

Thus, the Mountain Club of South Africa (MCSA) was established in 1891 in a context where the British government had a history of amenability to conservation intervention, and where the settler community had developed an affinity for the colonial environment. What was unique about this conservationist movement and which made it particularly urban, was that its mountaineering foundation advocated for human access to nature. The prime concern of the MCSA was mountaineering. However, their knowledge and observations of the ecological conditions on the mountains motivated the robust conservationist and preservationist discourse it presented throughout its existence (Goetze 2002). Mountaineering made the question of physical borders for a reserve inconceivable. Besides, at the time there were very few large animals left on the Cape Peninsula (these were not perceived in any case to be worthy of conservation). Mountaineering on the Cape Peninsula was no longer about the *sight* of wilderness but was "a *site* of elite learning about the natural world, but more importantly about the inner self" (Van Sittert 2002, 164, author's emphasis). Mountains, like the frontier, were repositories for sublime and romantic experiences (Voss 2003). The MCSA, like other cultural institutions in the city, was also a vehicle for mobilising a common "Englishness" that created class and ethnic boundaries in the city (Bickford-Smith 1999). It was especially their shared interest in the Cape Peninsula which led to the development of urban conservation. As described above, the government's main objective with respect to the Cape Peninsula, was to protect the reservoirs and streams from pollution. In recognising this objective, members of MCSA responded by suggesting that their presence was beneficial since they would be able to support the policing of polluters of the water. They also forged the notion of custodianship of the Cape Peninsula.

As the popularity of indigenous flowers in the city matured, so did the number of flower sellers in the city, a practice that had since the 1880s grown into a rich cultural heritage of Cape Town. The MCSA soon took notice of the dwindling effect these activities had on the numbers of the Cape flora. Thought to be the main source of flower pickers, the city's flower sellers were mostly coloured women and children and occupied the lower social ranks of the city (Rabe 2010)[1]. The second conservationist concern that occupied members of the club (and continues to do so) involved the eradication of alien trees. In order to prevent further erosion, soil conservationists had endorsed the planting of alien trees, and members of the club had taken it upon themselves to distribute pine seeds on the landscape during walks (Moll 1989). However, the seeds of conservation in the pockets of their hiking

[1]The term "coloured" refers to people with a mixed racial inheritance. The use of racial categories in this book is understood as categories inherited and produced by colonial systems. Their use underscores not their objective significance but the resilience of these social classifications in shaping the history of national parks.

trousers soon filled them with concern as they noticed the rapid colonisation of indigenous vegetation by alien trees (Van Sittert 2002).

By the time the MCSA was established the city's industrial and capitalist flows had grown to a considerable size. An urban geographical reordering took place in which the upper-class residents (of which Rhodes was one) came to prefer the pastoral eastern side of the Cape Peninsula where the rusty Dutch farms recalled the imperial garden idyll and where "neither [labour] nor violence intrudes" (Nixon 2011, 245). This was done in relation to the different "faces" of Table Mountain in which the less popularised and grandiose side of the mountain became a repository for situating English identity in the city. The nation, as it was presaged in the apartheid era, was divided along English/Imperial and Afrikaner lines, and nature provided a platform to build a common "white" nation. Therefore, eradicating pine trees and defending indigenous flora on the Cape Peninsula was a process by which the settler society was coding their white identity on the African environment. This analysis finds resonance in the contemporary period where the state's intention to eradicate aliens is a strategic priority to build a common nation (Comaroff and Comaroff 2001).

2.5.2 *Table Mountain: The Emblem for White National Identity*

The first nature reserve in Cape Town was founded in 1822 by Lord Somerset, and public game reserves were established in the province in 1903 and 1908 (Carruthers 2009). Prior to the 1920s, game reserves were authorised by those who had the right to proclaim crown-owned properties as game reserves for hunting practices, or to facilitate the wild animal trade market. It was only after this period that urban citizens started mobilising for reserves and national parks that were not intended for elite hunter-sportsmen but for the protection of wild flowers (Van Sittert 2005).

However, science on its own was not initially perceived to be enough to convince the city council to establish nature reserves, and tourism was highlighted instead (Lückhoff 1951). While large game sold the image of the country, during the 1920s and 1930s Cape flora and Table Mountain promoted Cape Town. Aided by the Cape Peninsula Publicity Association, Cape Town's natural features (especially the beaches) became the trademark of the city. Access to the praised natural sites of the city was radically improved by the construction of a cable car on Table Mountain in 1929, and the road to Cape Point. (In Rio, the funicular railway to Corcovado had been inaugurated as early as 1884). Cape Town actively promoted itself to tourists by endorsing nature as the heritage of the city. However, the influx of tourists to the Cape Peninsula soon caused considerable concern for the conservationists of the city. This led the Botanical Society and the MCSA to intervene and successfully pressure the council into establishing the reserves, albeit to sustain the tourist industry. Thus, between 1922 and 1939 the Cape Peninsula saw the formation of three floral reserves on its landscapes (Van Sittert 2002).

The unification of South Africa in 1910 facilitated the cross-pollination of conservationist ideas, and the notion of establishing state-proclaimed reserves was strongly influenced by the national park debates that occurred in the northern parts of South Africa. In the 1920s, the importance of aligning the imperial loyalties with Afrikaner nationalism became paramount, and creating a national park was about creating a white national identity and culture (Carruthers 2009).

Once Kruger National Park proved to be commercially successful in the 1930s, it changed the perception that game reserves were the exclusive domain of hunter-sportsmen. Representations and commodities of wild animals infiltrated the nation's visual simulacrum, and the national park experience embraced the modern motorcar, camping tent and camera made possible by the white middle-class salary. This development in tourism also contributed to the commercialisation of South African conservation. Parks became an integral part of South Africa's mainstream conservation movement, at least until the 1990s (Carruthers 2006).

For most of its existence, Cape Town has been the seat of British imperialism and initially the establishment of national parks was primarily about advancing Afrikaner nationalism. As a result, establishing a national park in the former British Colony's headquarters was deemed unwise. In sum, the likelihood of a national park on the Cape Peninsula was discouraged by a number of locally situated tensions rather than global ones. The issues around access, the lack of charismatic game to attract the attention of the National Parks Board and the political unsuitability of Cape Town for promoting Afrikaner nationalism, prevented the Cape Peninsula becoming a national park in the 1920s. The topography of the mountains and indigenous flora were brands and symbols of the city, rather than the country. The question of nationalising nature in Cape Town resurfaced in 1948, when a group of civil societies appealed to the central government to develop a management plan for Table Mountain, and declare the mountain a national asset under statutory legislation (Lückhoff 1951). This was also the year the National Party won the election, and the construction of an Afrikaner nation was being realised on symbolic, economic and social levels.

The announcement of Table Mountain as a National Monument in 1958 came therefore as no surprise, as was the demarcation of it as a White Group Area in 1957. In addition, the shaky legislation of the provincial Table Mountain Flora Reserve, from 1931, was replaced with national legislation in the 1960s. This introduced the mountain to a modern scientific conservation ideology, such as the controlling of veld fires and eradication of alien vegetation (Hey 1978). Moreover, from the 1950s to the 1980s, the image of Table Mountain was reproduced through a plethora of media that became material and symbolic references for white South Africa. Postcards, Christmas cards and newspaper images popularised the image of Table Mountain. Proclaiming the mountain as a national monument was thus also part of the process of creating a white national identity that symbolised a supposedly common white colonial heritage (Goetze 2002).

However, white custodianship of the Cape Peninsula was a contested process. Hostilities between the English and Afrikaner camps prevailed, and handing over ownership to the national authorities implied losing their regionally unique custodianship in the face of Afrikaner nationalism. Furthermore, while the white ideological ownership of Table Mountain officially denied non-whites a share in the mountain's symbolic and ideological ownership, it did not deter coloureds and blacks from visiting the mountains (nor could such ideological territorial pronunciation remain isolated from contradiction). Before the forced removals of the 1950s, many coloured families lived in the shadow of Table Mountain and fostered intimate relations with the urban mountains.

2.5.3 Nature Goes Global

Sixty-six years after their first attempt at establishing a national park on the Cape Peninsula, the Wildlife and Environment Society of South Africa (WESSA) established the Peninsula Mountain Forum that aspired to mediate between the state and civil society on the issue of nationalisation. The Cape Peninsula Mountain Chain was collectively subject to a legal border under the Environmental Conservation Act of 1982. This included the "entire Cape Peninsula mountain range above the 152 m contour", roughly 300 km^2 in size, and comprised of "land under the control of 14 different public authorities at national, provincial and local government level, as well as 174 private landowners" (Daitz and Myrdal 2009, 326). Given this multiplicity in ownership, a single authority (while recommended), proved unattainable and a Management Advisory Committee was implemented. In 1987, the responsibility of managing reserves again devolved to the provinces, which led to the formation of the Cape Peninsula Protected Natural Environment (CPPNE) in 1989. In general, the CCPNE's boundaries were "interpretable" and the cooperation between the different authorities was weak, which made conservation strategies ineffective (Daitz and Myrdal 2009).

The Cape Peninsula National Park was proclaimed in 1998, for the purpose of protecting the rare endemic *fynbos* (shrub) vegetation in particular. The property is included as part of the UNESCO Cape Floral Region World Heritage Site. The African National Congress (ANC) had embraced the proclamation of a national park in Cape Town, effectively endorsing a colonial inheritance in a period of political transition, when the model of parks could have been subject to radical revision (Ndebele 2007). Perhaps the potential for establishing a national park in Cape Town provided for the ANC a power threshold in the city that was politically held by the National Party, and later on by the Democratic Party.

In addition, the park was part of the establishment of a new South African identity, and doing so implied shaping a usable past that fitted into this discourse, even if it remained fraught with contradictions and unwanted legacies. The colonial history associated with the mountain mentioned earlier from the 1950s, was substituted with a pre-colonial heritage, in which the mountain became Hoerikwaggo:

the Khoi word for Table Mountain meaning “mountain in the sea”. Along similar lines, in 2004, the name was changed from Cape Peninsula National Park to Table Mountain National Park (TMNP). The former name was a legacy of the white custodianship of the Cape Peninsula, while the latter seemed to want to symbolise the new nation and its inclusion in global networks, as illustrated in the inscription of the Cape Floral Region as a World Heritage Site.

Secondly, despite Table Mountain’s capacity to capture (once again) the spirit of the nation, the government did not prioritise conservation in the transition process, and emphasised commercialisation in national parks. Realising this, SANParks (South African National Parks, formed in 1926) approached the Global Environment Facility (GEF) for financial support that allocated 56.7 million rands for the nationalisation of the Cape Peninsula (Daitz and Myrdal 2009). In 2004, the “Cape Floral Region”, a biodiversity hotspot that includes TMNP, was placed on the list of World Heritage Site. In sum, a number of circumstances and actors contributed to the nationalisation of the Cape Peninsula in which global—rather than local—factors seemed to carry most of the weight. In some way, the vitality embedded in protecting the Cape Peninsula transcended local custodianship and became encapsulated in a global environmental discourse.

2.6 Conclusion

Four parks, four urban environments, four nations and four specific cases: three of these parks were created to ensure, above all, the water needs of the city. Nairobi is not one of them. The protection of the flora is fundamental in Rio and Cape Town, but that of the fauna is just as fundamental in Nairobi and Mumbai. The nationalisation of protected areas took place rapidly in Nairobi, while in Mumbai and Cape Town protected areas went through more or less lengthy phases of local administration. Racial issues resulting from colonisation have made their mark in Nairobi and Cape Town, much less so in Mumbai and Rio de Janeiro. Fifty-two years separate the official creation of the national park of Nairobi (1946) from that of Cape Town (1998).

The remainder of this book intends to go further, by going beyond monographs so as to bring to light similarities and differences and reveal, beyond individualities, *types* of configuration and city-park interrelations. We will begin with an analysis in terms of social classes (cultural classes, political classes): Is environmentalism underlain by social determining factors, and if so, under which forms? To what degree do well-off urban populations truly care about the national park? Do poor city dwellers truly represent a threat to the national park?

References

Anderson, P., & O'Farrel, P. (2012). An ecological view of the history of the city of Cape Town. *Ecology & Society 17*(3) [online]. http://dx.doi.org/10.5751/ES-04970-170328.

Beinart, W. (1989). Part one: Conservation ideologies in Africa. In D. Anderson & R. Grove (Eds.), *Conservation in Africa: People, policies and practices* (pp. 15–20). Cambridge: Cambridge University Press.

Beinart, W., & Coates, P. (1995). *Environment and history: The taming of nature in the USA and South Africa*. London: Routledge.

Bickford-Smith, V. (1999). Leisure and social identity in Cape Town, British Cape Colony, 1838–1910. *Kronos, 25,* 103–128.

Boy, G. (Ed.). (2012). *Nairobi National Park*. Nairobi: Friends of Nairobi National Park.

Brown, K. (2003). 'Trees, forests and communities': Some historiographical approaches to environmental history on Africa. *Area, 35*(4), 343–356.

Carruthers, J. (2006). Tracking in game trails: Looking afresh at the politics of environmental history in South Africa. *Environmental History, 11*(4), 804–829.

Carruthers, J. (2009). National Parks in South Africa. In B. Child, A. Spenceley, & H. Suich (Eds.), *Evolution & innovation in wildlife conservation: Parks and game ranches to transfrontier conservation areas* (pp. 35–50). London: Earthscan.

CCN. (1966). *Floresta da Tijuca*. Rio de Janeiro: Centro de conservaçao da Natureza.

Coetzee, J. (2007). *White writing: On the culture of letters in South Africa*. New Haven: Yale University Press.

Comaroff, J., & Comaroff, J. (2001). Naturing the nation: Aliens, apocalypse and the postcolonial state. *Journal of Southern African Studies, 27*(3), 627–651.

Corrêa, M. S., et al. (2001). *Parque nacional da Tijuca: 140 anos da reconstrução de uma floresta*. Rio de Janeiro: Ouro sobre Azul.

Daitz, D., & Myrdal, B. (2009). Table mountain national park. In B. Child, A. Spenceley, & H. Suich (Eds.), *Evolution & innovation in wildlife conservation: Parks and game ranches to transfrontier conservation areas* (pp. 325–339). London: Earthscan.

Drummond, J. A. (1997a). *Devastação e preservação ambiental: os parques nacionais do estado do Rio de Janeiro*. Niterói: EDUFF.

Drummond, J. A. (1997b). *O Sistema Brasileiro de Parques Nacionais: análise dos resultados de uma política ambiental*. Niterói: EDUFF.

Gichohi, H. W. (2000). Functional relationships between parks and agricultural areas in East Africa: The case of Nairobi National Park. *Conservation Biology Series, 12,* 141–167.

Goetze, T. (2002). Table Mountain—South Africa's natural national monument. *Historia, 47*(2), 457–478.

Grove, A. (1989). Scottish missionaries, evangelic discourses and the origins of conservation thinking in Southern Africa 1820–1900. *Journal of Southern African Studies, 15*(2), 163–187.

Grove, R. H. (1995). *Green imperialism: Colonial expansion, tropical Island Edens and the origins of environmentalism 1600–1860*. Cambridge: Cambridge University Press.

Guerra, A. J. T., & Coelho, M. C. N. (Eds.). (2012). *Unidades de Conservação: abordagem e características geográficas*. Rio de Janeiro: Bertrand Brasil.

Hey, D. (1978). *Report on the future control and management of Table Mountain and the Southern Peninsula Mountain Chain*. Pretoria: Government Printer.

Homewood, K., Kristjanson, P., & Trench, P. (Eds.). (2009). *Staying Maasai? Livelihoods, conservation and development in East African Rangelands*. New York: Springer.

Huxley, J. (1961). *The conservation of wildlife and natural habitats in Central and East Africa*. Paris: UNESCO.

Juuti, P., Mäki, R., & Wall, K. (2007). Water supply in the Cape settlement from the mid-17th century to the mid-19th century. In P. Juuti, T. Katko, & H. Vuorinen (Eds.), *Environmental history of water* (pp. 165–172). London: IWA Publishing.

Khati, A. S. (2003). *Indian National Parks & Sanctuaries*. Noida: Pelican Creations International.

Knowles, J. N., & Collett, D. P. (1989). Nature as myth, symbol and action: Notes towards a historical understanding of development and conservation in Kenyan Maasailand. *Africa, 59*(4), 433–460.

Lückhoff, C. (1951). *Table Mountain: Our national heritage after 300 hundred years*. Cape Town: A.A Balkema.

MacKenzie, J. M. (1988). *The empire of nature: Hunting, conservation and British Imperialism*. Manchester: Manchester University Press.

Management Plan for nature reserve SGNP. (1974). Borivli: Government of Maharashtra.

Management plan SGNP. (2011), unpublished.

Matheka, R. (2005). Antecedents to the community wildlife conservation programme in Kenya, 1946–1964. *Environment and History, 11*, 239–267.

Moll, G. (1989). *Tafelberg: 'n natuurwonder*. Cape Town: WESSA.

Ndebele, N. (2007). *Fine lines from the box: Further thoughts about our country*. Roggebaai: Umuzi.

Nixon, R. (2011). *Slow violence and the environmentalism of the poor*. London: Harvard University Press.

Ogutu, J., Owen, S. N., Piepho, H., & Said, M. (2011). Continuing wildlife population declines and range contraction in the Mara Region of Kenya during 1977–2009. *Journal of Zoology, 285*(2), 99–109.

Pande, P. (2005). *National Parks and Sanctuaries in Maharashtra: Reference guide*. Mumbai: Bombay Natural History Society.

Paranjpye, V. (1997). *A comprehensive environmental assessment of the Sanjay Gandhi National Park (Borivali National Park)*. Mumbai: Commissioned by the Maharashtra State Division of the World Wide Fund for Nature.

Rabe, L. (2010). Living history—The story of Adderly street's flower sellers. *SA Tydskrif vir kultuurgeskiedenes, 24*(1), 83–103.

Rutten, M. (1995). The tragedy of individualizing the commons: The outcome of subdividing the Maasai pastoralist group ranches in Kajado District, Kenya. In *Reinventing the Commons, Fifth Biennial Conference of the International Association for the Study of Common Property*. Bodoe, Norway, (pp. 24–28).

Van Sittert, L. (2002). From 'mere weeds' and 'bosjes' to a Cape floral kingdom: The re-imagining of indigenous flora at the Cape, c. 1890–1939. *Kronos, 28*, 102–126.

Van Sittert, L. (2005). Bringing in the wild: The commodification of wild animals in the Cape Colony/Province c. 1850–1950. *The Journal of South African History, 46*(2), 269–291.

Voss, M. (2003). Wilderness domesticated: Nineteenth century perceptions of Table Mountain. *Historical approaches, 1*, 39–48.

Zérah, M.-H. (2007). Conflict between green space preservation and housing needs: the case of the Sanjay Gandhi National Park in Mumbai. *Cities, 24*(2), 122–132.

Zérah, M.-H., & Landy, F. (2013). Nature and Urban Citizenship redefined: The case of the National Park in Mumbai. *Geoforum, 46*, 25–33.

Chapter 3
Urban National Parks and the Rich: Friends with Benefits

Julien Dellier, Sylvain Guyot, Frédéric Landy and Rafael Soares Gonçalves

In Rio de Janeiro, almost all of the most beautiful views of the sea and city are from the *favelas* on the edge of the Tijuca National Park, whereas most of the well-to-do areas are at sea level, far from the park. In Mumbai, the rich live at a distance from the park when they can, further to the south towards the city centre (cf. Chap. 6). Even if this is less clear cut in Nairobi or in Cape Town, the situation in those places is far from being of the "Central Park"-type where the urban park is surrounded by rich populations exclusively. It is true that some of these cities' most majestic natural sites have not been incorporated into the national park (e.g., the beaches of Rio and Cape Town). It's necessary, too, to enquire as to whether the protection of the park and its surroundings, when efficiently managed, has not tended to tolerate only illegal occupation, primarily by indigent citizens who cannot afford legal housing. Also to be carefully considered is the attractiveness of the national park in the imagination of very many of the city's people: don't those who can afford to choose their residence wish to live in contact with a national park, relatively far from the centre of the city and not without various inconveniences? Does "nature" always make us dream?

The borders of the urban national park, taking into account space constraints, are, however, sometimes situated on the edge of distinctly residential quarters.

J. Dellier (✉)
University of Limoges, CNRS, GEOLAB, Limoges, France
e-mail: julien.dellier@unilim.fr

S. Guyot
UMR 5319 Passages CNRS, Pessac, France

F. Landy
University of Paris Nanterre (UMR LAVUE), Nanterre, France

F. Landy
French Institute of Pondicherry, Pondicherry, India

R. S. Gonçalves
Paris 7 University, Paris, France

F. Landy (ed.), *From Urban National Parks to Natured Cities in the Global South*,
https://doi.org/10.1007/978-981-10-8462-1_3

Moreover, if the rich don't often live within range of the park, they can always go there to frequent and use these natural spaces, and their buying power and sociopolitical power are a major factor in the evolution of these parks. Thus, the rich, being neighbours of the national parks (the park providing a scenery function) or regular visitors (usage function), have a distinct influence on their existence and on their future.

Those we are calling "rich" in this article are the actors who have the highest incomes and the capacity to influence the decisions of local powers. What thresholds are to be retained for defining them? This is a complex question in four cities with rather different socioeconomic profiles. A precise line of demarcation must be arbitrarily chosen. The 1% choice is highly symbolic but too restrictive considering the diverse logics of the implied actors. Hence, to be included in our definition of the rich are those residents whose economic and social capital ensures them a dominant position in local society, recognising themselves as such and being recognised as such: they make up part of a circle of local decision makers, often giving the impression of a "club" (Charmes 2011) amongst themselves. This definition of richness has to be discussed in relation to the specific contexts of the four case studies and to other factors associated with wealth (skin colour for example). Is being rich in Cape Town, in Rio, in Mumbai or Nairobi the same thing? Cape Town and Nairobi both have a white post-colonial bourgeoisie (Pape 2003), in Rio there is a well-to-do class considered white in Brazilian socio-racial representations, while in Mumbai there is an extensive well-to-do class not clearly related to criteria of caste or religion, even though some determinisms exist (Baviskar 2002; Mawdsley 2004). These rich people are part of an upper-middle and upper class. Whether bourgeois, high bourgeois or nouveau riche, they all have an individual house with land around, except in Mumbai where, given the population density, they most often own a more or less luxurious condominium apartment. They often have as many cars as individuals of driving age in their households, and they work in professional sectors (e.g., politics, law, intellectual and medical fields) or as senior business executives. Their representation of the natural sites in which they live is associated with the view (topography) or with amenities (beach) in Rio and Cape Town, forest in Mumbai or wildlife in Nairobi. Their perception of the national parks is mostly linked with the outdoor activities they engage in or with decorative scenery to be protected.

This chapter, therefore, poses the question of the inter-relation between the rich, nature and the national parks. In addition, it explores the possibility of the protection of nature embodying environmental capital (Richard et al. 2015) in which the rich may invest to protect their property rights, their social circles, their exclusivity of visiting and their influence on various local political networks (Guyot et al. 2014). In other words, this chapter wants to illuminate the role of the (upper) middle classes in the conservation of the urban environment. Now, an often militant scientific discourse, dominant in India but valid in other countries, commonly contrasts two types of environmentalism: the mostly rural "environmentalism of the poor" (Gadgil and Guha 1995), which opposes the construction of dams, major mining and industrial projects, etc.; and an urban "bourgeois environmentalism" (Baviskar 2011), often accused of associating sincere ecological concerns with agendas enabling the middle classes to enhance their property capital. The poor are ousted economically or physically from a city subject to "ecological gentrification" (Dooling 2013). However, such opposition must be qualified (Follmann 2016): on the one hand, many leaders of rural movements belong to the middle classes, while on the other hand, field studies sometimes show that "especially local environmental NGOs-although middle-class dominated-do not exclusively strive for middle-class interests, but rather endeavor for long-term ecological sustainability and social justice" (Follmann 2016, 49): the middle class environmentalism does "not simply follow the state-led bourgeois imaginary" (ibid.).

What is the influence of protected nature on the territorial strategies of the well-to-do population? Conversely, what is the influence of the rich on the national parks in terms of their creation, management and sustainability? To illustrate the answers to this question, the chapter ends by asking what the situation of the national parks would be without the urban rich, and what would become of the urban rich without the national parks: between geographic prospective and geography-fiction, these scenarios allow for an understanding of the value ascribed to the national park by the well-to-do population and vice versa.

3.1 The Proximity of the National Parks as a Factor in the Location of Well-to-Do Populations

3.1.1 The Rich and Nature, a Long-Term Dialectic?

Remarkable natural sites have their importance in the historic location of the rich in the city. It was not originally a question of potential conservation that caused these sites to be appropriated by the most prosperous populations, but most often facts such as topography, vegetation or the view belonging to particular local configurations. Obviously, these different facts were taken into consideration when the future perimeters of the protection of nature were put in place.

There is no historical connection in Bombay (Mumbai) between the location of the rich and the zone of nature that will be dedicated to a future park. In Nairobi, the connection exists throughout the Karen district (Fig. 3.1), a colonial residential area on the western boundary of the park, pre-existent at its creation in 1946. It is the same in Cape Town where the exceptional sites of the slopes of Table Mountain, oriented to the north, to the east or to the west, serve as the location of the initial residential areas of Cape Town Colony, such as at Constantia (Fig. 3.2) where the first vineyards in South Africa were planted. In Rio, the steepness of the slopes and the violence of precipitation due to the humid tropical climate made it difficult for the rich to install themselves in proximity to the coffee trees of the massif, then forest, of Tijuca, which serve as the framework of the national park (Lézy-Bruno 2014). All the same, some hills, such as at Cosme Velho (Fig. 3.3) or Santa Teresa, have been able to welcome the long-established cores of the privileged populations (villa de Chacara do Ceu).

Fig. 3.1 The house of Karen Blixen built in the 1920s (Karen district, Nairobi). *Photo* F. Landy, 2014

Fig. 3.2 Constantia Valley looking over False Bay towards Handklip in the south-east, by Johannes Schumacher 1776–1777, a watercolour on paper. *Source* http://www.zandvleitrust.org.za/images/zvthistorical%20maps%2010%20JF%20april%202013.JPG

The presence of the national park has supported the preferential settlement of the rich in its immediate surroundings in the cases of Nairobi and Cape Town. This is less the case in Rio even though the upmarket residential areas were developed on the southern boundary of the park. Finally, this is not at all validated in Mumbai, where the rich live in the city centre and don't move out unless there is a particular attachment to the proximity of the park. Consequently, the protection of a privileged way of life bestowed by the proximity of the park, in the form of an environmental buffer, is attested to in Cape Town, Nairobi and partially in Rio, but not in Mumbai. In the case of Mumbai, the view of the park seems to be appreciated, but at the same time it implies constraints for its beneficiaries, such as interactions with wild fauna that might become dangerous, the proximity of slums and barely regulated areas, not to mention the long distances to the workplace. In the other three cities, the process of "heritagisation" has served to maintain a status quo based on property: by identifying certain urban forms in the area of the park as elements to be consolidated, heritagisation tends to maintain and reinforce the concentration of coveted land possession in the hands of the elites.

Fig. 3.3 The Cosme Velho district in Rio, mid-nineteenth century at the foot of the Tijuca massif. *Key* Looking towards Zandvlei and False Bay from high up on the slopes above present-day Buitenverwagting farm, on the road to the top of Constantiaberg. *Source* https://upload.wikimedia.org/wikipedia/commons/5/50/Rio_de_Janeiro_Cosme_Velho_secXIX.jpg

Is the involvement of the rich in the park to the detriment of other urban environmental issues? This seems to be the case in Cape Town, where the rich ignore for the most part the environmental questions connected with Cape Town Flats where the poorest quarters are concentrated, and give all their attention to the Peninsula, which is mostly white and wealthy. Thus, the NGOs of the upmarket neighbourhoods are very little involved with environmental questions in the townships. This is less true in the other three cities. In Nairobi, the vision of the environment held by the residents' associations is close to being inclusive. And in Mumbai, the rich invest very little in the park compared with what might be expected: plenty of other subjects of debate and action exist in the urban agglomeration, from the safeguarding of mangroves, salt marshes or pink flamingos to more classically urban problems such as pollution. But those interested in the park are becoming aware, little by little, that the problems of the protected space are intrinsically linked to those of the city: for example, attacks by leopards are doubtless to be controlled by better urban management as much as by better park management; better garbage collection is required to prevent uncontrolled dumping attracting the leopards' potential prey. More cleanliness means fewer leopards? The park may, because of its positive image, turn out to be quite an efficient tool in better motivating the populations and the decision makers as regards other more specifically urban issues, equally important but less easily mobilised.

The presence of a park in proximity to upmarket residential areas may, however, also be seen as a factor of insecurity. This is the case in Mumbai because of the incursion of wild animals, and especially leopards, into inhabited areas with several mortal attacks recorded over recent years (Chap. 7) Paradoxically, the same cause produces the opposite results in Nairobi, where the proximity of big wild animals and of predators is often seen as much less of a threat, no mortal accidents having been recorded, and often as an extra protection against thugs. In Cape Town, the

Table 3.1 Influence of the national parks on the territorial entrenchment in upmarket districts

	Situation before the creation of the park	Current situation
Cape Town	– Well-to-do peripheries on the east slope of Table Mountain – Urbanisation regulated by apartheid laws	– An environmental buffer for the benefit of some rich residential areas – A limited involvement of NGOs issuing from high-end districts to work in the townships on environmental questions – The processes of heritagisation support a property-based status quo – Presence of an open space = potential risk
Nairobi	Cf. Karen Blixen: big plantations, (Maasai rangelands), reserve for white hunting, etc. → a cocoon in the west	– An environmental buffer for the benefit of some rich residential areas – The processes of heritagisation support a property-based status quo
Mumbai	The rich are far from the hills of today's park—for example on Malabar Hill at the tip of the peninsula	A place of relaxation for the well-to-do population, especially neighbouring communities. But the park attracts few well-off residents due to its peripheral location
Rio	Coffee plantations. The rich mostly lived in the centre and in districts far from the park. Even so, since the reforestation at the end of the nineteenth century, the area has been used for leisure activities particularly for the well-to-do populations	– An environmental buffer to the profit of some rich residential areas – The processes of heritagisation support a property-based status quo – Democratic use with the presence of the rich and the poor, but with signs of gentrification

proximity of the park is seen as a potential threat because of the absence of control over the comings and goings at the perimeter of the park in a high security residential context; the same is true in Rio because of the proximity of the *favelas* to the park. There, the more pronounced presence of the police in some *favelas* around the park had led to the reappearance of walkers in the protected space who entered from those points. Since 2015, however, the situation has again become dangerous and there are currently few walkers (Table 3.1).

3.1.2 National Parks Were Conceived for Frequentation by the Rich

Do the rich have privileged access to the national park, and if so for what usages? Accessibility of the parks may be a source of inequality between populations from the point of view of several criteria (e.g., social, spatial, etc.). Further, forms of degradation or of protection connected with the frequentation of these parks by rich populations need to be dealt with.

In Cape Town, the national park is highly frequented. This is due both to the neighbouring populations, usually well-to-do, and to international tourism (Donaldson et al. 2016). It is possible to differentiate the essentially touristic areas of frequentation, those with geo-symbolic value, such as Cape of Good Hope and Point Cape, or Table Mountain, where access most often requires payment, and areas of urban recreation, where access is most often free but where the practice of various activities requires a specific permit (e.g., dog walking, mountain biking, climbing, paragliding, etc.). The proximity of these places to rich localities (e.g., Constantia, Hout Bay, Noordhoek, etc.) is a supplementary factor promoting frequentation by the rich. Conversely, the popular areas are, generally speaking, often physically far from the edges of the park, and the co-presence of poor and well-to-do quarters on its borders brings about conflict, as in the sectors of Imizamo-Yethu, Masiphumelele and Red Hill (Guyot et al. 2015). Even though the "risky" areas inside the park may be at some distance from the well-to-do zones, a recurrent demand from the richer populations is for the securing of the approaches to the park, seen as the weak point in local residential vigilance arrangements.

In Rio, a social hierarchy in the usage of the park reflects the metropolitan geography. Thus, the slopes on the south zone are used by the well-to-do while those on the north zone are more often the destination of the working classes. Moreover, needing a vehicle in order to enter some areas of the park is a definite obstacle to the poor. Access to the park is free, with the exception of the Corcovado area, where the iconic statue of Christ stands; paid access makes visits here possible only for international tourists and the local well-to-do population.

In Nairobi, the park is essentially frequented by international tourists and rich Indian minorities, though pupils are certainly present there too. For the other privileged populations of Nairobi, the park mainly plays a role of decoration than of a place of activity, whose crossing is enforced by the need to avoid traffic congestion in the city. The presence of fauna in the upmarket areas, just as in Karen, puts a limit on specific interest in the park.

Lastly, the case of Mumbai demonstrates that there are numerous exceptions that make it possible for the rich to use the park without reference to the official rules. Thus, near Mulund, the park is the venue of visits without permission by the roadside dwelling well-to-do, who take advantage of a breach in the wall. Some luxury dwellings overhang the park or are situated on the so-called "private forest lands"—old private forests whose ownership remains contested today. However, this urban sprawl is limited compared with the invasion of the park by the slums which house more than 100,000 people—there were more than 500,000 in 1995.

3.2 The Influence of the Well-to-Do Populations on the National Parks

All the influences of the richer residents living near the national park must be analysed, particularly in terms of participation in the process of creation of the park and its governance.

3.2.1 The Part Played by the Rich in the Creation of the National Park

As regards our four case studies, there is no systematic involvement of the well-to-do populations in the process of creation of the parks. In fact, the rich develop an entire range of strategies in relation to a park or a park project, from ignorance to attempts at taking control. It is, however, possible to construct a hypothesis according to which the presence of remarkable natural elements (e.g., fauna, flora) was a favourable factor in involving the well-to-do populations.

Thus, in Cape Town, the prefiguration of the park is very much linked to the action of the residents' associations (Guyot et al. 2014), and its delimitation refers back to the limit of the urbanisation front during apartheid. The park is at once the fruit of the efforts of the well-to-do neighbouring population (as in the climbing clubs mentioned in Chap. 1) and the outcome of an environmental buffering strategy, underpinned by the preservation of endemic formations (*fynbos*). A similar logic is apparent in Nairobi, where the park corresponds historically to a buffer zone between the areas of the Kikuyu and Maasai populations that were occupied by the British. This function of buffer zone was subsequently continued between the Maasai reserve and the British district of Nairobi until the park was turned into military territory. It was during that period that its function of a game reserve for rich colonialists came into being. The creation of the National Park, effective in 1946, is thus partly due to the influence of the well-to-do (and British) populations interested in the management of the wildlife at the gates of the city. What one Maasai—himself fairly well-off and a member of the Friends of Nairobi National Park association (FoNNaP)—told us about white people, might be said of the rich:[1] "The Whites come from countries that no longer have wildlife. They know what it means to live without wildlife; and so they teach us about what could happen here." Many rich people and whites have forgotten what nature is, and therefore wish to find it again, in the name of "ecological imaginaries" (Gandy 2006) founded on conditions of urban wealth.

In Rio, the reforestation of lands that have now become part of the park was set in motion by the emperor Pedro II with the double objective of management of

[1]Interview, July 18, 2014.

water resources and creation of recreation areas. Indirectly, the creation of the park, nationalised in 1961, has also been to the advantage of real estate interests. The well-to-do populations are thus very much implicated in the control of the expansion of the *favelas* in the direction of the park, and especially in supporting the efforts of the public authorities: on the one hand, in the creation of eco-limits, physical barriers between the park and the *favelas,* and on the other, with the project of communal reforestation which, under the guise of protecting the *favelas* from landslides, also had the effect of fixing the boundaries.

Lastly, in the case of Mumbai, it was a matter as in Rio of responding to the imperative of protection of the water resource, but the process of creation did not involve the well-to-do populations at all. The territory of the park corresponds to an area of average elevation that has been through a period of forest clearing, both for the constructing of two urban water reserves and for cultivation by indigenous sharecroppers. Interactions between the park and the well-to-do populations appear to have been limited for a long time, apart from the extension of the park during the state of emergency of 1975–1977, when private forests were expropriated.

3.2.2 Clashes and Misfortunes in Co-management Between the Rich and the National Park

The presence of an urban national park, by its very nature, exerts an influence on the political agenda of the city. The running of the interface between the park and the city, the transition of protected nature to urban nature, and encounters between wild animals and residents are all debatable subjects. The concentration of issues makes the governance of the transitional areas between park and city very delicate. Three thematics are to be observed in particular: the creation of ecological corridors in the continuation of the parks, the management of the wildlife and the question of forums.

The creation of ecological corridors, aimed at ensuring the continuity of biodiversity outside the national parks is common to all four cities. However, the conditions of their creation and sustainability differ quite widely. In Cape Town and in Nairobi, an initial form of ecological corridor has materialised due to the presence of properties belonging to the rich in the buffer zones of the parks. The large associated "gardens", although often closed off by formidable fences, provide a habitat for a number of wild animal species and for varieties of flora, as is even more true at Cape Town since the residents were sensitised to the preservation of endemic species. More strategically, the city of Cape Town, in association with the residents of Constantia, works to establish ecological corridors along some watercourses, such as the Dieprivier, which gushes down Table Mountain to the ocean, going through the entire declination of the districts of Cape Town, from the high-end areas to the townships. In Rio as in Mumbai, the reality of the ecological corridors is quite different. In Brazil, the project of establishing a long ecological corridor, defined as the "Carioca mosaic", came up because of the decision of the local authorities to build a golf course for the Olympic Games on part of the Marapendi municipal park. The very notion of corridors seems to be ignored by the Mumbai rich, despite the unofficial presence of a corridor established by the Tungareshwar Wildlife Sanctuary to the north of the park. This is currently threatened by the reinforcing of the railway line that may in time physically cut it off from Sanjay Gandhi National Park (SGNP).

The other major subject of discussion and of co-management between park and city, over which the rich often take sides, is that of the wildlife. Wild animals are seen as dangerous in Cape Town, because of the damage to neighbouring properties caused by baboons; this attitude pertains even though the baboons are attractive to tourists. Wildlife is also seen as dangerous in Mumbai because of leopard attacks and, there too, damage caused by monkeys. In contrast, many rich residents of Karen district in Nairobi, are glad to have lions crossing the lawn since they discourage criminals. In Rio, the question of wildlife doesn't come up in specific discussions, given their low quantity, apart from some criticism regarding the presence of primates judged as alien: in fact, in Rio as in Cape Town, the problem of indigenous, even endemic, species is a subject, if not of dissension, then at least of lively discussion between some involved residents and the managers of the park.

This then raises the question of the park/inhabitants forums, whose objective is to ensure a form of co-management. Regarding Table Mountain National Park (TMNP), in exchange for their activism in having helped to create the park, the residents' associations took over the organisation of the forums between 1998 and 2011, but that was very soon found to be inoperative due to successive reorganisation. Observation of the evolution of these forums carried out by Didier and Swanepoel (2014) shows the relative loss of the influence of the elites with each change (Donaldson et al. 2016). In Rio too, forums came into existence from the creation of a consultation committee in 2002, according to the directives of the national Constitution (cf. Chap. 10). There are ongoing problems, especially the question of the weak representation of roadside *favelas*, but the reconstruction of the committee in 2012 seems to have led to a decrease in the amount of members from well-to-do populations and from big business.

The other two parks have always been less go-ahead as regards co-management. In Nairobi, a Management Committee was created only in 2014: among its ten members are three representatives from the business world as well as the son of a white proprietor, as against only two representatives from the "communities" (Maasai). This committee has budgetary autonomy, and no doubt the social skills of its members will quite easily be able to replenish its funds through events organised in and around the park (e.g., bicycle races). One of the members of this committee is the president of FoNNaP, an association of park users and environmentalists. They play a role of surveillance and support of park missions (e.g., counting of animals, conferences, lobbying, etc.). Amongst its members an overrepresentation of white or Indian Kenyans is noticeable.

The case is quite different in Mumbai, where there is hardly any willingness amongst the rich to join in the management of the park, not to mention the fact that there is nothing to encourage them to do so: no forum and no management committee etc. MfSGNP (Mumbaikars for SGNP) is, however, on the model of FoNNaP, an association whose members belong mostly to the higher social classes and for a few years have been involved by the park board in some expanding functions: social management of conflicts over leopards, scientific expertise and location of bicycles etc. (www.mumbaikarsforsgnp.com). In 2015, the director of the park talked to us about his wish to be able to count much more on the well-to-do classes, and his readiness to organise the space the better to attract them: they have more money, more political power and more interest in an environment policy corresponding to the model of a national park. Is this sheer cynicism with contempt for the poor, or realism when faced with threats to the park?

3.2.3 Multiple Influences of the Rich on the National Parks

If the rich cannot, in a general way, claim to hold control over the parks they have sometimes helped to create, this does not mean that they are lacking in the capacity for ad hoc interference, sometimes even parasitic behaviour.

Regarding park management, official regulations and practices often leave room for individual pressure or forms of collective lobbying which may raise micro-local problems (e.g., securing of a fence, access of domestic animals to the park, etc.), or more general ones (e.g., management of alien plants, securing of access and itineraries of the park's hikes, etc.). In Rio, given that there are coexistent upmarket areas and *favelas* around the park, there has been pressure over the last few years for the displacement of the poor quarters in the name of the protection of the environment, or sometimes under that pretext. An example is the attempts at resettling the *favelas* of Alta da BoiaVista, on the side of the road that crosses the park. We can also cite the conflict over the working-class quarter of Horto, situated in a relatively wealthy part of the city (Jardim Botânico) and accused of threatening both the national park and the botanical garden of Rio de Janeiro.

In terms of land, the proximity of the upmarket residential areas to the borders of the national parks is often seen as a stabilising factor and a guarantee of the existence of a buffer zone between nature and urban life as a result of a lower density of construction and particular attention being given to private gardens. This is the case in districts such as Constantia and Oranjezicht in Cape Town. However, the rich, through their financial means and their political interference, may reveal themselves to be formidable predators on the borders of the park. Encroachment by high-end residences is attested to in both Rio and Mumbai.

On the functional level, the rich areas, by their installation at the gates of the national parks, may also act as barriers, at least symbolically, to access to the park for the populations coming from working-class quarters. The absence of public transport services, and an urban atmosphere hostile to the lower classes (e.g., CCTV cameras, vigilantes, barbed wire, etc.), also act as barriers to the accessibility of the parks. It is possible to question the involuntary character of this type of spatial apparatus in some areas of Cape Town and Rio. Along these lines, the rich have had a tendency to privatise a slice of the national park to their profit, and in the case of Cape Town, to the profit of international tourists who stay at the numerous low capacity lodgings (e.g., guesthouses, lodges, etc.) situated in these quarters.

The relationship between the rich and the national park also has to be observed from the point of view of economics if the touristic dimension is taken into account. As central attractions of the cities of Rio and Cape Town, and in the case of Nairobi with its tremendous wildlife, the national parks are a key component of the touristic strategies of these three cities. Now, since a part of the touristic package—in the form of small to medium structures offering a high degree of comfort—is situated in the rich areas, close to the park, philanthropic commitment in the protection of a symbolic nature might be also the expression of a more down to earth economic interest.

3.2.4 Balance Sheet

In synthesising the totality of these observations, it is important to emphasise that the connections that hold together the parks, their managers and the rich, echo some of the original opposition between a model founded on the British heritage of a colony of settlement (landscaped aspects in Cape Town, big game hunting reserves in Nairobi) and a model introduced primarily for the preservation of a resource (water and secondarily forestry) in Rio and Mumbai. However, this is not an iron law for all that.

In terms of the co-construction of management and land, certain facts show the weight of the history of the setting up of society-city-park relations. In operation in Nairobi and Cape Town was a form of co-construction between the park's institutional actors (e.g., State services, political executives, etc.) and civil society. It attests to the weight of the local elites, especially those from long-standing colonial dominant classes, on the operating of the conservation policies directly connected with their living places. By extension, the rich have contributed substantially (except in Mumbai) to defending the geographical integrity of the park, especially with respect to the protection of the landscape and the security of the upmarket residential areas. It is, moreover, to be noted that the park-city closeness had led to the production of a specific urban aesthetic that encourages a low density of construction and the laying out of gardens, especially in the case of Cape Town (cf. Chap. 8). Regarding governance, in both Cape Town and Nairobi, that urban specific form is today being used by the rich as an asset with the putting in place of ecological corridors linking various areas of the park, or the park to the periphery of the city. And in fact, the setting up of ecological continuities by the executives of the city, including not only private properties but also open public spaces in the vicinity, results in the extension of the environmental buffer desired by the rich.

The situation is very different in Rio and Mumbai, where the construction of parks entails a complete shift in values that were previously linked to resources to be preserved and are now attached instead to environmental and ecological characteristics. The absence of communication relays within civil society as effective as those of Cape Town and of Nairobi is certainly one of the explanatory factors in the very great porosity of the park boundaries. The installation of poorer and richer residential quarters might have thus been facilitated by relative indifference outside the circle of park managers: this hypothesis does not, of course, take into account the differences in urban pressure which are much greater in Mumbai's agglomeration of twenty million inhabitants than in one much less dense and populous such as Cape Town. There is, in any case, evidence that the rich can do much more harm to the park than the poor as they have the means and capacity to mobilise networks of influence, unlike the poor who have less social and political capital.

Yet the weight of history on contemporary interactions between the urban rich and the parks must be qualified.

With regard to accessibility, the specificity of the parks of Table Mountain and Tijuca is in their form of archipelagos resulting in a significant increase of points of contact between city and park. Paradoxically, the great number of entrances to the park do not make it much more accessible: the access points positioned within the rich areas or the installation of toll booths in some zones are yet more factors that favour the accessibility of the rich and reduce the perceived risk of coming into contact with the less-privileged public. Cape Town and Rio appear, moreover, as touristic metropolises where the parks are frequented as much by tourists as by local residents (Donaldson et al. 2016). In contrast, the use of the park of Nairobi seems to be primarily for tourists while some rich locals envisage a simple zone of passage to avoid bottlenecks. Finally, in the case of Mumbai, the more disorganised and scarcely controlled access to the park makes all sort of usage possible.

The matter of wildlife is another marker of the relations between the rich and the parks. As regards Mumbai and Cape Town, wildlife is a great source of tension between the rich and the park's managers because of leopard attacks in Mumbai and damage caused by baboons in Cape Town (Chap. 7). These two cases certainly reveal the similarity of these two species in adapting their behaviour to uncover fresh resources: dogs and rodents for the leopards and the food stores of the residences bordering the park for the baboons. This tension, however, casts light on the gap between those resolutely urban dwellers who have lost any awareness of the rules of co-presence with wild animals, and the animals themselves reduced to a nuisance (Table 3.2).

3.3 Where Would the Park Be Without the Rich or the Rich Without the Park? Geography-Fiction

In the preceding two parts it has been possible to establish intersecting influences between the rich and the parks in the urban framework of our four emerging metropolises. In this final part we intend to present a semi-fictional geographical

Table 3.2 Evolution of the influence of well-to-do populations on the national parks

	Situation before the creation of the park	Current situation
Cape Town	– Impact of apartheid laws on the proximity of the richer residents to the Peninsular natural spaces – Lobbying well-to-do residents to establish a national park	– Interference of well-to-do roadside dwellers in the management of the park (fauna and security) – Environmental buffer coming into being through the properties of the rich – Residential vigilance and the demand by the rich for security
Nairobi	– Natural space as a buffer zone between the Kikuyu and Maasai peoples, occupied by the British as a projected game reserve and then by the national park	– The FoNNaP association marginally collaborates with the park management – The Karen ecological corridor to the west of the national park represents a very bourgeois space
Mumbai	– Natural exploited forested space, protected for the water needs of the city	– The rich frequent zones of the park prohibited without authorisation – Residences overlapping the park: the specific case of "private forest lands". But generally no positive impact of the park on purchase prices – The rich do not involve themselves in the running of the park in spite of their recurrent complaints about leopards. However, they play an increasing role in some associations (MfSGNP)
Rio	Coffee plantation in the nineteenth century; ban on coffee and reforesting from the second half of the nineteenth century as a protected forested place for the water needs of the city	Substantial presence of the rich in the management spaces of the park, especially through the Friends of Tijuca Park Despite a large number of *favelas* around the park, there are several upmarket areas, notably on the south slope, very near the sea Significant economic impact of tourism, especially on the site of Christ the Redeemer The park's landscape constitutes the city's main tourist attraction (principal reason for its World Heritage status as of 2012) Catchment of water in the park's sources by the *favelados*, but also by upmarket residences around the park

sketch: what would happen to the park if there were no rich people? Reciprocally, what would become of the rich if there were no park? To reason in a dystopic vein ("what would happen if …?) has illuminating heuristic values. This innovatively throws light, in negative, on the pressure of different actors and factors on the changes, whether in the past or in progress.

3.3.1 Where Would the Park Be Without the Rich?

The influence, not to say interference, of the rich in the creation, management and perspectives of the park is very varied as our case studies show. The potential consequences of the absence of a rich class, very active in Cape Town, but little interested or present in Mumbai, can be gradually described. In Mumbai, the changes would have been the less dramatic. True, the presence of the rich has certainly helped to stop the expansion of the slums there: an ecological association of middle and upper classes laid a complaint against the Forest Department (Zérah 2007). Without this complaint, the situation in 1995, with more than 500,000 people living in the park, would no doubt have deteriorated much further. Yet, the role of protector played by the rich also concerns the scale of the agglomeration and other aspects (e.g., the movement for conservation of mangroves and pink flamingos): their role is rather limited in proximity to the park where their influence is slight and, if the officials of the park are to be believed, sometimes harmful due to the overcrowding of certain touristic areas and their amenities. This lack of interest in the park on the part of the rich as an element of mobilisation is explained to a great extent by the national dimension, disconnected from the local elites, of the management of SGNP.

Nairobi offers a milder version based on this same logic of a disconnection between local issues and park management responding to national imperatives. Here, the absence of well-to-do populations would no doubt lead to a limited extension of the lower-class quarters, especially in the vicinity of the Kibera slum, and the disappearance of the Karen buffer zone to the west of the park. The absence would have very little effect on the management of the park as such, committed as it is to a national cause of protection of the biodiversity; but it would have contributed to substantially changing the surroundings, causing, with the disappearance of the Karen district, a loss of ecological continuity between the national park and the Ngong Road Forest Nature Reserve. Another potential change with significant impact on the ecological functioning of the park would have been the opening of a motorway in the northern part, envisaged at one time by the municipality and eventually abandoned following strong involvement, especially by the rich of Karen, that would certainly have had a different outcome in their absence. The influence of the rich, who are enthusiastic defenders of the park for its ecological and recreational value, is nevertheless ambivalent as it seems at the same time to be at the origin of the purchase of some Maasai lands which puts a squeeze on the park between the south-west and south-east. It is therefore possible to imagine a shifting of the park's limits, encroached upon by the city in the northern part, towards the Maasai lands in the south that might have been purchased by the State had the rich not desired these resources.

As a third case of dominance of the national upon the local, Rio de Janeiro offers a more contrasted situation in the juxtaposition of rich districts, disadvantaged quarters and the park. In the absence of both the residential proximity and frequentation of the rich, the park ought to be weakened by the extension of the 46 neighbouring *favelas*, and the possible creation of new *favelas* (*favelas* paradoxically contribute to the protection of the park by limiting the appetites of the well-to-do). This slum extension would certainly flourish in the southern and most dense part of the metropolis with a non negligible accentuation of environmental challenges and risks. Notably, however, with the statute of Christ of Corovado, which has taken on the status of a national symbol and represents one of the country's most important touristic resources, the issue is certainly a national rather than local one. In this light, the absence of richer populations would not have much impact on the sustainability of the park and its management.

Unlike the other three cities, it is tempting in the case of Cape Town to suggest a scenario that gives a more important place to local issues, on the one hand because of the central role of the elites in the creation of the park (Guyot et al. 2014), and on the other because the geo-symbol of Table Mountain is overtaken, in terms of its level of national outreach, by that of Robben Island, where, notably, Nelson Mandela was imprisoned: in this light it doesn't constitute a sufficient marker for national issues to predominate forcefully over local ones. The most plausible scenario is that of a more disputed limit to the park, weakened by being submitted to the growing pressure from disadvantaged populations to settle on the edges of the park. Thus the surroundings of the main roads crossing the park would be progressively doubled in a precarious urban linearity. The city of Cape Town would be faced with the same slum extension syndrome previously mentioned, which tends towards the park no longer resisting the city, but in a more virulent form because of the absence of the rich from environmental activism.

All in all, the consequences of the absence of the rich would seem to be essentially negative for the parks.

On the economic level the rich by their presence in proximity to the parks and their habits of frequentation of them, participate in creating the conditions for the parks as tourist attractions, especially in Cape Town. Some districts near to the park are integrated into the touristic image of the park and are today sold as part of the park atmosphere, with characteristic architecture, as in Constantia. In addition to an influence on national and international touristic frequentation, the rich also generate economic resources for the park through permits for activities or rights of access to the various amenities.

At the ecological level, the worst offence by the city against the park through the process of slum extension would have at the very least a triple consequence: a reduction in the park's area, of variable magnitude; a loss of the richness of flora and fauna linked to the deterioration of the environment in the areas affected by brutal urbanisation as well as in the buffer zone of the shifting border between the city and nature; and the isolation of the park due to urban densification leading to the closing of ecological corridors connecting to peripheral natural spaces. Furthermore, the disappearance of property walls, which serve the integrity of the park as much as that of the residences of the well-to-do quarters, increasingly poses the question of the interactions between city and nature. In the framework of urban densification in the surroundings of the park, the hardening of conflicts over wildlife between roadside dwellers and the parks might become a much more charged issue, especially in Nairobi where there would be a reflection of the leopard situation in Mumbai. It is very much to be feared that one of the solutions suggested would be to put fences all along the borders of the park, which would not be without consequences at the level of ecological continuity for the fauna.

At the crossroads of these two negative consequences, it is useful to consider whether in the event of the realisation of these scenarios in their most dramatic form, these parks would preserve their national labeling. If there is little doubt about this in terms Rio, Mumbai and Nairobi, the question remains open for TMNP.

If, however, these environmental aspects could possibly lead to disaster, it is necessary to provide a balance by emphasising that the absence of the rich could lead to a more efficient taking into account of the social aspects. In the context of an emerging city, these scenarios would in fact lead to much greater use of the park by the disadvantaged populations. On the one hand, these populations would no longer be relegated geographically, economically and socially to the exterior of the parks, while on the other hand, above and beyond the issues of social justice, it is the question of urban identity that must be looked at: rather than think of the working-class quarters and specifically the slums as a form of habitation incompatible with the formalisation of nature in the city embodied by the national park, this geography-fiction elicits our re-examination of the city-park relation outside of social, political and scientific norms.

3.3.2 Where Would the Rich Be Without the Park?

Turning the question around does not produce a simple negative. The pure and simple suppression of a spatial object, such as a national park, entails territorial consequences for the rich that are difficult to predict precisely, but whose major traits may be sketched out here with regard to the pre-existing local dynamics.

In Mumbai, the lack of interest in the park by the rich in general suggests that its absence would not cause a major upheaval in their relationship with their natural environment. Much more polemical: despite some harmful consequences (e.g., disruption of the microclimate, loss of recreation grounds, etc.) the balance sheet of this disappearance may even be seen in a positive light by the Mumbai rich. As regards the urban dynamic of the megalopolis, it may in fact be imagined that with the declassification of the park, the city might experience a scrap of fresh space allowing for some loosening of the densification and providing the opportunity to create infrastructures aimed at easing traffic. A policy of construction of social housing might result in the reduction of the presence of slums, which concerns almost half the population of Mumbai. This would indirectly please the rich who complain about this kind of urbanisation. Such a fictional geography, however, brings up the question: for how long? Urban pressure is such that it is hard to imagine this breath of air lasting for more than a short time.

In the case of Nairobi, the park is nowadays little used as a place of amenities by the rich of the city. The absence of the park would thus provide an opportunity for the rich to extend their land properties, even if this goes along with the progression of slums into the contact areas of the park. As regards the habits of Nairobi elites, the loss of the national urban park would, moreover, not be an obstacle to their search for amenities as several parks of comparable, even superior, attraction are only a few hours away by road. Consequences would not be more significant in the economic and touristic sectors either, the park being without value as an urban icon: its absence alone would not be likely to disturb the international image of Nairobi. In the longer term, however, the consequences might be more problematical as it is not only the progressive overcrowding of the area, particularly by slum populations, that causes a loss of heritage held by the rich. More importantly, the increase in contact with the Maasai lands would entail conflicts over questions of land and access to pastures. Currently the park plays, to some extent, the role of a buffer zone.

Unlike Mumbai and Nairobi, the identities of the cities of Rio de Janeiro and Cape Town are deeply connected with the presence of areas of nature such as the parks of Tijuca and Table Mountain. The absence of the park would also have very striking consequences, not least from the point of view of urban marketing, which is in both of these cases one of the keys to the attractiveness of these metropolises.

In the case of Rio de Janeiro, the loss of national park status for Tijuca would doubtless lead to an upsurge of appetite for the taking over of these lands in the form of *favelas* by the disadvantaged populations and perhaps especially through the monopolising by the rich of this property reserve for urban development. It would result in potentially violent conflicts over control of the property. On the metropolitan scale, we may question the possible realisation amongst the rich of the necessity to preserve natural spaces, especially around the peaks (sugarloaf), considering the appetite for real estate investment and the necessity for metropolitan infrastructures. Finally, the disappearance of Tijuca Park would reinforce the process of segregation already in progress in Rio de Janeiro.

In Cape Town, in the absence of the national park, some neighbouring areas would still have the status of a municipal or provincial natural reserve. These levels of management, that are more accessible than the national level, would allow the rich to strengthen their influence on the management of the reserves. In addition, it is very probable that some of the rich would, on their own or through collective structures, acquire a part of the Peninsula. In fact, the park today extends into lands that belong to different owners: local and provincial authorities, domain of the State and erstwhile military territories. Without their protective classification as national parks, these parcels of land would become the object of greed. The stranglehold of the rich on the environmental buffer of the Peninsula would therefore be more direct than in the actual situation. It is, however, difficult to imagine that these rich people would succeed without the park, in preserving the integrity of their maintained private turf in the face of the needs of development (especially for housing) supported by the provincial and even national authorities. The absence of the park and a policy favouring the construction of social housing, at the expense of the preservation of a natural space, would be prejudicial to the landscape of some rich districts on the edge of the park. Such a situation might support the strategies of circumvention and "clubbisation" (Charmes 2011) already present in Cape Town. On the one hand, the multiplication of gated communities with private environmental amenities would add to the socio-spatial fragmentation of the metropolis. On the other hand, despite the resistance of some geo-symbols, the attractiveness of the city to rich migrants in quest of environmental amenities might be weakened. Some positive aspects might, however, emerge as a result of the decline of protected areas, thanks to a possible report of the involvement of the rich in favour of the environment on the metropolitan scale, especially regarding the articulation between rich and poor quarters, particularly those of Cape Town Flats. In an urban configuration, where the rich topographically would no longer systematically dominate the poor, it is appropriate for the former to occupy themselves much more energetically with the functioning of the ecosystems beyond the single connection between the park and the rich quarters.

It would be tempting to suggest an iconoclastic conclusion to the end of this part: the rich would do better without the urban national parks. The disappearance of the park would greatly encourage the growth of the space allocated to the city. Transportation infrastructures, residential developments and urban decongestion are among the possibilities offered by this breath of fresh air which would help in improving living conditions in these emergent metropolises, particularly for the rich. But this is only a puff here whose primary condition is ephemerality. Very quickly, the partial or total filling up of the park's space would lead to escalating conflict between the rich and poor quarters, accentuated by segregation and intensified by the logics of real estate speculation by the rich. The backlash would then be particularly bitter.

Hence, it makes sense to conclude with the pacific qualities of the urban national park that benefit the metropolises containing them. With a governance more or less detached from local interests, the park supplies the rich with an opportunity to profit from its proximity as an environmental buffer preserving their heritage interests as much as their security. As such, and certainly in a paradoxical manner, it stands as guarantee of a share, at minimum, in environmental amenities amongst all the constituents of urban society.

3.4 Conclusion

The rich usually share values of environmental preservation with the managers of the park. In this sense, they are allies of the park in the emergent metropolis. They participate by their proximity to the park, their frequentation and sometimes beyond their investment in participative management committees by constituting and disseminating the social and environmental norms of good practice and the good management of nature in the city on the scale of these metropolises.

Beyond this convergence of ideas, the rich are capable of thoroughly integrating the park into their experience of the emerging metropolis. Be it through environmental amenities, social distinction (by the proximity of their place of residence or by frequentation of touristic hotspots) or economic opportunities, the rich are the principal beneficiaries of the presence of urban national parks, probably even in Mumbai.

Far from halting at a logic of consummation of the park and at the valorisation of its economic potential, especially for tourism, the rich exhibit a capacity to manipulate the existence of the park to their profit. With their historical lobbying for the creation of the park or with their mobilisation for the defence of its spatial integrity, the rich can put all their influence into raising the park as a guardrail against urban growth. In this way, they preserve, as a by-product, a landscape and heritage that is part of the desired exclusive urban identity.

The rich are not, in fact, often genuinely interested other than by the positive externalities of the park in their favour. The inconveniences and restrictions that concern them are seen as being sheer dysfunctions needing to be corrected, rather than as major structural problems caused by the presence of a natural space open to the vicinity of their residence. Management of the wildlife and of social diversity are two flagrant examples. Yet the sharing of values by managers and the rich remains fragile: because of their dominant position, the rich are reluctant to have their legitimacy discussed. Thus, as their part in the mechanisms of co-management is reduced little by little, or a decision is detrimental to their practices, initiatives aimed at fixing their legitimacy in the new spaces will come into being. It is possible here to cite examples of the battle against invasive plants in Cape Town which give rise to animated exchanges between park managers and rich environmentalists: the latter often criticise a management deemed too weak or vague about protected spaces, even though the park develops awareness campaigns for controlling invasive species close to the protected area.

However, the rich may also turn out to be downright harmful to the park by participating in its predation, especially in terms of land, by the installation of buildings within its enclosure. The economic, political and social capital of the rich allows them, in urban contexts scarcely or badly regulated and susceptible to corruption, to be top-level actors in investment where the park is bordered by the city. This is most obvious in Rio de Janeiro, followed by Mumbai.

Urban national parks are for the rich, therefore, both a tool that allows them to consolidate, if not their domination, then their privileged position in the city and a glass ceiling at least, given that the governing of a park with relevance to the national scale tends to reduce their range of influence. Thus the capacity of the rich to influence the policies of the park in their favour tends over time to balance out between shared interests and breaking points, particularly around questions of usage. It is in this precarious balance that room to manoeuvre is to be found in urban national parks for reaching towards social and environmental justice regarding access to nature for all the populations of these metropolises.

As for deciding whether the rich are benevolent neighbours of urban national parks, it must be concluded that the rich can be as much of a nuisance as an asset for these parks. Nor is this ambivalence specific to them since it is also to be found at the other end of the social scale: the case of poor populations is analysed in the following chapter.

Bibliography

Baviskar, A. (2002). The politics of the city. *Seminar, 516*, 40–42. www.indiaseminar.com/2002/516/516%20amita%20baviskar.htm.

Baviskar, A. (2011). Cows, cars and cycle-rickshaws: Bourgeois environmentalism and the battle for Delhi's streets. In A. Baviskar & R. Ray (Eds.), *Elite and everyman: The cultural politics of the Indian middle classes* (pp. 391–418). New Delhi: Routledge.

Charmes, E. (2011). *La ville émiettée. Essai sur la clubbisation de la vie urbaine.* Paris: PUF. English long summary on http://www.metropolitiques.eu/Beyond-dreams-of-village-life.html.

Didier, S., & Swanepoel, J. (2014). A park for all... and by all? Identifying scalar traps in the building of participatory mechanisms for urban parks: Evidence from Cape Town City case, Conference *BiodiverCities. Rescaling Natural Parks and City*, Cape Town, April 7–9.

Donaldson, R., Ferreiras, S., Didiers, S., Rodarys, E., & Swanepoel, J. (2016). Access to the urban national park in Cape City: Where urban and natural environment meet. *Habitat International, 57*, 132–142.

Dooling, S. (2013). Ecological Gentrification: Re-negociating justice in the city. In S. Fol, S. Lehman-Frisch, & M. Morange (Eds.), *Ségrégation et justice spatiale* (pp. 167–184). Nanterre: Presses universitaires de Paris Ouest.

Follmann, A. (2016). The role of environmental activists in governing riverscapes: The case of the Yamuna in Delhi, India. *South Asia Multidisciplinary Academic Journal, 14*. http://samaj.revues.org/4184.

Gadgil, M., & Guha, R. (1995). *Ecology and equity: The use and abuse of nature in contemporary India*. London: Routledge.

Gandy, M. (2006). Urban nature and the ecological imaginary. In N. Heynen, M. Kaika, & E. Swyngedouw (Eds.), *In the nature of cities. Urban political ecology and the politics of urban metabolism* (pp. 62–72). London & New York: Routledge.

Guyot, S. (2011). The eco-frontier paradigm: Rethinking the links between space, nature and politics. *Geopolitics, 16*(3), 675–706.

Guyot, S., Dellier, J., & Caillot, A. (2015). 'Our rural sense of place'. Rurality and strategies of self-segregation in Cape Town Peninsula (South Africa). *Justice spatiale/Spatial Justice, 7* (online).

Guyot, S., Dellier, J., & Cerbelaud, F. (2014). L'environnement au profit des plus riches? Construction et hybridation d'un front écologique métropolitain dans la Péninsule du Cap. *VertigO, 14*(1). http://vertigo.revues.org/14660.

Lézy-Bruno, L. (2014). Usages et conflits sous fond de protection de la nature en ville. Le cas du Parc national de Tijuca à Rio de Janeiro, in Laslaz L. et al., *Les espaces protégés. Entre conflits et acceptation* (pp. 209–227), Berlin, Paris.

Mawdsley, E. (2004). India's middle classes and the environment. *Development and Change, 35*(1), 79–103.

Mawdsley, E., Mehra, D., & Beazley, K. (2009). Nature lovers, picnickers and bourgeois environmentalism. *Economic and Political Weekly, 44*(11), 49–59.

Pape, J. (2003). The struggle against encroachment: Constantia and the defence of white privilege. In D. A. McDonald & J. Pape (Eds.), *The 'New' South Africa in cost recovery and the crisis of service delivery in South Africa* (pp. 123–142). Pretoria: HSRC Press.

Richard, F., Saumon, G., & Tommasi, G. (2015). De l'hégémonie des valeurs environnementales à l'affirmation d'un nouveau capital? Paper for the seminar *Capital environnemental*: *Pratiques, dominations, représentations, appropriations spatiales*, Limoges, November 18–21.

Zérah, M.-H. (2007). Conflict between green space preservation and housing needs: The case of the Sanjay Gandhi National Park in Mumbai. *Cities, 24*(2), 122–132.

Zérah, M.-H., & Landy, F. (2013). Nature and urban citizenship redefined: The case of the National Park in Mumbai. *Geoforum, 46,* 25–33.

Chapter 4
The "Poor", the Park and the City: Policies of Social Stigmatisation Rather Than Inclusion

Frédéric Landy, Pauline Texier and Sylvain Guyot

Academic research was making the connection between the environment and politics well before the arrival of "political ecology". Human geography, for example, was one of the many disciplines that spoke early about dominance relations vis-a-vis the environment between social classes and ethnic groups, though rarely, it is true, between genders (in the field of French tropical geography alone, see Antheaume et al. 1989; Pélissier 1966; Rossi 2001). However, the environment was for a long time no more than a minor component of research, as it was of policies, so there was little emphasis on the issues. Since the 1990s, however, with the triumph of the "environmental paradigm" and the emphasis on ecological priorities at both local and global levels, in both political and scientific debate, dominance relations around the environment have emerged into the light. Perhaps the (deserved) success of political ecology is therefore more linked to the development of ecological awareness and studies focusing on the environment, than on a genuinely fundamental difference in the approaches it proposed (see Laslaz 2016; Baviskar 2008).

F. Landy (✉)
French Institute of Pondicherry, Pondicherry, India
e-mail: frederic.landy@ifpindia.org

F. Landy
University of Paris-Nanterre/UMR LAVUE, Nanterre, France

P. Texier
Department of Geography, Jean Moulin University Lyon 3/UMR EVS, Lyon, France

S. Guyot
University of Bordeaux-Montaigne/UMR Passages, Bordeaux, France

F. Landy (ed.), *From Urban National Parks to Natured Cities in the Global South*,
https://doi.org/10.1007/978-981-10-8462-1_4

This volume demonstrates this clearly enough for there to be no need to labour the point (see Chap. 9 in particular): the urban milieu multiplies the potential for conflict and domination, in terms of both diversity and intensity. The differences between the objectives and interests of the actors are often huge, the socioeconomic and cultural differences between social groups substantial, and the tensions between multiple scales of political power often great. More than ever, therefore, interpretation of the environmental question needs to take into account the role of sociocultural diversity in the emergence of conflicts, conflicts of resource appropriation, and also, more profoundly, conflicts of *Weltanschauungen,* of world views. In particular, the vulnerability of the poorest populations demands that we consider environmental policies that can enjoy real democratic legitimacy.

The countries and cities studied in this volume are characterised by immense poverty. This is not at all incompatible with economic growth. In fact, it is a feature of the "emerging economies" that they retain a considerable mass of poor populations (and spaces), alongside well-off middle classes and chic urban neighbourhoods. For some analysts, this poverty is a legacy that will disappear through economic growth. For others, it is one of the very factors of growth, since it allows cities and countries to be internationally competitive: the fact that a quarter of India's population still earns less than $1.25 a day makes labour cheap, raising uncertainties about whether all the country's decision makers are keen to see poverty vanish. In Mumbai, 42% of the population live in slums (as of 2001, the last credible census): this makes the city's economy run cheaply, since almost half its people have to rely on themselves to find accommodation. It is therefore an abuse of language to speak of social "exclusion" (Saglio-Yatzimirsky and Landy 2013), given that these individuals, who are rarely unemployed, are an integral part and essential cogs in the urban economy and society: we prefer to speak of "marginality" and "domination". The coexistence of these groups and areas, with their great contrasts between wealth and poverty, means that emerging countries and their cities can be spoken of as "iceberg countries" and cities, in which only the tip is visible. In this view, marginals are all those social groups or spaces that remain below the surface.

This chapter begins by outlining the complex and disputed links between poverty and environmental protection/destruction. It then goes on to show how park management policies, though claiming (at least in Rio and Cape Town) affiliation with the inclusive paradigm pursued in the great international directives, continue to maintain processes of exclusion and environmental injustice that are reflected in the marginalisation of poor populations. The case of informal housing is also examined, in order to show how the populations of the four cities considered continue to use natural resources despite the fencing of protected spaces, and yet do not constitute a real factor of environmental deterioration, except in Mumbai. In its final part, the chapter will discuss the management choices in the four national parks, on the spectrum between entrepreneurial management and nationalisation, assessing the extent to which poor populations are allowed access to nature and the management structures.

4.1 Can Poverty Be Defined by the Environment?

Being poor in Mumbai is not the same thing as being poor in Nairobi or in Rio. We therefore need to define what we mean by “poor”, just as the previous chapter briefly discussed the notion of “rich”. We know that defining these categories simply in terms of an income threshold is reductive: it ignores the problems of economic or land insecurity, the question of education, of stigmatisation, whether sociocultural (e.g., Dalits in India, skin colour everywhere) or spatial (having a slum address can be a major factor in discrimination) (Saglio-Yatzmirisky and Landy 2013). Being primarily qualitative, however, our research can be content with a very generic definition specific to our environmental perspective, founded on two components—income and education—that are intrinsically linked, even if this combination of variables is open to discussion.

4.1.1 A Problem of Resources?

We will use the term “poor” to refer to groups and individuals who *may* be prompted by scarcity of income to see the natural urban environment as a resource or a livelihood. The possible implications of this are twofold: it *may* either provide an incentive to protect national parks in order to conserve this vital resource, or conversely a temptation to use them to a maximum as an economic coping strategy, resulting in involuntary damage through overexploitation. Here, the term “may” is used to signify that there is nothing inevitable or universal about these characteristics, and that they often correspond to sometimes imaginary representations on the part of wealthier social groups or park managers. Indeed, the latter often believe, erroneously, that the populations living on the urban fringes near protected areas are invariably responsible for multiple forms of damage, above all because of their ignorance of environmental impacts: they are perceived as unaware of the need to preserve environmental resources and incapable of recognising the wider public interest (for Cape Town, see Guyot et al. 2014). In India, for example, “the consonance between dirt and disorder, pollution and poverty, is a resonant one in higher caste and class consciousness” (Mawdsley 2009, 248). This view of matters stretches back a long way, and in the countries of the global South developed in particular during the period of European colonisation. Even then, however, it was at least as much to do with social as racial prejudices, directed as much at the poor as at indigenous populations: in the same era, Europe was experiencing the enclosure movement, which had a negative impact specifically on poor populations who depended on the commons and on “nature”. Three years before the French conquest of Algeria, the ferocious Forestry Code of 1827 in France prohibited the picking of acorns, even in woodland commons (Dardot and Laval 2014).

This vision of poor populations as dangerous, or at least powerless, explains why they were left out of decision-making processes and resource management plans such as the national parks. The ways in which the poor used resources were seen as based on their perceived ignorance of the universal need to protect nature; measures were therefore recommended to expel and exclude the people living in the areas designated for protection. These strong-arm external political measures, carried out without consultation with the populations concerned, are experienced by the latter as "development aggression" (Heijmans 2001; Oliver-Smith 2001).

In the 1970s, however, this technocratic vision was profoundly challenged by the research community and local actors, leading to the emergence of a new paradigm which, after many years, came to be embodied in big international directives (Hyogo Framework for Action 2005–2015, Millenium Development Goals 2000–2015, Nagoya Protocol 2010 and Aichi Biodiversity Targets 2011–2020). In this view, it is the lack of access rights to resources that drives populations to adopt environmentally dangerous or damaging behaviours, not their lack of environmental awareness: their practices are coping strategies, the outcome of a quest for the least bad options (Davies 1996; Scoones 2009; Texier 2009). "A livelihood is environmentally sustainable when it maintains or enhances the local and global assets on which livelihoods depend, and has net beneficial effects on other livelihoods. A livelihood is socially sustainable which can cope with and recover from stress and shocks, and provide for future generations" (Chambers and Conway 1991, 1) (cf. Benson and Twigg 2007). The emphasis is then placed on the capacity to develop strategies in order to prepare for, cope with and recover from a damaging event; populations are no longer considered to be powerless victims, but as agents endowed with the ability to act (Texier 2009). Policies that fit within this paradigm therefore tend to favour social and economic measures (e.g., reducing social inequalities, combating poverty, etc.).

The same period saw the emergence of a new eco-centric ethic, which argued for a symbiotic form of development that includes the protection of human productions (Rodary et al. 2003; Depraz 2008). In the concept of "integrated" conservation, protection is incorporated into the principles and objectives of sustainable development. The problems arising from the forced evictions carried out on the pretext of protecting nature, common in South Africa (Guyot 2006; Dellier and Guyot 2009), gradually led to a new awareness of the need—in the management of protected areas—to take into account the social, economic and cultural facets of nature as embodied respectively in spaces of habitation, production and identity. Within this new and still sparsely implemented conceptual framework, uses of nature by local populations are accepted without automatic recourse to expulsions, and are considered as potentially protective of biodiversity. If they are not, this fact can be explained by a lack of access to alternative resources.

In consequence, the current practices and habits of the "poor" in relation to nature or risk are to be understood from a holistic temporal and spatial perspective, in terms of trajectories and adaptation to socioenvironmental events. What access did they have to natural resources in the past? What sociopolitical events changed their rights of access? The use of a park's natural resources can thus be the outcome of a lack of access to certain resources not provided by public services (e.g., water), rather than the expression of a failure to perceive the need to protect those resources. From this perspective, the hypothesis of a lack of education seems highly open to challenge.

4.1.2 A Problem of Education?

Educational levels among the “poor” are often low: 10% of the population of Greater Mumbai are illiterate. More specifically, they may have no exposure to the types of ecological values spread through schooling (The political issues of environmental education are discussed in Chap. 13). Here again, however, this way of defining the “poor” as a category is reductive. Low levels of education can be balanced by other channels of access to environmental knowledge. In Rio, for example, environmental education projects run by the voluntary sector in concert with the park’s management structures are bringing specific and appropriate knowledge into the *favelas* adjoining Tijuca Park, drawing on inclusive and participatory methods of resource management. Moreover, in the older *favela* communities, young people have access to collective and intergenerational legacies of natural resource use. These are not strictly speaking conservation practices, with the exception of a few examples, such as in the Horto district in Rio, where former employees of the Jardim Botanico Park living in the area pass on their knowledge to their children. But at the least, they entail smart ways of using resources or managing waste (e.g., recycling, waste limitation, etc.), which can be seen as environmentally beneficial social and cultural practices that compensate for low levels of education or schooling (e.g., Parque Da Cidade). It is recognised that the original inhabitants have a strong emotional connection with the park and protect its resources, unlike newcomers who have moved in for economic reasons and have developed no such relationship. More than ever, “practices” and experience of resource management seem crucial to reinforcing ecological awareness and creating “environmental subjects” (Agrawal 2005). In passing, it might be asked whether this ecological awareness would not develop further if the local populations were more frequently invited to participate in one way or another in the governance, or even the management, of the national parks studied. This explains why, in Cape Town, initiatives of this kind are being backed by the national park (see Chap. 13).

Criticising the deep ecology movement, Guha (1989) noted that it “believes that environmental preservation is a ‘full stomach’ phenomenon that is confined to the rich, urban, and sophisticated”. Yet in the countries of the global South, poor populations often have “a finely balanced relationship with nature” precisely because of their dependence on it, and they are not the last to mobilise to protect the environment. Guha explains: “Two features distinguish these environmental movements from their Western counterparts. First, for the sections of society most critically affected by environmental degradation—poor and landless peasants, women, and tribals—it is a question of sheer survival, not of enhancing the quality of life. Second, and as a consequence, the environmental solutions they articulate deeply involve questions of equity as well as economic and political redistribution. Highlighting these differences, a leading Indian environmentalist stresses that 'environmental protection per se is of least concern to most of these groups. Their main concern is about the use of the environment and who should benefit from it'" (p. 81).

The two points of view—deep ecology and Guha's style of social ecology—agree on at least one point: environmentalist approaches in the rich countries and in the poorer social classes of the South are different. Guha does not say that the poor have the same conceptions of environmental protection as the rich, just that they have conceptions. A sharper opposition characterises the sense of the link between poverty and environment. For not only can poverty have an impact on the environment (positive or negative), but the environment can also affect poverty. Living near a national park can have positive day-to-day consequences, whether on health or on potential income from tourism. This should not blind us to the fact that in most cases, the creation of a national park deprived local populations of resources and habitat. For the adherents of political ecology, the creation and sequestration of protected spaces usually corresponds to a real transfer of resources from the poor to the rich. There is undoubtedly "spatial" and environmental injustice (Landy et al. 2017), for example in the creation and subsequent expansion of Mumbai park: even if it may be considered that conservation priorities could legitimately restrict the rights of local populations, often Adivasi ("tribals"), it was not necessary to try to expel them all (a goal never, in fact, fully achieved) or to prohibit absolutely all activity. Can this initial injustice be remedied today by an environmental policy of spatial justice? As we will see, this is notably one of the components of the post-apartheid rhetoric on facilitating the access of black populations to Table Mountain National Park (TMNP).

4.1.3 Poverty and National Parks: Spatial Promiscuity

By contrast with the situation in many big cities in the countries of the global North, where the environs of urban parks are reserved for the homes of a social elite (e.g., Central Park in New York, Bois de Boulogne in Paris, etc.), our four parks are characterised by small distances (sometimes even overlaps!) between protected spaces and poor areas. In Nairobi, the Bangladesh slum adjoins the national park to the south-west. And the biggest slum in Kibera (though probably not more than 250,000 inhabitants, despite the legend) (Desgroppes and Taupin 2011) is located a few hundred metres north of the national park. In Cape Town, the peninsula is riddled with villages and urban neighbourhoods, both poor and rich (Guyot et al. 2015). In particular, there are two formal coloured townships in Hangberg (Hout Bay) and Ocean View (Kommetjie), and three squatter camps populated by Africans ("local" and African migrants): Imizamo Yethu (Hout Bay), Masiphumele (Kommetjie-Noordhoek) and Red Hill (Simons Town). These five formal and informal districts are essentially home to populations that are modest in means or

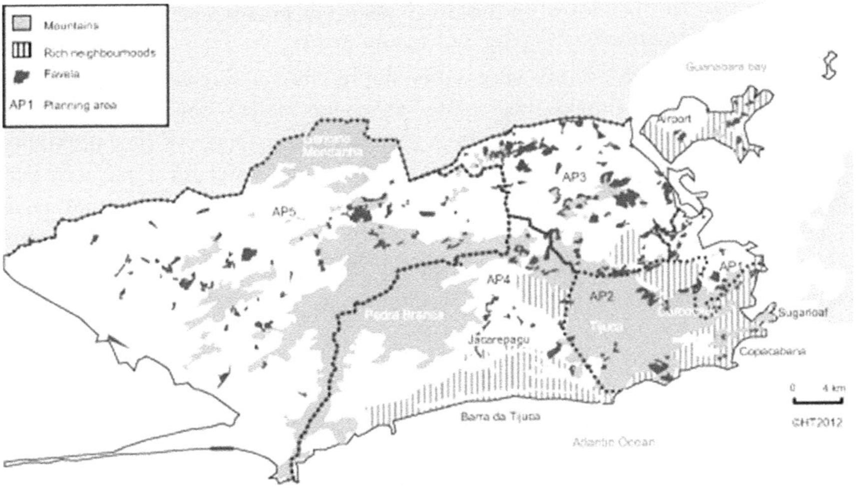

Fig. 4.1 Location of the *favelas* surrounding the protected areas of Tijuca and other urban parks in Rio de Janeiro. *Source* Google Earth collection, zoning by photographic interpretation

very poor (e.g., recurrent cholera epidemics). They have the particularity of adjoining the boundaries of the national park, and sometimes overlapping (very modestly in the case of Hangberg, associated with cases of abalone poaching in the protected marine area). In terms of spatial extension, however, they bear no comparison with the chic neighbourhoods that take up almost all of the interface between the park and the city, and cause more day-to-day problems for the park authorities than the poorer districts—invasive plant species, security, access to the mountains or the beach and, in particular, mountain bikes and dogs (interview with M. Slayen). Organic pollution mainly occurs downstream from these poor neighbourhoods and is therefore a matter for Cape Town's municipal management.

In Rio de Janeiro, the *favelas* tend to be located high up, on the steep slopes of the *morros* (hills), and therefore surround the parks of Tijuca, Pedra Branca, the Jardim Botanico or Parque da Cidade, given that these are to some degree defined by the topography (Fig. 4.1).

It is also the topography that accounts for the existence of Mumbai national park, by enabling it to survive urban growth. However, the expansion of this park has brought it dangerously close to the urban frontline, itself very active, with the result that by the late 1980s it had begun to experience the incursion of slums. These were home to more than 500,000 people in 1995, the year in which an environmental organisation, the Bombay Environmental Action Group, lodged a complaint against the Forest Department (with the latter's discreet blessing) in condemnation of these invasions. In 2016, there were probably more than 100,000 people still living within the boundaries of the park. The Adivasi villages are small in number compared with the mass of the slums, which are concentrated on the southern edges, as these are the closest to the urban centre of gravity and are therefore the most attractive (Saglio-Yatzimirsky and Landy 2013).

Does this spatial proximity necessarily imply close interrelations between the poor and the park? This is what many environmentalists object to, citing the environmental deterioration, or at least the risks, generated by this proximity. However, we should be wary of blindly accepting such a claim. First, it is very often a front for ulterior motives, dreams of socio-spatial discrimination that would enable the rich populations to live together, or quite simply strategies of real estate development. Second, there is no evidence that poor populations necessarily damage the environment, or indeed that they do so more than the rich.

4.2 From Conservation to Environmental (in)Justice

4.2.1 Parks as an Instrument of Justice and Spatial Inclusion

Almost everywhere in the world, the spread of the paradigms of sustainable development and integrated conservation has led to sometimes quite profound changes in the objectives and "governance" of national parks (see Chap. 10 and Introduction). These paradigms have clearly been transcribed into local conditions in varying ways, since they differ in terms of the actors involved, the conservation priorities and the pressures on the park, etc. However, it can be said that in response to urban demands, all the four national parks have undergone—with varying degrees of willingness—an increase in the flexibility of their management, in the direction of greater participation by at least a section of "civil society".

In Brazil, the 1988 Constitution provides for consultative committees to be established even in the national parks. So Tijuca park, where a management priority is "to promote participation of civil society", opened its consultative committee to civil society in 2008, notably through the Friends of Tijuca Park Association. Inhabitants of the neighbourhoods local to the park (notably in Horto) participate in this committee, though they are not genuinely poor populations, but rather community leaders and neighbourhood "representatives".

In Kenya, it is the recent decentralisation which—in granting more power to the counties—may ultimately contribute to the needs of poor populations being incorporated into the management of the protected areas (see Chap. 9). In the case of Nairobi National Park (NNP), there was already a Community Warden responsible for relations with the city and the rural areas south of the park. The rich neighbourhoods of Karen-Langata, as well as the most powerful of the Maasai, have the ear of the Kenya Wildlife Service (KWS) administration, but many rangers try not to give them too much weight at the expense of the needs of the poor populations.

In India, to the great displeasure of many environmentalists, even the national parks are covered by the Forest Rights Act of 2006, which allows people to acquire up to four hectares of land in a protected area: this applies to tribals, but also to any "other traditional forest dweller", provided that they have been on the land for 75 years. This law is only applicable to rural areas, so only a small part of Sanjay Gandhi National Park (SGNP), north of Vasai Creek, is affected. Nonetheless, numerous applications have already been entered under the provisions of this Act.

In South Africa, finally, the aim is to use a "social ecology", in the words of South African National Parks (SANParks), to generate "social integration" and erase the traces of apartheid.[1] Since 2003, advisory boards have been compulsory, in the effort to avoid conflicts with the local populations. In TMNP, which is located in an area designated "white" in the 1960s, public works programmes designed to give employment to the poor have been used to develop the park, remove non-native plant species and also to familiarise these populations with the park, in order to encourage them to buy entry passes at heavily discounted prices. Coaches sometimes pick up children from school to take them to the park for certain classes that are part of the curriculum: lessons in ecology, combined with a strong civic education component, since the park is presented as a national asset that belongs to all citizens of South Africa. Ultimately, the park is seen as a symbol of democratic transition, a vehicle of environmental and social justice (Belaidi 2015).

The coloureds who lived in the Peninsular (and still live in the two townships cited above) do not have the same links to the park territory as the Africans, of whom only a tiny minority lived in the environs of Simons Town. These coloured populations use the park in ways that are often tied to family legacies. However, it is obvious that the "racial" aspect of the TMNP slogan, "A park for all for ever", tends to mask the social component of the problem (Donaldson et al. 2016). Among blacks and coloureds, it is the richest who most frequently visit the park. A study conducted in Johannesburg (Butler and Richardson 2013) showed that half the people interviewed in Soweto had never been into a national park. True, Table Mountain is a particular kind of national park, without significant wildlife and close to the city. Nonetheless, it is not hard to guess the level of park use by the inhabitants of Khayelitsha, Cape Town's enormous slum, much poorer than Soweto. In addition to marginality in terms of distance and cost, there is the factor of cultural marginality.[2] It is perhaps quite revealing that one of the few examples of active collaboration between the city and SANParks is a forum on security in the park, which includes certain city dwellers but also members of the Police. There is a high incidence of rape and a temptation to stigmatise certain types of poor populations.

---

[1]Nadia Belaidi and Sophie Didier, papers at the UNPEC workshop, lab Mosaiques-Lavue, University of Paris-Nanterre, 27.9.2013.

[2]Likewise, it is known that very few Latinos visit Los Angeles' Santa Monica Mountains National Recreation Area (Byrne et al. 2009).

In Rio, a large majority of the *favela* inhabitants questioned had never been to Tijuca Park, for reasons of geographical distance and cost: the trip involves several bus rides and takes at least 90 min, unless you enter via Guararapes, the starting point for the rack railway or for cars with a tourist guide, which is too expensive. On the other hand, according to C. Junius, director of the Mosaico Carioca programme, which aims to complete a green corridor connecting all the protected areas of Rio, the poor are more likely to visit nearby natural areas, where entrance is free: for example, Parque Da Cidade for the inhabitants of the *favela* of the same name, or the Babylonia forest zone near Chapeu Mangueira *favela*, which is a private forest, although maintained with the help of the local population (e.g., a replanting programme based on paid community work and managed by Rio municipality). As with the other parks, Tijuca reveals discrimination against poor populations in access to nature. And the effect of the efforts being made to develop tourism in the park is not to make access more democratic: paid entry, costly activities reserved for an elite, compulsory guides and prohibition of barbecues and picnics.

4.2.2 *Parks, Factors of Injustice and Marginalisation*

4.2.2.1 The Original Sin of Expulsion and Its Consequences

A first injustice is embedded in the very genesis of the parks. Indeed, in the case of three of them, it is as if they remain marked by an original sin that attended their creation: the removal of their poor populations (for its part, the future Tijuca Park was created in the nineteenth century, with the expulsion of the relatively well-off population of planters). Far from establishing a certain level of justice, they often help to maintain, or even reinforce, the marginality of the poor. The typical cases are of two kinds:

- marginalisation is an *indirect* consequence of conservation policies. This is generally the case when a national park is created or extended: amongst the populations displaced, the poorest are obviously the least mobile and find it hardest to obtain new homes and sources of revenue. Chap. 7, which focuses on the—globally poor—indigenous populations, shows the difficulties that conservation policies have created for them, not only historically but still today. Nor should we forget that any park inherently places limits on intra-urban mobility. In Rio, for example, Tijuca forms a barrier that tends to isolate the city's broadly working-class western and northern districts from the more central and middle-class south and east, thereby introducing a degree of social segmentation;
- environmental conservation can, however, be *directly* exploited for the real purpose of social exclusion. "In the context of the Cape Peninsula, rurality and urban sprawl result in a conflict of representations between several forms of intensive house building in the outlying suburbs, for non-white and poor people

versus white and well-off people, which invariably refers back to the legacies of apartheid urban policies” (Guyot et al. 2015). True, opposition to the national park and SANParks often comes from the well-off populations of the peninsula, who alternately accuse them of laxity on matters such as urbanisation and invasive vegetation, or of excessive ecological zeal when the administration rejects household projects. However, the existence of TMNP and the type of environment that it maintains are used, in the discourse of the well-off populations, to try to prevent suburban expansion—especially when it is black and poor—ostensibly on the grounds of protecting nature and rurality. Similar processes are found in Rio, where *favelas* are demolished in the name of “modernity” and hygiene—or to coincide with big sporting events. These demolitions are not simply physical: they also smash social structures and disrupt the coping strategies of the populations, which lose their social networks and capital following expulsion. Even in Nairobi, the GreenLine project to plant trees and create a green corridor along the boundary of the park can be seen as having little to do with the needs of the poor. The picnics included in the project may perhaps have something to offer, but what about the jogging route that seems likely to be largely used by the well-off populations?

4.2.2.2 The Poor as Destroyers of Nature? Reality of the Practices and Their Impacts

A second injustice towards the poor is perpetrated through words and stigmatisation. Park management principles foster certain practices and criminalise others, particularly those originating in lower-class populations. Often, they simply reflect the widespread perceptions of the middle and upper classes, who see the poor as beings with no environmental awareness or hygiene (cf. the notorious *gande bachche*, the “dirty children” condemned by the middle classes in India, or the views expressed by a white municipal councillor in Cape Town Peninsula: “toilets are not part of their culture, those people prefer to pee and poo in the bushes, in the mountain”). In Rio, similar attitudes place a stigma on the *favelas* in the eyes of visitors to Tijuca Park Museum (Fig. 4.2).

Are the poor as destructive of nature as is claimed by the well-off classes and park managers? The surveys we carried out in the *favelas* located on the edges of Tijuca Park showed that their inhabitants perceived nature as a resource that is necessary both to their well-being and survival, and were aware of the need to preserve it. Moreover, as we will see, their day-to-day practices relating to Tijuca Park and its natural resources were limited, with no serious impact on biodiversity or environmental quality.

Among the 100 or so *favelas* near the boundaries of Tijuca Park, to the south-west, is Parque Da Cidade, which has the particularity of bordering a municipal park of the same name (Fig. 4.3). This neighbourhood is made up of 342 homes (in reality more like 2000) plus 14 households located in the municipal park

Fig. 4.2 Exhibition panel at Tijuca Park Museum. *Source* Photo by P. Texier, August 2012

(mainly park employees or their families who have lived there for 60 years, whom the government has been unable to persuade to leave): a total of 6000 people. The authorities accuse them of a variety of practices: picking fruit and plants for their own consumption or for sale, allowing their pets to foul the environment during daily walks, parking cars, illegally building houses in the park, criminality (since 2005, two people have in fact been murdered in the municipal park), fly tipping and drawing spring water from the park. In fact, our interviews and direct observations showed that the fruit picking was trivial (confined to jackfruit, mangoes and melons, which are invasive, non-native species), was practised by just a minority of people for their own personal consumption, that leisure activities had no impact, since the visitors invariably remained on the marked tracks and that the parking took place in a designated and permitted zone at the entrance to the park, whereas the cars travelling through the park in the direction of Tijuca belonged not to the inhabitants but to outside visitors (since most of them never went to Tijuca). The presence of dog faeces is indeed a problem, partly dealt with by the few employees occupied in cleaning the park. However, it is collected on a daily basis, and has no real impact on the environment. No houses are now built within the boundaries of the park, whereas the 14 old houses that accommodate the families of public servants working for the park are substantial; these households are not among the poorest, and in fact enjoy more comfortable living conditions than in the *favela*, as well as an exceptional natural landscape. There is nothing that would persuade these families to leave their neighbourhood. According to the local people we spoke

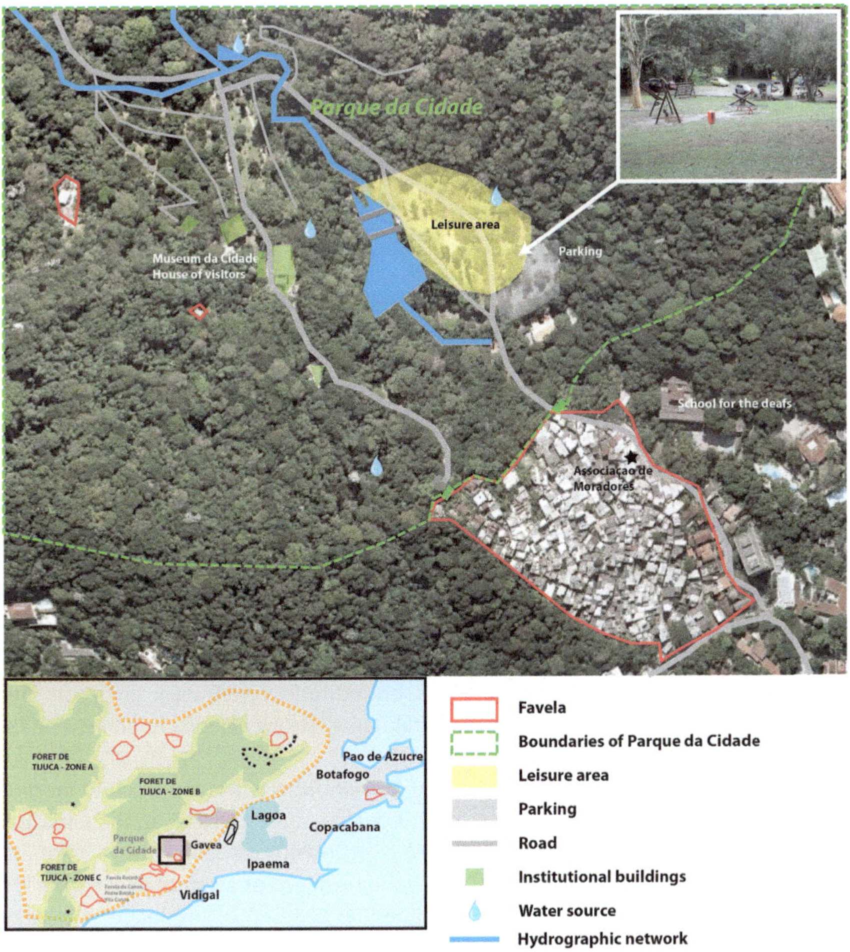

Fig. 4.3 Layout and zoning map of Parque Da Cidade *favela*. *Source* Photo by P. Texier, 2012

to, the criminality of the past has reduced, and the enquiry conducted after the two murders had concluded that it was a double homicide perpetrated by somebody from outside the district.

The only activities that may be considered to have an impact on natural resources can be interpreted as strategies for coping with a lack of access to resources, such as the absence of a drinking water system. As a result, populations have no choice other than to construct a system of pipes to bring water from the park—which in fact does no harm (Fig. 4.4a). These families are unfairly accused of damaging nature, whereas it would seem logical to inculpate the government for failing to provide access to water. Moreover, some areas were only connected to the network in 1983, prior to which, since the 1960s, the inhabitants had built a

Fig. 4.4 Informal pipe system bringing water from the park (**a**); accumulated waste at the bottom of the Parque Da Cidade district (**b**). *Source* Photos by P. Texier, August 2012

network of pipes to connect the houses to springs in the park, and have therefore naturally continued to pursue the same strategy. Formiga *favela* has its own "water companies", alternative networks using water captured from springs in the park. The inhabitants also receive public service water, but these alternative networks play a role that goes beyond providing a service: they influence the forms of organisation of local life. These companies are managed as community entities, with no profit motive: they are just one of multiple community environmental projects in this *favela*, which shows the inhabitants' strong interest in preserving the resources of the park.

Similarly, the temporary storage area for household waste is too small to meet all the needs of the Parque Da Cidade district. Located at the entrance to the park, it is often marked by an accumulation of refuse overflowing its containers, a sight and smell that are completely incompatible with the presence of a sumptuous park a few metres away (Fig. 4.4b). But can the families be accused of producing too much waste, given that they reuse all their recipients and sort their plastics? Or is it the municipal collection services that are to blame, for failing to collect often enough or to distribute containers of sufficient capacity for the demography of the district—in circumstances of a steadily rising population?

Finally, the surveys showed that most of the inhabitants of this *favela* maintain a particular relation to the nature around them: it is important for them, for their day-to-day well-being, their health and their children; it is an essential amenity, reflecting a real attachment to their home territory: "nature means everything, if there was no nature in my life…" (Woltern, aged 45, a 25-year resident). Many go to relax, to walk in Parque da Cidade or visit the picnic areas and playgrounds. They appreciate the proximity of nature: "Yes it is a very good thing to live here, it's beautiful, pleasant, to be close to nature: it is a fantastic thing, with the birds, the trees, it's very beautiful, the air is pure…But I don't use the plants" (Irene, aged 72, a 50-year resident, who thinks that the *favelas* are more an "aid" to the forest than a threat). They would like to go to Tijuca more often, if the park were accessible.

The park guards we spoke to (local agents, not high officials) noted the good behaviour of the *favela* inhabitants when they come to relax, and specify that they take part in community work ("*mutirao*") to keep the park clean.

Yet Afro-Brazilian religious celebrations, together with certain evangelical services, are often criminalised and forbidden in the national park (e.g., risk of fire from candles, food residues that disrupt animal diets, etc.). However, the rhetoric condemning the environmental practices of the poor is far from reality. Likewise in Mumbai, the slum dwellers who have cleared the inner boundaries of the park cause scarcely any damage beyond them, since they are afraid of the forest and of leopards and rarely venture there. As for Nairobi, our surveys in the Kibera shantytown produced very contrasting accounts, ranging from a love of nature to fear.[3] One man said to us: "There are sometimes so many monkeys here, they are like schoolboys. They are our neighbours! I don't close my door, I play [*imitating monkeys*], they are not harmful, they just want to do what you do, eat what you eat, etc. I give them bananas. I appreciate this place: in the town centre it is a forest of stones, whereas here there is shade, we can eat bananas free, have free water, fetch firewood, whereas in town everything must be purchased." But another couple did not agree: "The park should be locked, fully fenced, and animals should be fed inside. They are a danger for the community, they are a threat. Once a lioness killed a goat in our cemetery. After goats they will attack humans! There are not enough rangers. And sometimes power cuts occur in the fence [bordering the park], which may also be broken by street kids."

4.2.3 How Far Should Informal Housing Be Accepted in the Parks? The Limits of the Integrative Paradigm

In the world of social science research, the dominant position is one of tolerance towards shantytowns, which represent a city built by its inhabitants with their own hands, with no help from the authorities, which are thus saved a great deal of investment. The blame lies with the failures of urban policies, which have resulted in a shortage of housing for low-income populations. Researchers therefore tend to favour the legalisation of these neighbourhoods, or at least their recognition by the authorities in the form of basic public services (e.g., electricity, toilets, schools, etc.) (Zérah et al. 2011). In Cape Town, there are a few squatter camps within the boundaries of the park at Hangberg, but they seem to cause no major problem to the authorities. The case of Tijuca in Rio is emblematic of the need to legitimise the poor populations of the nearby *favelas*, which need the state to provide essential services that will prevent them employing coping strategies that involve the natural resources of the urban parks. The solution is, above all, political and technical.

[3]Interviews, July 19, 2014.

However, not all national parks are the same: in some cases questions need to be raised, even if it means playing devil's advocate. Of course, it is impossible not to be scandalised by the situation of the Bangladesh slum in Nairobi, which is so close to the national park that accidents caused by the electric fence are frequent—a charge designed to deter a rhinoceros can be fatal to a child. At the very least, a wall should be built to isolate the neighbourhood from the fence. In our view, a different attitude is needed to the slums of Mumbai. Can a national park accommodate more than 100,000 people (not to mention the 500,000 who lived there in 1995)? Isn't there a risk that the park will be irremediably damaged, not necessarily by the practices of the inhabitants (if basic amenities such as water, toilets or energy are someday provided), but simply by the loss of land in the protected area? Some 200 hectares were occupied in 1995. Then there is the potential knock-on effect: given the housing pressure in Mumbai and the efficiency of certain mafias in profiting financially or electorally from the distress of households, other slums could quickly spring up. It has to be said: yes, in this case, poverty has had a negative impact on conservation. Here, perhaps, we reach the limits of the critique of environmental injustice. On the other hand, two points need to be specified: first, the problem in Mumbai was exacerbated by the sudden expansion of the national park, which brought it dangerously close to the fabric of the city. If this expansion had not taken place, the slums would have grown in more urban areas. Second, and above all, it needs to be emphasised that the problem of housing is caused by the failure of upstream urban policies. It is they who deserve condemnation, rather than the "poor".[4] Poverty in itself is not the cause of the deterioration of the national park: both are attributable to inadequate urban management.

4.3 Between Entrepreneurial Management and Nationalisation of Nature: What Place for the Poorest in the Tourist Economy?

The "nationalisation" of nature embodied in the creation of national parks was, as we have seen, hardly favourable to the poor. This is a fact regularly cited by advocates of a more entrepreneurial form of park management, who are happy to argue that disadvantaged populations have little to lose through a change in the management models and objectives of protected areas. This does not necessarily mean full privatisation, but often a "contamination" of public conservation institutions by private sector goals (profit generation), on the grounds of financial autonomy and economic efficiency: hence the rise of subcontracting, the casualisation of labour and a short-term cost-benefit approach.

---

[4]It should be recalled that not all of India's slum dwellers are poor, and that many poor city dwellers in India do not live in neighbourhoods considered to be slums (Saglio-Yatzimirsky and Landy 2013).

In this approach, there is a strong temptation to favour, for example, upscale tourism which will bring in new revenues, but is likely to restrict access to the park for the poorest populations, or indeed to reinforce negative representations of their practices. This new form of management drives long-term elitist trends in the park and runs counter to any hope of real appropriation by the most disadvantaged. The park would no longer be seen as a collective heritage, but as a state asset managed by private interests, with nature viewed as a commodity. Privatisation, public-private partnerships, outsourcing and subcontracting: what public service missions can remain if the terms of reference prove lax in this respect?

In Tijuca, the concession granted to the Paineiras-Corcovado area has reinforced elitism in the use of this sector and limited tourist business for residents of the neighbouring *favelas*, generating certain conflicts. For example, the desire to give tourist guides official status by requiring them to receive training and sign a charter can boost their profession by making it more serious and attractive to international visitors, but can also discriminate against the young, informal guides who have no schooling and may not be able to do the training. Similarly, the quality criteria for activities conducted within the park have become stricter, and businesses now have to pay a fairly high tax for permission to operate, which has already driven out a local restaurant owner and prevents the small informal businesses carried on by the residents of the *favelas* in big tourist areas like Corcovado. The park now operates by issuing calls for tenders from private companies, which must provide an elaborate technical project corresponding to the multiple economic and technical criteria imposed. While these projects are theoretically open to all, in reality only private contractors of significant size are eligible to tender, whereas poorer inhabitants wishing to open a small business lack the resources to pay the amount required for projects (40 million reals). In any case, in the absence of training, they will fail to meet certain quality criteria, such as having the ability to speak foreign languages as guides, or knowledge of hygiene in the food sector. For the moment, there are no projects designed to develop the capacities of the local populations so that they can take advantage of this tourist market, leaving them marginalised (interview E. Botelho, TNP, August 2012).

Only a few exceptional initiatives show that the populations of the *favelas* possess skills that they can exploit for micro-projects in ecotourism, a long way from the dominant paradigm of mass tourism. One such example is the ecotourism project in Vale Encantado (Enchanted Valley): located inside the forest of Tijuca (but outside the national park), the community of natives from Vale Encantado created a cooperative in 2006, successively pursuing a number of economic ventures (e.g., coffee, horticulture, stone extraction, and, since the Earth Summit in Rio in 1992, reforestation and tourism). This cooperative aims to develop community scale ecotourism, such as discovery trekking (flora, fauna and history), or local organic food (Fig. 4.5), in order to generate employment and income, limit daily commuting and encourage people to protect their environment through appropriate practices such as the development of water sanitation projects. This ecotourism project shows how it is possible to enhance local knowledge on environmental protection and to optimise resources while involving people in nature conservation.

(a) (b)

Fig. 4.5 A discovery trek on the mining history and endemic plants of the Enchanted Valley (**a**); community kitchen that employs three women, cooking with local produce (**b**). *Source* Photos by P. Texier, August 2012

It would appear to constitute a major argument for keeping people in the area, despite the pressure coming from government and economic stakeholders. This type of sustainable project brings environmental, social and economic benefits, and its only weakness is a lack of institutional support to attract a continuous influx of tourists.

In Nairobi, while management of the park itself remains in the hands of the Kenya Wildlife Service (KWS), the dispersal area south of the park, so necessary to wildlife migration, is increasingly managed in consultation between KWS and the owners, in the perhaps somewhat vain hope of slowing down the process of plot fragmentation and fencing. This embryonic joint management does not specifically concern the poor: in fact, it is an elite that sits on the management committee formed by the KWS in 2014 (e.g., major landowners, businessmen, etc.). On the margins, however, the poor are perhaps able to make their voices heard and even benefit, notably through support for the *conservancies* gradually being created in the south of the park. These conservancies are genuine shared assets, in the sense that these avatars of the "group ranches" formed after Independence seek to maintain hundreds of hectares undivided within a family clan. The Oleroi Conservancy, for example, involves 3500 ha and 800 adults: the properties are officially private, with deeds dating back to 1984, but a memorandum of understanding has been signed between the members of the community for the purpose of collective management. The objectives of the two projects currently underway in the dispersal area are to combine ecotourism and conservation, and pastoralism and free wildlife migration. Will these conservancies be able to work despite land speculation, such a big temptation for some members of the community who dream of selling their shares? Will support by the KWS and the state—more, it would seem, through regulation than through subsidies—and earnings from tourism be enough to make the project viable? And will not any profits be captured by the most influential members of the community?

In Nairobi, it costs 500 shillings—almost 5 euros—for an entrance ticket to the national park: social selection is already at work. Compare this with Mumbai, where entrance costs only 30 rupees (less than 0.5 euro), whereas a daily wage labourer in the city earns 200–300 rupees a day. Affordability in India is the upside of the still strongly top-down state management of Sanjay Gandhi National Park (SGNP): the Nehruvian model persists, with its claim, at least in theory, of nationalising resources in the public interest, i.e. including the poor. No private enterprise has a concession in the park and everything is managed by the Forest Department, with the recent exception of bicycle hire, which has been conceded to local “tribals” by the Mumbaikars for SGNP association. That being said, the model is perhaps in the process of being challenged, on the grounds of conservation itself. The management of the national park, but also members of Mumbaikars for SGNP, are now adopting the same language as Friends of Nairobi National Park (FoNNaP) or GreenLine: undoubtedly, the poor must have access to the park, but it is not they who will be able to defend its existence, given that they are politically and economically too weak. It is better to rely on the rich, who have political contacts, who have financial clout, and who, if attached to the park, will be better able to protect it against uncontrolled urbanisation (see previous chapter). SGNP is therefore in the process of converting its recreation area, in a sense with the aim of making it less popular: there are plans to limit attractions such as pedal boats on the lake and the small train, in order to reduce the “theme park” aspect and to make it a place of initiation into the environment and the landscape. This could attract better educated and wealthier visitors from India’s increasingly influential middle class, whose attention is so important, even if it means further exclusion of the poor.

Finally, in Cape Town, with the exception of the souvenir sellers located near the Cape Point gates or Masiphumelele craft market, the poorest populations seem to be absent from the environs of the park and to obtain very little direct benefit from tourist flows. On the other hand, SANPark pursues a fairly dynamic policy of providing employment for the poor, whether for work in the park or in the implementation of different projects.

4.4 Conclusion

The path is therefore narrow on the one hand, official policy in favour of facilitating access to the park for disadvantaged populations, or even using the park as a tool of social and environmental justice (Table Mountain); on the other, processes that tend in the opposite direction, either because the pressure is too strong (the slums of Mumbai), or because of a growing trend towards entrepreneurial management (Tijuca), or because a certain social elite is seen as better able to defend the park (Nairobi and Mumbai). All this despite the fact that our surveys, notably in Rio, show that poor populations are not in themselves a factor in the deterioration of the park. What should we conclude from this with regard to the environmental awareness of the poor? Is it only possible to pursue sustainable practices and hold environmentalist representations on a full stomach, to employ Guha's formulation?

First, it is not true that the poor use the national park and more generally nature itself extensively as a source of livelihood. One major exception is of course the indigenous populations, such as the Maasai or Adivasi (Chap. 7), but for the rest it would seem that poor city dwellers tend rather to turn their back on the resources of the national park, often of course because those resources are prohibited, but also because they do not match their needs. In Mumbai, slum dwellers are afraid of the park, and often go there only to fulfil certain natural needs, not to fetch firewood (gas is subsidised in India) or water (there are a few public hand pumps). The park is used as building land, but not as a day-to-day private or commercial resource. In Tijuca, we have seen how parsimonious the inhabitants are in their use of park resources. In Nairobi, it is true that the inhabitants of Kibera often use firewood gathered illegally in the neighbouring Ngong Road Forest and sold to them by neighbours. There are also reports of street children who get into the national park to steal wood or wire from the electric fence, but there is no evidence that the poachers come particularly from the shantytowns.

Conversely, it seems fairly clear that environmental awareness mirrors a class culture that corresponds to the national park model (i.e. fencing to protect nature and to attract visitors for leisure and educational purposes). The original model comes straight from the United States, and more generally from what in the postcolonial approach is referred to as "the West": however, the westernisation of practices and cultures is socially highly selective in emerging countries, and it is therefore certain that the middle and upper classes adhere more readily to public park objectives and procedures than do the lower classes. This is one of the aims of environmental education, whose highly political aspects are highlighted in Chap. 12.

However, diagnosis is particularly complex given that while the idea of the national park may seem to epitomise modernity and efforts of purification through the separation of nature and culture, today's ecological consciousness is increasingly driven by a postmodern vision that challenges this concept of purification as in fact synonymous with energy waste and extreme urbanisation. In this ecological perspective, the tendency is to seek to rearrange national parks by breaking this dichotomy and bringing nature conservation more into line with social priorities. However, the poor are mostly very "modern". In Cape Town, an activist working for an NGO that trains people in gardening and environmental education told us that some migrants living in the poor Gugulethu township had refused to take part in its activities, arguing: "Hey, if I left Eastern Cape for Cape Town, it's not for gardening!" (interview, April 4, 2014). The clearest case is India: sometimes newly arrived in the city as part of the rural exodus, the poor are often proud to live in the concrete jungle rather than in the natural jungle, so to speak. They are often happy at no longer having to frequent the natural spaces of the forest and common land where their fathers in impoverished villages used to graze their animals or fetch wood. Their cultural roots remain too close to the rural world for them to miss it, to visit it, to exaggerate or mythologise it. We love nothing so much as that which we no longer have, because loss creates desire. This sense of loss does not exist yet amongst most poor people, whereas it is present among some of the rich (cf. previous chapter).

To finish, in comparing this chapter with the one that precedes it, one may wonder whether the issue of the poor is not akin to a photographic negative of the issue of the rich. It is true that the questions seem similar. However, a systematic analysis of the two reveals significant nuances, at least on three points.

The first of these concerns the opposition between "nature as livelihood" (for the poor) and "nature as recreation and conservation" (for the rich). This would seem too stereotypical a view. As we have seen, with the exception of indigenous populations, the poor scarcely use protected areas as a source of economic resources, whereas many rich households earn their money from conservation or tourism. Moreover, it is not true that the poor almost never use national parks for leisure: if not the poorest, at least the least poor of the poor sometimes take a pedal boat ride in the recreation area of Mumbai park, or hold a *brai* (barbecue) with family or friends in TMNP. But it is true that in this case, the focus is usually on nature as an environment conducive to leisure and not on nature as biodiversity in action.

The same is true for the land issue. There is no doubt that in Mumbai it is the slums that pose the greatest threat to the perimeter of the park, and have even penetrated it already. However, middle-class houses or chic hotels have not been slow to make incursions into no-build areas either (e.g., Royal Palms, a housing estate built illegally on a golf course). The wall built to protect the national park is also sometimes interrupted to allow an apartment building or temple to be inserted into protected space.

Let us conclude with representations. It is probably true that ecological awareness, in the Western sense, is fairly rare among the poor, but Chap. 7 shows how the holistic view of humanity as a simple component of a vast ecosystem is shared by indigenous populations and middle-class environmental activists. Moreover, as the previous chapter showed, environmentalist sentiment is not necessarily dominant in many rich households, while those who profess to love the national park do not necessarily demonstrate, among themselves, practices that are particularly environmentally friendly, whether in the park or in their day-to-day lives. Many, for example, show their appreciation of Nairobi National Park above all as a venue for corporate events or wedding celebrations.

Ultimately, the simplistic distinction between "rich" and "poor" is less than ever justified in terms of relations to the environment: it is a distinction that needs to be investigated, since it is present in many accounts, but only in order to deconstruct it more effectively.

References

Agrawal, A. (2005). *Environmentality: Technologies of government and the making of subjects*. Durham; London: Duke University Press.

Antheaume, B., Blanc-Pamard, C., Chaleard, J.-L., et al. (1989). *Tropiques, lieux et liens*. Paris: ORSTOM.

Baviskar, A. (2008). Introduction. In A. Baviskar (ed.), *Contested grounds. Essays on nature, culture and power* (pp. 1–12). Oxford University Press.

Belaidi, N. (2015). Théorie du droit et front écologique: apport à la (re)définition de la justice environnementale. *Développement durable et territoires, 6*, 1. https://developpementdurable.revues.org/10806.

Benson, C., & Twigg, J. (2007). *Tools for mainstreaming disaster risk reduction: Guidance notes for development organizations*. Genève: Provention Consortium.

Butler, G., & Richardson, S. (2013). *Observing the social and economic barriers to visiting South African national parks: A study of residents from Soweto*. University of Johannesburg-STH-SANParks.

Byrne, J. A., Wolch, J., & Zhang, J. (2009). Planning for environmental justice in an urban national park. *Journal of Environmental Planning and Management*, 38 *52*(3): 365–392.

Chambers, R., & Conway, G. R. (1991). Sustainable rural livelihoods: Practical concepts for the 21st century. *IDS Discussion paper*, p. 296.

Dardot, P., & Laval, C. (2014). *Commun. Essai sur la révolution au XXIe siècle*. Paris: La Découverte.

Davies, S. (1996). *Adaptable livelihoods: Coping with food insecurity in the Malian Sahel*. Basingstoke: Macmillan Press.

Desgroppes, A., & Taupin, S. (2011). Kibera, the biggest slum in Africa? *Les cahiers de l'Afrique de l'Est, 44*, 23–34.

Depraz, S. (2008). *Géographie des espaces protégés*. Paris: Armand Colin.

Donaldson, R., Ferreira, S., Didier, S., Rodary, E., & Swanepoel, J. (2016). Access to the urban national park in Cape Town: Where urban and natural environment meet. *Habitat International, 57*, 132–142.

Guha, R. (1989). Radical environmentalism and wilderness preservation: A Third World critique. *Environmental Ethics, 11*(1), 71–83.

Guyot, S. (2006). *Rivages Zoulous: l'environnement au service du politique en Afrique du Sud*. Karthala–IRD: Paris.

Guyot, S., & Dellier, J. (2009). Rethinking the Wild Coast (South Africa): eco-frontiers versus livelihoods in Pondoland, VDM Verlag: Saarbrücken.

Guyot, S., Dellier, J., & Caillot, A. (2015). 'Our rural sense of place'. Rurality and strategies of self-segregation in the Cape Peninsula (South Africa). *Justice Spatiale/Spatial Justice, 7*. https://www.jssj.org/wp-content/uploads/2015/01/guyot-JSSJ7-ENG.pdf.

Guyot, S., Dellier, J., & Cerbelaud, F. (2014). L'environnement au profit des plus riches? Construction et hybridation d'un front écologique métropolitain dans la Péninsule du Cap. *Vertigo, 14*, 1. http://vertigo.revues.org/14660.

Heijmans, A. (2001). Vulnerability: A matter of perception. *International conference "Vulnerability in Disaster theory and practice"*

Landy, F., Belaidi, N., & Gaudry, K. H. (2017). Urban protected areas: Forces of justice or injustice for indigenous populations? The cases of Xochimilco and the national parks of Mumbai and Cape Town. *Justice Spatiale/Spatial Justice, 11*. http://www.jssj.org.

Laslaz, L. (2016). Avide d'espaces. *HDR dissertation*, Chambéry: University of Savoie.

Mawdsley, E. (2009). 'Environmentality' in the neoliberal city: Attitudes, governance and social justice. In H. Lange & L. Meier (Eds.), *The new middle classes: Globalizing lifestyles, consumerism and environmental concern* (pp. 237–251). Springer.

Oliver-Smith, A. (2001). Displacement, resistance and the critique of development: From the grass roots to the global. *Final report prepared for ESCOR R7644 and the research program on development induced displacement and resettlement*, p. 123.
Pélissier, P. (1966). *Les paysans du Sénégal*. Fabrègue: Saint-Yrieix.
Rodary, E., Castellanet, C., & Rossi, G. (2003). *Conservation de la nature et développement, l'intégration impossible?* Paris: GRET-Karthala.
Rossi, G. (2001). *L'ingérence écologique*. Paris: CNRS Editions.
Saglio-Yatzimirsky, M. C., & Landy, F. (Eds.). (2013). *Megacity slums. Social exclusion, space and urban policies in Brazil and India*. London: Imperial College Press.
Scoones, I. (2009). Livelihoods perspectives and rural development. *Journal of Peasant Studies, 36*(1), 171–196.
Texier, P. (2009). *Vulnérabilité et réduction des risques liés à l'eau dans les quartiers informels de Jakarta, Indonésie. Réponses sociales, institutionnelles et non institutionnelles*. PhD dissertation, University Paris Diderot—Paris 7.
Zérah, M. H., Dupont, V., & Tawa, Lama-Rewal S. (2011). *Urban policies and the right to the city in India. Rights, responsibilities and citizenship*. New Delhi: UNESCO-CSH.

Chapter 5
Categorisation of People and Places, Indigenous Peoples and Urban National Parks: Between Eviction, Instrumentality and Empowerment

Nadia Belaidi, Karl-Heinz Gaudry and Frédéric Landy

In 2000, the High Court of Mumbai, India, received a petition from a group of indigenous people (called Adivasis or tribals in India) asserting their right not to be evicted from Mumbai's Sanjay Gandhi National Park (SGNP) since they had been living there for about "500 years"—unlike the 3–400,000 slum dwellers living in the national park at that time. The order, issued in 2003, dismissed the petition but admitted that "Adivasis and tribals are 'wedded' to the forest and they preserve, protect and propagate forest. Their case, therefore, had to be dealt with separately".[1]

The ambiguity of this order points to the difficulty decision makers have in categorising peoples and places. National parks are areas governed by specific rules, which generally prohibit human beings from living within their borders: their categorisation in terms of both spatial boundaries and management rules can either label local dwellers as "encroachers" or legal inhabitants. Similarly, the category of "indigenous peoples" is the product of a political decision that gives or denies certain rights to the groups concerned.

This paper is about the links between these two categories of space and people, urban national parks and indigenous peoples. It highlights the twofold uncertainty regarding the status of indigenous peoples living in or near urban national parks: the "usual" controversy over the definition of "indigeneity" and the specific rights possibly associated with it adds a further dimension to the complex issue of the

[1]Writ petition no. 925 of 2000, judgement September 15, 2003.

N. Belaidi (✉)
CNRS, MNHN - Dept Man and Environment
(UMR Eco-Anthropology and Ethnobiology), Paris, France
e-mail: nadia.belaidi@mnhn.fr

K.-H. Gaudry
University of Freiburg, Freiburg im Breisgau, Germany

F. Landy
University of Paris-Nanterre (UMR LAVUE), Nanterre, France

F. Landy (ed.), *From Urban National Parks to Natured Cities in the Global South*,
https://doi.org/10.1007/978-981-10-8462-1_5

original status of national parks located in megacities, which are protected areas under particular threat because of their urban location, population pressures and multiple activities. "Indigeneity" is a notion considered by many—and not just anthropologists—to be highly controversial. The meaning of "indigenous" becomes even more complex given the diversity of interpretations in different countries. This underlines the breadth of the indigeneity narrative and signals a rhetoric of "insiders versus outsiders" or "us versus them".

In India, the official word "tribe", or the more politically correct *adivasi* (literally "aborigines"), are the most commonly used terms when referring to indigenous peoples. The word "natives" has fallen into disuse for political reasons: no prior claim is officially recognised among the various peoples that constitute the Indian Union for fear that it might open the Pandora's box of historically rooted claims.

The situation in Kenya is virtually the opposite: apart from the small white minority, all the country's ethnic groups are classified as "indigenous" in most international documents. The legal situation of the Maasai who live near Nairobi National Park (NNP) does not therefore differ from that of the majority ethnic groups such as the Kikuyu or the Luhya. However, their economic situation is generally more precarious because of their pastoral identity, their small numbers (2% of the population), and a history in which the colonists "granted" them a reserve on the edge of the future park and the city of Nairobi.

Though on the same continent, the South African case is very different because of the legacy of apartheid. The question of the priority of the Khoisan in the Western Cape is contested only by extremist Afrikaners; and Jacob Zuma, as Deputy President of the Republic of South Africa, has defined the Khoisan as "the first indigenous peoples of our country"[2]; yet they are not recognised as such.[3] The *Traditional Leadership and Governance Framework Amendment Act* (2003) states that the "indigenous peoples of South Africa consist of a variety of cultural communities" and covers the country's entire black African population (DTA 2013).

Out of the four countries under study, Brazil is the only one that has ratified the International Labour Organization's (ILO) Convention 169 on "Indigenous and Tribal Peoples" (1989). Fairly substantial reserves have been granted to the Amerindians (more than 13% of national territory). Yet the recognition of their identity is equally complex (Le Tourneau 2015). The colonial system caused a demographic disaster across the Americas. In Brazil, the Amerindian population is

[2]Opening ceremony of the National Khoisan Consultative Conference, March 29, 2001. The term "Khoisan" is a relatively recent invention, coined in 1928 by Leonard Schultze as a collective category for early hunter-gatherer and herding peoples in southern Africa. It increasingly prompts tension and controversy.

[3]The reinstatement of traditional authorities (Chap. 12 of the Constitution) did not originally refer to Khoisan authorities, because of their "non-survival" to colonisation and apartheid (Mukundi 2009). Yet the Khoisan Renaissance led to the creation of the National Khoisan Council in 1999. Its negotiations with the South African government resulted in the Traditional Leadership and Khoi-San Bill. Revamped repeatedly, the Bill presented to Parliament is strongly disputed.

estimated at around 900,000, or only 0.4% of the population.[4] This figure needs to be qualified, since population counts are based on the number of people living on land officially declared to be indigenous and on people who declare themselves as such. Above all, the language and culture of Brazil's very mixed society are strongly marked by their Amerindian roots. While blacks and Amerindians continue to be overrepresented in the most disadvantaged categories of the population, the populations that consider themselves as white are more proud to claim their Amerindian heritage than the small drop of African blood that might "leave a stain".

Controversies about the rights of indigenous peoples are well documented, particularly in relation to national parks, since their creation has often coincided with the expulsion of local indigenous peoples who had settled there (Dunlap 1999; Harris 2002; Randeria 2007; Héritier and Laslaz 2008). The example par excellence is the first national park ever created, Yellowstone National Park in the United States, whose establishment led to the forced eviction of the Blackfeet tribe. Less well-known are the disputes that arise in urban settings, which is the topic of this paper. Whereas in many cases, parks were originally created by excluding people or activities, they may later be used as tools of political inclusiveness to foster social or political integration, a process in which urban parks are sometimes at the forefront. How can conservation practices, often seen as antagonistic to social development, be remodelled to foster not only better links between society and nature, but also better social links between human beings? Are policies that advocate this new paradigm nothing more than rhetoric and media hype, or do they have a real impact on the management of the protected areas and the empowerment of indigenous peoples?

This chapter is not about how indigeneity is experienced and perceived by the indigenous peoples themselves. It does not deal with how they conceive and shape themselves and their territory, or how they use biodiversity, nor with their demands for greater social and spatial justice (Landy et al. 2017). It is not an anthropological paper that seeks to critically deconstruct identities and visions (Kuper 2003) (on the Indian case, see Karlsson 2003; Rousseleau 2003). We do not try to define what indigeneity is: our aim is rather to highlight the spatial and political impacts of the categorisation of peoples. The first part of this chapter introduces the parks and ethnic groups concerned by this study. The second part employs the analytical concept of eco-ethnicity to assess how the identities claimed by indigenous peoples may be recognised and instrumentalised. Ethnicity is explored as a marker of environmental governmentality—what Agrawal (2006) calls "environmentality". As such it can be used, recognised or refused in various commercial, political or social enterprises.

[4]http://www.funai.gov.br/index.php/indios-no-brasil/o-brasil-indigena-ibge, visited on 26/04/2016.

5.1 Four Urban National Parks, Four Types of Indigeneity

In the case of Mumbai, some indigenous peoples live within the boundaries of the park. In other cases, such as Nairobi, they inhabit the outer fringe, while in the case of Cape Town, indigenous populations are not visible. The city of Rio de Janeiro officially has an indigenous population of only 12,000, but the presence of indigenous peoples remains visible through the culture and practices of the inhabitants of many *favelas*, including around the park. These various situations raise questions about indigeneity in these cities and its degrees of domination (Sect. 5.1.1). The functions assigned by park managers and decision makers to the protected areas are another clue to the position allocated to indigeneity (Sect. 5.1.2). Until recently, a former director of Mumbai national park considered that it should be managed like any other remote Himalayan national park. By contrast, Table Mountain National park (TMNP) in Cape Town has announced it is to pursue other objectives, notably social inclusion for the impoverished African population adjacent to the park. Whether a park is perceived as a wilderness enclave or a social entity shapes the different approaches to indigeneity.

5.1.1 Parks and Peoples, Degrees of Domination

Since the word "indigenous" means "dwelling in the place of origin", indigeneity implies the idea of original settlement, a specific link to a place, to a territory, as well as the sense of an identity forged over time (Bellier 2015). However, while priority in the occupancy of a territory remains a fundamental aspect of the definition of indigeneity, this definition and its evolution are heavily influenced by political factors. Dominance relations, discrimination and marginalisation complicate the definition of "indigenous peoples": they can be described as groups that have in common an experience of historically unequal power relations with the nation-states in which they live. Applied at local level, this definition raises the question of the extent to which indigenous populations are "dominated" by the city and by the park: what rights do they have to the city? What rights to the park?

Whilst indigeneity itself is a debatable notion, in our four case studies the blurred identity of the indigenous peoples is further distorted by their location, since they live in an interstitial space between a managed nature and the city. Within their national entities, these peoples remain a minority, marginalised to varying degrees, but locally they may represent a rather large population. Yet their access to the protected areas is highly regulated or even prohibited. We will now look at the situations of the indigenous peoples in the four countries and regions under consideration. In what follows, the order in which the parks are discussed reflects the diminishing importance of the indigenous peoples in and around them.

Nairobi

While NNP is kept uninhabited, numerous Maasai pastoralists live on its southern side and use the huge grasslands extending from Nairobi to the Tanzanian border to raise their cattle. This land is a legacy of the colonial "Maasai Reserve" that constituted the sole territory left to the pastoralists by the British in the 1904 treaty (Rutten 1995); NNP acts as a buffer zone between the capital city and this former "extra-provincial area" (see Chap. 2). Today the Maasai coexist more or less peacefully with large numbers of wildlife that migrate seasonally through the dispersal area south of the park. In the 2009 census they represented 2% of the Kenyan population. This is a small minority, but the country as a whole is ethnically fragmented, since no ethnic group accounts for more than 17% of the population. More even than in South Africa, whites and Asians are small in number compared with the dominant African tribal population, which makes the situation of the Maasai and other Kenyan indigenous peoples unique amongst our case studies: Nairobi is situated in a country where the great majority of the population consists of officially indigenous peoples. The Maasai occupy an ambiguous position, a (tiny) minority among other minorities, but enjoy—as this paper will recall—iconic status across the globe as herdsmen who live in harmony with nature. Their cattle are not allowed to graze in the NNP, but there is no legal threat to their pastoralist activities since with some exceptions they do not claim the right to live within the park. However, their traditional way of life is threatened by rampant urbanisation and industrialisation, the privatisation of collective pastureland and the introduction of fencing.

Mumbai

In India, indigenous peoples essentially belong to the category of "Scheduled Tribes", which are the beneficiaries of affirmative action policies. Officially, reservations and quotas in their favour are not justified in terms of their historical presence, but by their poverty and marginality, such as Dalits (formerly Untouchables). They constitute a minority in India's immense population (8.6% in 2011), but with more than 104 million people they nevertheless represent over a fifth of the world's indigenous population. Since India is a democracy, until recently the "tribals" argued their claims for greater rights in the national arena only, which explains why their numbers are often underestimated in global forums and public opinion. They were the first victims of India's major colonial and postcolonial projects (e.g., dams, mines, factories, etc.). More recently, economic liberalisation has attracted domestic private companies and foreign multinationals to the tribal areas, with their abundant raw materials and energy, leading to the displacement of indigenous peoples.

The mushrooming of protected natural areas in India is another factor that has triggered tribal impoverishment. The case of SGNP is an example of this process, as since its registration in 1950 and its subsequent expansion, Adivasis living in the area have been considered to be "encroachers". Today, 2,415 indigenous households (according to the 2011 Management Plan), largely Warlis and Kohlis, most of them hinduised, live within SGNP.

Rio de Janeiro

In the sixteenth century, when the Europeans landed in Guanabara Bay, this territory was inhabited by thousands of people. Most of the Amerindian populations were Tupinambá and Tamoio, members of the greater Tupi family, but also Temimino or Maracaja, along with Tupinikin or Margaia. They lived by hunting and fishing, by cultivating manioc and by gathering fruit. In interaction with their environment, they practised the selection of useful species and food storage, and contributed to the floristic composition of the forest (Hladik et al. 1996). According to a map drawn by André Thevet, a member of Villegagnon's crew, towards 1556 there were 36 *tabas* (Indian villages) on today's Governador Island. Around Guanabara Bay, the French Calvinist missionary Jean de Léry, in what is considered to have been a partial assessment made around 1558, identified 32 Tupi *aldeias* (villages). The first *aldeia* identified in Léry's list is the one called Kariok or Kariog, located in the mouth of the River Carioca, a river sacred to the Tupinambas.

The Tupinambá and Tamoio populations lived on the seashore and in the valleys. A new transformation of the landscape began with the arrival of the European colonisers, the establishment of the urban nucleus and farms. This process accelerated, resulting—between the end of the eighteenth century and the second half of the nineteenth century—in the destruction of almost all the tropical forest. The Amerindian populations were gradually decimated or displaced, after a fruitless attempt to reduce them to slavery. However, in Rio as elsewhere in Brazil, the Amerindian heritage remains present in the population's ethnic mix, in place names (like Tijuca), in the syncretism of the religions, in music and dance and numerous cultural practices, such as local food, constituting a backdrop to modern times.

Cape Town

In its report on the registration of Table Mountain National Park (TMNP) as a World Heritage site, the IUCN noted the importance of its cultural qualities, although the South African State did not request it. This was due, in particular, to the connection between the endemic *fynbos* (shrub) vegetation and the precolonial history of the Cape: San people (hunters-gatherers) are said to have been present until 1978 (IUCN 2004).

By the end of the 1670s, the indigenous peoples of the Cape Peninsula and the immediate interior, mainly Khoikhoi pastoralists who had migrated into the region about two thousand years earlier, had been subjugated and dispossessed of their land and livestock by Dutch-speaking interlopers. Some dispossessed Khoikhoi resorted to hunter-gathering, like the San people. While others migrated beyond the reach of colonial influence, many Khoikhoi were taken as labourers by farmers. Initially valued by the white *trekboers* because of their intimate knowledge of the natural environment and skills in animal husbandry, by the end of the eighteenth century most Khoikhoi were in effect forced labourers, little better off than serfs (Elphick 1985; Elphick and Giliomee 1989; Penn 1986). A number became stock raiders, at times in collaboration with the San, putting up fierce resistance to further colonial incursions (Adhikari 2011). In addition to this, epidemics, in particular the smallpox outbreak of 1713, took a huge toll on Khoikhoi society, wiping out an estimated 90% of the population (Mitchell 2012).

Today, the KhoiKhoi and the San are known as the Khoisan. Despite differences in their livelihoods, culture and languages, this ethno-linguistic group remains a recognised identity in South Africa.[5] "San" as a term came to be used to differentiate the hunter-gatherer communities that did not speak Khoi languages—known to white settlers as "Bushmen"—from the Khoi-speaking herders. Under the apartheid government, Khoisan peoples were forced to register as Coloured, a label that subsequently came to be widely resented, especially from the 1980s, for its neglect of their distinct identity.[6] Such policies of classification, or self-classification without official recognition, along with land dispossession that came as a result of forced relocation policies under the apartheid era, further fractured Khoisan identity (Mitchell 2012). The Khoisan are now classified into five groups, which include the "Cape Khoi"—the Khoisan revivalists of the Western Cape (Verbuyst 2015), a very small group compared with the larger Coloured community living in Cape Town, who define themselves as the descendants of the Khoisan and are working to revive their identity in the region.

In the four case studies, the indigenous peoples are minorities who enjoy a (claimed and generally accepted) status as first settlers while suffering from low demographic, socioeconomic and political status (Bellier 2009). Yet our four case studies show huge differences. Whereas the Tupis had settled well before the arrival of Europeans, the Maasai did not arrive in Kenya before the seventeenth century, and probably most of the Adivasis in Mumbai national park did not settle there before the twentieth century. In short, the Maasai and the Adivasis are probably more "indigenous" (etym. "born here") than "aboriginal" ("here from the origin"). Their cases differ historically from those « clearly native » and world famous autochtonous, such as the Australian aborigines or the North-Amerindians who lives in countries with a dominant white « recently » settled population (Fritz and Porteilla 2005). Yet the point made by Barnard (2004, 19) is valid for the four case studies: "The acceptable point about the classification of peoples as indigenous is not that they pre-date other peoples, but that they pre-date state formations established to protect such other peoples." Be they Maasai, Adivasis, Tupis or Khoisans, they all predate the formation of the modern nation-state.

---

[5]Some San communities, claiming that their people have been especially marginalised, even in comparison with the Khoikhoi, are beginning to assert a distinct identity, encouraging the use of "Khoi and San" as opposed to "Khoisan" or "Khoi-San" in official references.

[6]The issue of Coloured registration and identity is a complex one. Within the Coloured category, there existed various subgroups, including Cape Coloured, Cape Malay, Griqua, Nama, and "other Coloured". The Khoisan community was not neatly categorised within one of these groups, but instead individuals with slightly different heritages were classified in different subgroups. Those of Khoikhoi and Afrikaner descent, for example, often classified themselves as Griqua, whilst those with a stronger slave heritage tended to be classified as Cape Malays (Mitchell 2012).

5.1.2 *Different* raisons d'être *of the Parks, Different Regulations*

In terms of legislative frameworks, we identify a sharp contrast between the protected areas of Tijuca and Cape Town, as parks covered by several legislative regimes, ranging from the local to the international, and those of Mumbai and Nairobi, which are parks with a single designation status that are managed in a way often described as fortress-like.

5.1.2.1 The Dominance of the Conservation Objective: Mumbai and Nairobi

Eastern Africa and India were two core areas of the "radical paradigm" of conservation (Laslaz 2012). Today the Kenya Wildlife Service (KWS) and the Indian Forest Department (IFD), which respectively manage the Nairobi and Mumbai national parks, are still marked by this legacy and only reluctantly open the protected areas to uses other than conservation. Their senior officers are trained in premier institutions, and the sense that many of them have of belonging to an elite contributes to making the KWS and the IFD "a state within the state". The conservation policy implemented by the British was maintained by independent India and Kenya, with more and more areas being defined as protected, at the expense of the local inhabitants. A major difference is that in Kenya—unlike in India—tourism is a key resource for both the KWS and the government. As a result, the ability of the Maasai to attract tourists puts them in a strong position to promote their rights. In India, by contrast, not only are the national parks a limited tourist resource, but the Adivasis themselves also hardly constitute a tourist attraction. There is therefore little incentive for the IFD to be open-minded in supporting the rights of the Adivasis in and around national parks. Moreover, anthropogenic pressure in Mumbai is so high and conditions so bad that the IFD fears that any rights or privileges granted to the Adivasis may be exploited by other groups and that another wave of slum development might once again overrun the park boundaries.

5.1.2.2 Multiple Functions, Multiples Rules: Rio and Cape Town

In the Brazilian and South African cases, the conservation priority is accompanied by varied objectives. One reason for this is the multiple levels of park governance, a legal intricacy specific to Cape Town. Traditionally, the logic of conservation in South Africa's national parks saw surrounding communities as potential poachers and competitors for land and water, and their poverty was seen as an embarrassment to tourism. The SANParks authorities began in the mid-1990s to be sensitised to community issues, especially after holding a series of community forums. They saw the need to systematise community relations for what was called "social ecology":

a new conservation ethic that promotes community-based development through conservation and tourism (Cock and Fig 2002). Now, the significance of national parks is not limited simply to their capacity to preserve biodiversity and promote ecotourism, they are a base to promote empowerment strategies: hence the official objective of TMNP is to be "A Park for All, Forever".

In Brazil, the racial context is less tense, and less political responsibility is given to national parks such as Tijuca. Yet conservation policies must comply with education to environment (cf. Chap. 13), and that supposes to open the park to visitors and students as much as possible. Moreover, Tijuca National Park (TNP) plays a crucial role in funding conservation institutions thanks to its high number of paying visitors. A former director of TNP made the open choice to favour education and revenue at the risk of attracting too many visitors. He was even accused to create "a green Disneyland".

As can be seen, therefore, the parks under consideration vary in their main objectives. Yet, whether during colonial times (Nairobi, Cape Town) or in the postcolonial era (Mumbai), the establishment of the protected areas was a top-down process, entailing the removal of people (Nairobi) or rights (Mumbai).[7] In this respect, the creation of a national park entails the same process of historical modernisation as the administrative division of ethnic groups, into non-porous spatial or social categories: the processes of categorisation of space as well as groups were contemporary and to a large extent were the mark of reinforced state control on territories and populations. Today, in the era of postmodernity, these definitions and classifications are increasingly disputed and there is growing recognition that boundaries, whether spatial or social, are actually blurred or fluid. Hence, while there are few challenges to the existence of national parks in themselves, the nature of their management is attracting more and more criticism, and the issue of indigenous peoples is often making the situation even more complex and debated.

5.2 The Paradoxical Power of Ethnicity: From Claims to Recognition, Instrumentality … or Ignorance

5.2.1 A Gradual "Eco-ethnicity"? Various Nature-Culture Relationships

Can the groups with an environmentalist image make use of it as a tool of soft power? This is what we explore through the notion of eco-ethnicity. By "eco-ethnic

[7]Tijuca and Table Mountain are intermediary cases: their original population was evicted during colonial times, as happened in Nairobi, but that process took place long before the creation of the national park, even though, in Tijuca, conservation was already the reason given for the eviction of the local (rich non-Indian) population between 1845 and 1860.

identity", or eco-ethnicity, we mean an identity made up of two components. The first is the "ethnic" aspect: a strong, visible ethnicity, marked by a vibrant (often reconstructed) folklore and culture, may bring many advantages. Every foreign tourist in Kenya knows about the Maasai; hence this group enjoys some bargaining power in the national and international arenas due to the support of local or global NGOs and general opinion. On the contrary, very few foreigners coming to India are aware of the existence of Adivasis. In terms both of notoriety and soft power, therefore, the Adivasis come up short—their identity is not "ethnic" enough and lacks visibility.

The second component of eco-ethnicity is ecological identity, in other words the types of interrelations between an ethnic group and its environment. What counts is not only the interrelations as they are in reality, but also as they are perceived by the Other. These interrelations may be genuinely sustainable but not generally recognised as such in the country itself or abroad; or conversely, some communities may enjoy a mythical and romanticised aura of "ecological wisdom" while the reality is very different. In any case, if the group possesses an eco-friendly image, given the increasing dominance of the environmental paradigm in most national and global narratives, this can in many cases provide soft power, be it in negotiating with a government, attracting investment in ecotourism, acquiring funds from an environmental NGO, or obtaining rights of use in protected areas.

5.2.1.1 Mumbai and Nairobi: Opposite "Eco-ethnicities"

The Adivasis who lived in the area now called Sanjay Gandhi National Park were mostly farmers, with some livestock for ploughing (oxen), milk (buffalo) and meat (goats). They also used to collect numerous non-timber forest products (e.g., game, fish, medicinal plants, etc.). Today, farming and livestock are prohibited, and the collection of forest products is tolerated only for family use (Fig. 5.1). Though the Adivasis do not always obey these rules, they have lost much of their livelihood and have to rely on urban incomes (day labouring for men in construction sites, domestic service for women, etc.). They are not allowed to obtain any income from the park. The only exception is the few Adivasis engaged by the IFD for menial tasks, and the handful of tribal women allowed to sell fresh cucumbers to tourists enjoying the toy train that passes near an Adivasi hamlet.

In Nairobi, the Maasai pastoralists are similarly forbidden to use the national park. The difference is that, by contrast with Mumbai, there exists a huge savannah area, south of the protected zone, where their cattle can graze in coexistence with wildlife (Lézy 2012; Fig. 5.2). Many of the Maasai have been able to maintain their traditional activity, although pastoralism now accounts for less than two-thirds of the income of Maasai households in Kitengela (61% according to Kristjanson et al. 2002).

In India, the 2006 Forest Rights Act gives every Adivasi household living in a protected area the right to claim for up to four hectares of land, provided that it can prove that it depends on the forest for its livelihood. In Mumbai, even before this

Fig. 5.1 Washing dishes without tap water in Taleopada village (Sanjay Gandhi National Park). On the right, the Warli house has walls built traditionally with bamboos and mud. *Source* Photo by F. Landy, January 2014

Fig. 5.2 At dusk, an old Maasai takes his goats and cattle back to his family settlement, one kilometer from NNP. *Source* Photo by F. Landy, March 2012

act, in order to ensure a hearing from activists and the government, associations defending Adivasis contended that their income is forest-based. Yet most households have a variety of sources of income, and the fact is that most Adivasis are both urban workers and park dwellers. They live in hamlets in the inner fringes of the park, many of them invisible to visitors, and they commute daily on foot and by bus to their place of work. Their dress has nothing specific, and all the men and most of women can speak Marathi, the regional language. In fact, in the park, Adivasis are conspicuous by their transparency. By contrast, in Nairobi visitors are welcomed by a group of dancing and singing Maasai dressed in traditional red robes (Bruner 2005). Sometimes these dancers are not even from that ethnic group but are instead only acting as if they are Massai. In NNP the Maasai are fully part of the picture—in all the senses of the term, since tourists appreciate the opportunity to photograph them in the savannah wildlife habitat. Despite representing a smaller percentage of the national population than the Adivasis, the Maasai are perceived as attached to famous reserves (e.g., the Maasai Mara, or Serengeti in Tanzania), and enjoy a prestigious status. Their shield and spears are in the centre of Kenya's national flag. While the Maasai have succeeded in linking themselves with protected areas in the international psyche, the Adivasis are often seen as practising environmentally harmful agricultural methods and lacking environmental awareness because of their low literacy rate and "backward" condition.

5.2.1.2 Rio: An Eco-ethnicity of Intermingling

Tijuca forest is undoubtedly a very fine cultural production, in spite of painter Debret's (1834) vision of a forest that was still virgin before the arrival of the Europeans. In fact, like the Amazonian rainforest, it had probably been a "garden" cultivated by the Amerindian populations to meet their requirements for food, medicines, constructions and tools, etc. (Hladik et al. 1996). Today, the forest is home to Afro-Brazilian myths and legends: Yara, the queen of fresh water; Curupira, protector of the forest, etc. Swarming with spirits, it is a veritable open-air temple for followers of *Candomblé* and other religions associated with the cult of nature. Tijuca Forest is a cultural space which has, in reality, been reconstructed by the hand of man.

Tijuca's name, like other place names in the national park and the city (Ipanema, Copacabana, Corcovado, etc.), is taken from a Tupi Guarani word meaning mud, marsh or swamp. The presence of Amerindian culture is also revealed in the operation of those urban tribal areas—the *favelas*—which are home to the descendants of the Amerindians and black slaves. In Rio today, *favelados* combat and oppose the established order just as their ancestors did in colonial times. They are the ones who manage their environment, not so much in the ecological sense of the term, but in their control of numerous entrances to the forest, their understanding of the pathways and codes of the territory, including those relating to hunting (see Chap. 4). However, clandestine religious events that attract large numbers of people have had a significant impact in certain sections of the park (sector D—Covanca/Pretos Forros). Other Afro-Brazilian religious practices involve leaving candles and food in the forest as offerings: these practices—like those in the Hindu festival of Shivaratri in—represent a significant danger, both in terms of fire and impact on animal diets.

5.2.1.3 Cape Town: Eco-ethnicity as Revival

In Cape Town, Table Mountain is the central landmark for every member of the population. Even before the city existed, the mountain exerted power over the landscape. The KhoiSan called it *Hoerikwaggo*, "the mountain of the sea", the name used since 2005 for guided mountain hikes organised by SANparks, with the result that Table Mountain is now identified with its Khoisan name. The guides are expected to be aware of the mountain's sacredness to the indigenous peoples and how they used its biodiversity (Tourism manager interview, 2010). The San

Fig. 5.3 KhoiSan Rights Now! Cape Town residents protesting for their Khoisan rights as part of a larger movement against a new development project on the scenic Chapman's Peak drive showcasing parts of Table Mountain National Park. *Source* Photo by J. Swanepoel, January 2012

had extensive knowledge of the resources of their ecosystems, particularly with regard to food, medicinal plants and habitats (Fritz and Porteilla 2005; Belaidi 2005). On its website, SANParks stresses this connection with indigeneity, immediately emphasising the temporal and cultural roots of the mountain. It is now accepted that the recognition of indigenous knowledge is part of the national reconciliation process (Landy et al. 2017; Fig. 5.3).

As TMNP is registered as a World Heritage site, it is illegal to damage any indigenous flora, but the park is surrounded by a community of traditional users, ranging from Rastafarians and herbalists to traditional healers (*sangomas*), including people who define themselves as Khoisan.[8] A solution to the tension between conservation initiatives and traditional uses of plants today protected by TMNP, was proposed with the establishment of medicinal herb gardens offering access to a larger population. The idea is that once the garden is established, healers are allowed either to collect what they need or, even better, to take cuttings and seeds to grow in other areas of Cape Town. Yet this trend should not be overemphasised. True, the granting by the civil authorities of a "traditional garden" to traditional healers and other African stakeholders constitutes a recognition of their knowledge of the environment. However, it is also a way to keep people out of the park, by giving them incentives to obtain their resources away from protected areas. Eco-ethnicity is a soft power with strong limitations indeed.

All in all, in protected areas as well as in our four case studies, the ethnicisation of the local populations is a key factor in the process of empowerment (Guyot 2011). But this ethnicity has to include some environmental component, unless the indigenous peoples are refused rights and recognition. This is highlighted by the difference in the fates of the Maasai and the Adivasis, since the former possess

[8]For an overview of the tensions between the park and traditional healers over plants, see http://livelihoods.org.za/projects/informal-economy-of-biodiversity.

much greater eco-ethnicity than the latter. Whereas in Nairobi the resources around the park are usually considered to be used sustainably by the Maasai, this is not the case in Mumbai where the image of the Adivasis is negative for most citizens. Furthermore, even when the group is endowed with a quite high degree of eco-ethnicity, the case of the KhoiSan proves that it may be far from enough to provide important rights on the protected area.

5.2.2 *Is Ethnicity a Marker of "Environmentality"?*

5.2.2.1 Environment, Class and Nation

The environment can become a technique of government if, in a way analysed by Foucault (2004), environmental policies and environmental education work as a government's tools of control over its population (see Chap. 13). This is what Agrawal (2006) calls "environmentality". "Environmentality refers to the knowledges, politics, institutions and subjectivities that come to be linked together with the emergence of environment as a domain that requires regulation and protection" (Agrawal 2006, 226). Since "subjectivities" and inner perception relate directly to the institutions and techniques of government (i.e. to Foucault's "governmentality"), the various social relations to the environment cannot but trigger conflicts between groups with differing environmental identities.

In the conflicting environmental views of the middle classes, the Adivasis and the slum-dwellers have harsh political consequences, which could be softened only by a fragile hybridity, the components of which are yet to be clearly defined. For the moment, there remain huge gaps between social groups that are defined culturally as much as socioeconomically. For example, the "bourgeois environmentalists" (Baviskar 2003) are doing their best to enforce the judicial decision to evict the tens of thousands of slum-dwellers living in the park—and possibly the Adivasis as well. But many "bourgeois" are not environmentalist at all: far from being fascinated by the urban forest, they are the first to complain to the authorities about leopards roaming in the suburbs (see Chap. 7 and Landy 2017).

Unlike the three other case studies, the issues at stake around are mostly local: they cause conflict between social and cultural groups along class and ethnic lines made even more complex by metropolisation and globalisation—and by the desire for power and money. This is less the case in Nairobi, where there is some national component to relationships with the Maasai. In Kenya, Maasai identity is highly instrumentalised by environmentality as an iconic ethnic group, both touristically (for foreigners) and environmentally (for all). It is a way to reinforce the fragile unity of a nation around a common goal: protecting and exploiting parks and reserves that are immense reservoirs of biodiversity and currency.

In Cape Town, behind a genuine concern with conservation, there are also very significant cultural and identity factors at work. Here biodiversity supports a culture and identity based on the process of nation building. Registration on the UNESCO list reinforced the site's cultural and natural values and tightened their links and interdependences. TMNP, and environmentality processes and policies around it, are directly involved in this process of building the new South Africa (Belaidi 2012).

5.2.2.2 Political and Commercial Uses of Identity

In the case of SGNP, there is no official or commercial use of indigeneity: hence there is no instrumentality in the fullest sense of the word, let alone empowerment, since until recently the Adivasis were not supposed to have any claim over the park, and therefore no grounds for compensation to be granted. Adivasis are almost invisible to outsiders, not only because they live inside the park and because those of them who live outside it are a small minority lost in the mass of Mumbai's population. There exist associations (*sangathnas*) who fight for Adivasi rights, but they are small, factionalised and often led by non-Adivasis.

Some qualification is needed, however. First, there have recently been some changes. The IFD now employs many male Adivasis, often throughout the year. They are recruited for menial task (sweepers, etc.) and unfortunately their environmental knowledge is hardly used. Yet this employment can be seen as a de facto recognition of their dwelling within the park. Furthermore, in 2012, the new park director ordered Warli paintings to decorate the Education Centre. A young local Adivasi was paid for this, in both a (partial?) recognition of his right to live in a forest hamlet and, equally importantly, as an acknowledgement of the importance of Adivasi culture in the perception of nature. If indigenous representations of nature can be painted on official buildings, if animism gets some right of presence within the boundaries of a national park, isn't it an indication that the paradigm of a nature-culture dichotomy is slowly breaking down? (Landy 2017).

Second, some instrumentalisation of ethnicity may be encountered, but it does not come from stakeholders in the environmental or touristic spheres, but rather from real estate. Strategies based on the hunger for land in such a crowded metropolis explain some applications for land grants within the framework of the Forest Rights Act: builders and developers hoping to obtain land in and around the are encouraging some Adivasis and *sangathnas* in their quest for land, with the aim of subsequently buying it from the tribals.

The situation is different in Nairobi, where the Maasai are a very visible ethnic group. It may seem surprising that such a small tribe has become emblematic not only of Kenya's natural heritage but of the whole nation. In fact, the Maasai possess this status because—not in spite—of their small numbers. They are instrumentalised to maintain a (fragile) national harmony. Putting the spotlight on the Maasai is a way to limit the often bloody rivalry between Kikuyus, Luos or Luhyas. True, at the local level some Maasai with large land holdings may possess sufficient power to demand more prerogatives, especially in land issues. Moreover, in Kenya "ethnicising nature" may turn out to be politically dangerous: if the national park is too closely associated with specific ethnic groups, it may gradually lose its identity as a resource for the whole nation, not to speak of the status of its biodiversity as a common global good. The can of ethnic worms would be difficult to close. A national park considered mostly Maasai property would lose its iconic status as an inviolable public resource, and become the target of inter-ethnic rivalry and political agendas.[9] But demands remain in a low tone: being pastoralists, the Maasai are less attached to land than traditional agriculturists would be. Furthermore the area in and around the present NNP was more a no-man's land between Kikuyu and Maasai territories than a core area for the latter, which incidentally was an incentive for the British to choose the Nairobi site as their main settlement.

Are the Maasai such big beneficiaries after all? Maasai presence in the tourism industry is very small in terms of investment. Using Maasai eco-ethnicity as a commercial and political tool leads to some limited income for them, to some even more limited empowerment, incommensurate with the scale of instrumentality (Homewood et al. 2009; Kristjanson et al. 2002). In Mumbai, the Adivasis' situation is even worse: they receive no empowerment. The painter in the park may perhaps earn some rupees, but a sword of Damocles always hangs over the group: expulsion from the park, especially if a new, more rigid director is appointed. Claims pursued under the Forest Rights Act may eventually provide a few Adivasis with some access to natural resources or land in the Nagla block north of the park, but this might come at the expense of other Adivasis, since the IFD is becoming tougher out of fear that granting privileges may be harmful to conservation (which is, indeed, quite possible).

A similar ambiguity can be found in the South African case study. At the opening ceremony for the National Khoisan Consultative Conference on March 29, 2001, then Deputy-President Zuma declared it a "powerful demonstration of the enduring strength of the Khoisan people. It was, after all, the Khoi-Khoi in the Cape who waged the first wars of resistance against the colonial onslaught of the seventeenth century … It is of historical significance that the descendants of those who were cruelly victimised, repressed, exploited, driven from their homes and suffered worse injustices and inhuman treatment, are today joining together to participate in building a better and stronger South African nation". And in his State of the Nation address on February 9, 2012, President Zuma made a promise that provisions would be made "for the recognition of the Khoi-San communities, their leadership and structures". This address held special significance as one of the first official recognitions of Khoisan communities in the history of modern South Africa.

A central aspect of these concerns relates to the use of Khoisan languages. Not one of South Africa's eleven official languages is Khoisan. Yet indigenous Khoi and San languages are recognised in the Constitution. Interestingly when compared with the Maasai weapons on the Kenyan flag, the South African coat of arms features a phrase in Xam, a now extinct Khoisan language, potentially implying the

[9]True, the Maasai Mara National Reserve, not very far from Nairobi, has its name linked with the Maasai tribe. But it is a "national reserve", not a "national park", with a status that allows more local management by the county and the neighbouring communities.

growing perception of the importance of Khoisan history and culture in wider South African identity[10] (Mitchell 2012).

In this context, the mountain's Khoisan name could be interpreted as a pointer in the direction of Khoisan claims. It was chosen over the Xhosa term, Umlindi Wemingizimu ("the watcher of the south"), despite the fact that the Xhosa language is South Africa's second most common domestic language, and many Xhosa-speaking people live in Cape Town. But the restoration of the name Hoerikwaggo is just a symbolic restitution while the goal of the Khoisan peoples is to achieve recognition as the country's first occupants and, constitutionally, for the cultural genocide they underwent (Walker 2008). The recognition in the law of Khoisan indigeneity and rights will be difficult to obtain, as these are perceived, for the moment, as a threat to the nation-building effort.

5.3 Conclusion

The cultural legacy of the Khoisan starts to be emphasised, in particular through the management policy of TMNP. But the recognition is more political than effective. As TMNP promotes conservation values, the city of Cape Town benefits from appearing to be a green, modern and ecologically responsible city. The emphasis on these values, at international and national level, is a way to transform the image of a city known for its profound racial and social disparities. Hence heritage and conservation priorities are instrumentalised as part of an urban marketing strategy to promote a city project.

A rather similar situation can be found in Rio de Janeiro. The national park has a Tupi name—Tijuca—despite the fact that the Tupi Indians are now extinct in the region. Moreover, the city as a whole is marked by Amerindian identity in its place names, down to the very name given to Rio's inhabitants, the Cariocas. Yet, as in Cape Town, the global visibility of the city is not due to the mark of indigenous peoples. Table Mountain, not the Khoisan heritage, is well known abroad. Similarly, this is TNP—with the enclave of the Corcovado Christ—not the Tupi legacy, that directly participates in the construction of the global, green image of Rio de Janeiro bringing investments, tourists and international conferences to the city.

By contrast, thanks to their stronger eco-ethnicity the Maasai play a role as important as the protected area to enhance the global image of Nairobi. Locally, however, where they might represent a threat to conservation policy, they are

[10]Recent developments show some schools beginning to revive the Khoi or San language use, and new books in these vernaculars are being published. In Schmidsdrift, there is a Khoisan radio station, XK-FM, with an estimated 5,000 listeners concentrated in the Northern Cape. Programmes are broadcast in the !Xhu and Khwe languages. Furthermore, the Pan South African Language Board currently claims to promote the development and use of Khoi, Nama and San languages (Mitchell 2012).

carefully kept out of the decision-making process, and away from the grasslands of the park. In Mumbai, the situation is very different, since there are thousands of Adivasis within the park. In consequence, they can be granted no real recognition by the IFD without breaking sharply with the entire paradigm of Indian national park management. It would open the door, officials fear, to too many claims, by too many groups more powerful and numerous than the Adivasis.

The fact that the eco-ethnicity of the Adivasis is very weak is one more disadvantage, since they lack the powerful environmental image that would give them political clout. Until recently, their knowledge of the environment of the MNP was not even recognised by the IFD, which prevented their involvement in park management and watered down the pledge that they would be allowed to live in the park. This is a further contrast with Nairobi, where some Maasai work as rangers. Nonetheless, in all four parks the idea that indigenous knowledge should be fully integrated into the management of protected space continues to make little headway. Yet would not such a development bring benefits for both conservation and indigeneity, with greater efficiency for the former and greater rights for the latter?

References

Adhikari, M. (2011). *The anatomy of a South African genocide. The extermination of the Cape San peoples*. Cape Town: UCT Press.

Agrawal, A. (2006). *Environmentality. Technologies of government and the making of subjects*. New Delhi: Oxford University Press.

Barnard, A. (2004). Indigenous people. *Comment, anthropology today, 20*(5), 19.

Baviskar, A. (2003). Between violence and desire: Space, power, and identity in the making of metropolitan Delhi. *International Social Science Journal, 175*, 89–98.

Belaidi, N. (2005). Apport de la vision du monde des peuples autochtones à la question environnementale sous l'angle juridique. In J.-C. Fritz, F. Deroche, G. Fritz, & R. Porteilla (Eds.), *La nouvelle question indigène, Peuples autochtones et ordre mondial* (pp. 401–424). Paris: L'Harmattan.

Belaidi, N. (2012). Le Patrimoine mondial pour créer une identité commune dans l'Afrique du Sud post-apartheid? Exemple de la ville du Cap. In L. Bourdeau, M. Garavari-Barbas, & M. Robinson (Eds.), *Tourisme et patrimoine mondial* (pp. 269–284). Quebec: Presses de l'Université Laval.

Bellier, I. (2009). *Autochtone*. EspacesTemps.net: www.espacestemps.net/articles/autochtone.

Bellier, I. (Ed.) (2015). *Terres, territoires, ressources: Politiques, pratiques et droits des peuples autochtones*. Paris: L'Harmattan.

Bruner, E. M. (2005). *Culture on tour. Ethnographies of travel*. Chicago: The University of Chicago Press.

Cock, J., & Fig, D. (2002). From colonial to community-based conservation. Environmental justice and the transformation of national parks (1994–1998). In D. A. Mc Donald (Ed.), *Environmental justice in South Africa* (pp. 131–155). Athens: Ohio University Press.

Debret, J. B. (1834). *Voyage pittoresque et historique au Brésil*. Paris: Firmin Didot.

DTA (Department of Traditional Affairs). (2013). Select Committee on Cooperative Governance and Traditional Affairs presentation. 2013/2014 Performance Plan (Including progress made on targets set in the 2012/13 Annual Performance Plan).

Dunlap, T. R. (1999). *Nature and english diaspora, environment and history in the United States, Canada, Australia and New Zealand*. New York: Cambridge University Press.
Elphick, R. (1985). *Khoikhoi and the Founding of White South Africa*. Johannesburg: Ravan Press.
Elphick, R., Giliomee, H. (Eds.) (1989). *The shaping of South African Society*. Maskew Miller Longman.
Foucault, M. (2004). *Sécurité, territoire, population. Cours au Collège de France 1977–1978*. Gallimard-Seuil.
Fritz, J.-C., Porteilla, R. (2005). Les Khoisan d'Afrique australe. In J.-C. Fritz, F. Deroche, G. Fritz, R. Porteilla (Eds.), *La nouvelle question indigène. Peuples autochtones et ordre mondial* (pp. 169–203). Paris: L'Harmattan.
Guyot, S. (2011). The instrumentalization of participatory management in protected areas: The ethnicization of participation in the Kolla-Atacameña Region of the Central Andes of Argentina and Chile. *Journal of Latin American Geography, 10*(2), 9–36.
Harris, C. (2002). *Making native space. Colonialism, resistance and reserves in British Columbia*. Vancouver: UBC Press.
Héritier, S., & Laslaz, L. (Eds.). (2008). *Les parcs nationaux dans le monde*. Paris: Ellipses.
Hladik, M. C., et al. (Eds.). (1996). *L'alimentation en forêt tropicale: Interactions bioculturelles et perspectives de développement*. UNESCO: Paris.
Homewood, K., Kristjanson, P., & Trench, P. (Eds.). (2009). *Staying Maasai? Livelihoods, conservation and development in East African rangelands*. Livelihoods: Springer.
IUCN. (2004). *World heritage nomination—Technical evaluation, The Cape Floral Region (South Africa)*, ID n°: 1007 REV.
Karlsson, G. G. (2003). Anthropology and the "indigenous slot": Claims to and debates about indigenous peoples' status in india. *Critique of Anthropology, 23*, 403–423.
Kristjanson, P., et al. (2002). *Valuing alternative land-use options in the Kitengela wildlife dispersal area of Kenya. An ILRI-ACC report.*
Kuper, A. (2003). The return of the native. *Current Anthropology, 44*(3), 389–399.
Landy, F. (2017). Urban leopards are good cartographers: Human-non human and spatial conflicts at Sanjay Gandhi National Park, Mumbai. In A. Rademacher, K. Sivaramakrishnan (Eds.), *Places of nature in ecologies of urbanism* (pp. 67–85). Hong Kong University Press.
Landy, F., Belaidi, N., Gaudry Sada, K. H. (2017). Urban protected areas: Forces of justice or injustice for indigenous populations? The cases of Xochimilco and the national parks of Mumbai and Cape Town. *Justice Spatiale|Spatial Justice*, n° 11. Online: http://www.jssj.org.
Laslaz, L. (2012). *Atlas mondial des espaces protégés*. Paris: Autrement.
Le Tourneau, F. M (2015). En marge ou à la marge? Les populations amérindiennes du Brésil. *Espace populations sociétés*, 2–3. http://eps.revues.org/5859.
Lézy, E. (2012). 'Out of the park', le Parc National de Nairobi dans la géodynamique Maasaï. In L. Bruno & E. Lézy (Eds.), *BiodivercCités. Les aires protégées urbaines, des laboratories grandeur nature* (179–196). Paris: Le Manuscrit.
Mitchell, F. (2012). *Khoisan identity*. SAHO Public History Internship: http://www.sahistory.org.za/topic/khoisan-identity.
Mukundi, W. G. (2009). *Kenya: Constitutional, legislative and administrative provisions concerning indigenous peoples*. Geneva: ILO & African Commission on Human & Peoples' Rights.
Penn, N. (1986). Pastoralists and pastoralism in the Northern Cape frontier during the eighteenth century. In W. James & M. Simons (Eds.), *The South African Archaeological Society Goodwin Series*: Vol. 5. *Prehistoric pastoralism in Southern Africa*.
Randeria, S. (2007). Global designs and local lifeworlds: Colonial legacies of conservation, disenfranchisement and environmental governance in postcolonial India. *Interventions: International Journal of Postcolonial Studies, 9*(1), 12–30.
Rousseleau, R. (2003). Entre folklore et isolat: La question tribale en Inde, de Mauss à Dumont. *Social Anthropology, 11*(2), 49–73.

Rutten, M. (1995). The tragedy of individualizing the commons: The outcome of subdividing the Maasai Pastoralist Group Ranches in Kajado District, Kenya. Conference paper: http://dlc.dlib.indiana.edu/dlc/handle/10535/5237.

Verbuyst, R. (2015). Claiming Cape Town. Ethnographic interpretations of Khoisan activism and land claims, Master Thesis - Research Master African Studies Leiden University/African Studies Centre, 2013–2015.

Walker, C. (2008). *Landmarked: Land Claims and Land Restitution in South Africa*. Cape Town: Jacana Media.

Part II
Beyond Local Challenges: The City-Park Relationships

Chapter 6
Urban National Parks and the Making of the Housing Market in Emerging Cities: Places of Exclusiveness, Land of Opportunities

Julien Migozzi, Abdul Shaban and Rafael Soares Gonçalves

6.1 Introduction

Mumbai, Rio de Janeiro, Nairobi and Cape Town, as emerging (GaWC 2012, 2008) and postcolonial cities of the Global South (Parnell and Oldfield, 2014; McFarlane 2008) share some common trends with regards to the local property market. As epicentres and recipients of the strong economic and demographic growth that characterised these emerging countries, these cities display new trends in the property market, which reshape the urban fabric and question patterns of social stratification (Michel et al. 2011; Kopper 2016; De Neve and Donner 2015).

Over the last two decades, the urban fabric of these metropolitan areas has been reconfigured by the emergence of homeownership as the dominant form of tenure and political agenda (de Soto 2000), the rising influence of the middle class, depicted as inherently urban, upwardly mobile and aspiring homeowners (Kharas 2010; Nijman 2006), the greater interconnectedness of property markets at the global scale (Aalbers 2016) and the influence of private property developers on the production, and fragmentation, of urban space (De Neve and Donner 2015; Nainan 2008; Bénit-Gbaffou et al. 2008; Tomlinson 2003). Mega events, such as the 2015 World Cup and the 2016 Olympic Games in Rio, or the 2010 World Cup in Cape Town, were also instrumental in the remaking of the urban landscape through the

J. Migozzi (✉)
Department of Geography, Ecole Normale Supérieure,
48 boulevardJourdan, Paris 75014 France
e-mail: julien.migozzi@ens.fr

A. Shaban
Tata Institute of Social Sciences, Mumbai, India

R. S. Gonçalves
Pontificale Université Catholique de Rio de Janeiro (PUC-Rio), Rio de Janeiro, Brazil

F. Landy (ed.), *From Urban National Parks to Natured Cities in the Global South*,
https://doi.org/10.1007/978-981-10-8462-1_6

promotion of mixed-use, market-oriented urban developments (Steinbrink 2014; Alegi and Bolsmann 2013; Samara 2010; Sánchez and Broudehoux 2013; Gaffney 2015).

Urban dwellers are confronted with a phenomenal increase in property prices, which unfolded along a similar timeline: from 2002 to 2008 for Cape Town, Mumbai and Nairobi, and more recently for Rio (2008–2014). The spectacular rise of property prices and the completion of intermediate to high-rise, speculative, real estate developments, whether on available land or in renovated urban areas, left a large part of the urban population with limited housing opportunities, especially in the low-income categories (Brown-Luthango et al. 2013; Searle 2013; Dupont et al. 2016). Formal and informal property markets coexist in these cities (Birch et al. 2016). National urban parks are precisely one of the sites where formal housing and informal settlements are juxtaposed, sometimes only separated by a few metres (Abramo and Cravino 2012; Gunter and Scheepers 2011).

Despite the rise of land and property values, and the intensity of urban and social change (López-Morales 2015), these cities remain quite neglected by the large scholarship on property market and housing prices. Southern cities have been historically marginalised in the production of economic geography and dominant urban economic theories (Robinson 2006). Due to the opaque nature of the property market (Harris 2015) and a general lack of reliable data for Southern cities (Parnell and Oldfield 2014), few studies analysed the evolution of these cities through the lenses of real estate markets with a quantitative perspective.

Using original data on property prices, this chapter questions the presence and the role of a large urban park in terms of market dynamics. The housing crisis and the spectacular increase in real estate prices had clear implications for urban national parks in these cities: the land and resources provided by the park are claimed by the urban poor, in the form of informal settlements and squatting strategies (Roy 2015) while the park, as a major amenity and a potential reserve of land, attracted developers and investors. Urban national parks are arguably depicted and understood as a valuable amenity for city dwellers, but little is actually known about the relationship between urban parks and housing prices in emerging cities. For the sake of data availability and accuracy, we limited our analysis to the formal property market, even though informal markets are dominant in many areas of these urban agglomerations. As less data is available for Rio de Janeiro and Nairobi, this chapter mainly focuses on here Mumbai and Cape Town.

This chapter is divided into three sections. The first one presents the methodologies used for gathering the relevant data and building a database for the study of two emerging cities. Tracing the spatial dynamics of housing prices and the local market trends both in terms of supply and demand, the second section investigates the interweaving of the park within the urban housing market. Findings highlight how the interplay of market forces and residential contexts led to a process of homogenisation in Cape Town and to a process of differentiation in Mumbai. Finally, focusing on marketing strategis and discourses, we discuss the making of the park as an amenity and investigate its role in market trends and the determination of property prices.

6.2 Searching and Mapping Housing Prices in Emerging Cities

Accurate, timely, reliable and geo-referenced data is hard to obtain or generate, especially for the sake of comparative research. The challenge is even more complicated when conducting research on cities of the Global South, where public institutions are confronted with a lack of resources which hampers the collection and the reliability of data (Arfvidsson et al. 2016). To overcome this issue, this chapter combines therefore primary and secondary sources to build databases on housing prices.

6.2.1 Collecting Data on Housing Prices

For both cities, we use administrative boundaries as spatial units for measuring and mapping the dynamics of property prices. We selected the metropolitan area of the City of Cape Town, as delimited by the boundaries of the 2011 census. For Mumbai, we refer to the Municipal Corporation of Greater Mumbai (which despite its name does not include recent suburbs). Its its various "sections", extracted from the 2001 Census of India, are used for mapping ratio and growth rates.

For Mumbai, the data on housing prices was collected from secondary sources, in this case the *Times of India.* This newspaper regularly publishes the selling prices of residential flats, for major localities in the city, and focuses on the market values as assessed by corporate and real estate agents. Prices have been collected on three different dates from the *Times of Indias*'s supplementary paper *Property Times* on September 6, 2003, December 25, 2010 and September 15, 2012. These prices have then been extrapolated, using a polynomial inverse weighting method, and adjusted to neighbouring areas (Chen et al. 2015). A property value for each tax ward was calculated. The data on Mumbai focus therefore on a particular market product (flats), from which an aggregated variable was derived.

For Cape Town, this chapter exploits the original and primary data sourced from of the City of Cape Town. For the 1990–2015 period, a sample of 600,000 residential sales has been extracted and georeferenced. This is a primary and original source, as in South Africa most existing research on the real estate market relies on databases constructed by banks, such as the Absa House Price Indices (Simo-Kengne et al. 2013; Das et al. 2010), or data originally derived from the records of the deeds office and then made available by private companies (Steedley 2014; Du Preez and Sale 2013). Information on housing prices at the neighborhood level and cartographic visualisations are scarce for most Southern cities: this chapter provides empirical evidence from original, fine-grained data.

For Cape Town, prices have been adjusted to the 2014 rand value, using the Consumer Price Index as calculated by the South African Reserve Bank and Statistics South Africa, in order to mitigate the effect of inflation when analysing the spatial dynamic of prices over time.[1] Furthermore, we excluded transactions located in the central business district (CBD). The CBD might be close to the park but constitutes a different, specific segment of the metropolitan real estate market, especially since the late 1990s and the renovation of the CBD through public-private partnerships targeting mixed-use, commercial and residential developments (Didier et al. 2012; Pirie 2007). Using the real estate agencies' own practices and spatial subdivisions, we delineated four main residential areas that circle Table Mountain National Park (TMNP) (Fig. 6.1).

- the Southern Suburbs, which extend on the eastern side of the park, from Rondebosch to Muizenberg;
- the Peninsula, which encompasses a large portion of the park and all the residential areas from Hout Bay to the most southern part of the peninsula;
- the Atlantic Seaboard, the upmarket residential area bordering the western side of the park and stretching along the Atlantic coast;
- the City Centre, circled by the northern section of the park.

Finally, for Mumbai, the polygon centre of each tax ward was used to measure the distance to the nearest boundary. Similarly, for Cape Town, a distance variable was calculated for each transaction, using the Euclidean distance to the nearest park's administrative boundary.

6.2.2 Computing and Mapping Property Prices

For Cape Town, using a Stewart method (Giraud and Commenges, 2016), we computed potential housing prices on a 100 meters grid, with a span of 200 meters. These potential prices are then mapped using a decile discretisation, which allows to compare the distribution of housing prices across time. This modeling of housing prices, based on spatial interactions and a distance function, allows a better grasping of spatial dynamics by overcoming the issue of the modifiable areal unit problem (Fotheringham and Wong 1991), and helps to capture and retrace the evolution of the housing market at a fine-grained scale (Le Goix 2016). Finally, using aerial pictures, we excluded the transactions in the large commercial and industrial areas when computing prices. One main challenge regarding the data was to eliminate transactions resulting from the transfer of government houses, for which prices are recorded but are actually disconnected from market values. These transactions are simultaneously registered (on the same day), display absurd prices and tend to be concentrated in previous black and coloured townships. Various filters (e.g., deviation to the area's median price; number of transactions recorded the same day with a similar amount, etc.) were applied to identify these outliers and eliminate them. The lowest house value remains difficult to define, but field research and

[1] In 2014, 1$ USD = 11 rands (ZAR).

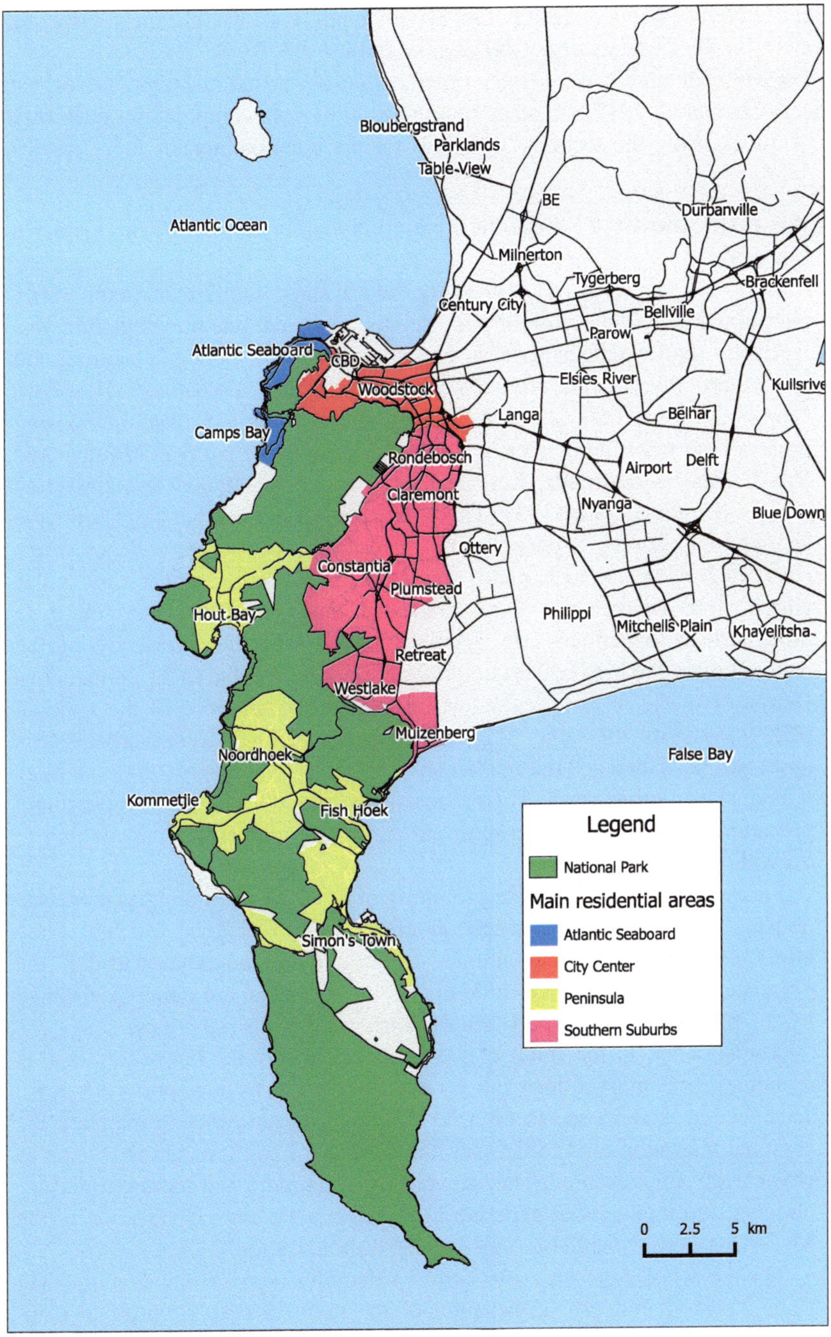

Fig. 6.1 Localities around Table Mountain National Park

online advertisements indicate that 80,000 rands was the cheapest price for a property in the Cape metropolitan area in 2014.

For Mumbai, an inverse weighted method has been used to interpolate real estate prices (Chen et al. 2015). Average real prices have been computed for each section of Mumbai, using the sections' centroids for the interpolation.

Box 6.1: Contrasted Situations Around Nairobi National Park (Frédéric Landy)

Nairobi has experienced a property boom since the 2000s (temporarily interrupted in 2008 because of post-election violence): the price inflation was fueled by rural exodus and urban growth, but also by money laundering and real estate investments from South Sudanese and Somalians. However, Nairobi National Park is unable to generate a general increase of housing prices in the surrounding area, a similar scenario to the parks in Mumbai and Rio. On the western side, Karen Blixen's former plantation was converted into residential lots after 1931. This green and posh area, that neighbours the Ngong Road Forest, is occupied by many institutions (e.g., colleges, churches, etc.) endowed with parks and gardens as well as upper-class villas where white Kenyans are overrepresented among the owners. Yet, even though these inhabitants are mostly satisfied with the proximity of the national park, the high real estate prices are here more due to the endogenous factors of the local historical situation than to the presence of the national park just across the large Magadi Road. Furthermore, other urbanisation processes have changed the market environment, with the growth of the huge Kibera slum further north. There, more than 250,000 inhabitants have found precarious homes through an informal real estate market that has nothing to do with the national park.

On the north of the park, the proximity of downtown Nairobi has increased housing prices, while further east, along the Mombasa Road which is a key artery leading to the international airport and an industrial corridor, the presence of the park has very little impact on prices that are rather determined by the proximity to transport infrastructures and industrial zones.

The situation in the south of the park is probably the one where market dynamics are most influenced by the proximity to the protected area. That section is supposed to be a no-building zone in order to maintain the seasonal migration of wildlife. Yet the Kitengela Land Use Master Plan was never really implemented. This savannah land is more and more coveted for housing developments or agricultural farms, and the individualisation of the Maasai properties facilitates mostly illegal urbanisation.

On the whole, the case of Nairobi resembles more Mumbai more than Cape Town. It reflects the isolation of the national park from urban planning frameworks and the lasting effect of a "conservation fortress" view that contributed little to the integration of the park within the urban fabric.

Box 6.2: Tijuca National Park and the Rio Housing Market (Rafael Soares Gonçalves)

The real estate market plays a central role in the production of space in the city of Rio de Janeiro, especially in the recent context of preparing the city for the 2016 Olympic Games. Some relevant work aimed at analysing how the interventions for the Olympic Games removed many *favelas* and benefited the areas with higher real estate values (Brum 2013; Magalhães 2013). Similarly, many analyses underlined the negative impacts of the social housing project of the Labor Party (PT) government (*Minha Casa, Minha Vida*). The completion of large housing stock located in the periphery was instrumental in removing and further marginalising the poor (Cardoso 2013; Leitão and Araujo 2013). Few researchers focused on the issue of informal markets, excepting Abramo, who researched the particularities of the informal market without losing sight of the interdependence between the formal and informal areas of the city (Abramo 2011).

It is complex to define the different factors that influence the housing market in Rio, but the distinction, or rather the division, between the North and the South of the city, is critical to understanding the dynamics of the urban market. The mountains of the Serra da Carioca, in Tijuca National Park (TNP), act as natural boundaries between these two areas of the city. As Abreu describes, Rio de Janeiro was consolidated between the forest, the mountains and the sea. There is therefore a clear distinction between the districts close to the coast (the south zone), and those located in the northern zone, further away from the beach (Abreu 1992). The majority of working-class neighbourhoods are located far from the centre. On the western side of the park, the new, highly valued, speculative and mediatised neighbourhood of Barra da Tijuca, nested between the ocean and the Olympic Park constitutes the new frontier of the real estate market.

We seek to identify, by analysing the discourses produced by property professionals on two real estate websites,[2] the role and the importance of TNP when marketing a property or a neighbourhood. To encompass this north-south division, we chose three neighbourhoods: the Jardim Botânico, situated in the south zone, the Tijuca neighbourhood in the north zone and the Alto da Boa Vista, a neighbourhood nested in the park between the two areas of Serra da Carioca and Tijuca forest zones.

[2]Zap Imóveis (http://www.zapimoveis.com.br/) e Vila Real (https://www.vivareal.com.br/).

The neighbourhood of Tijuca is a traditional middle-class neighbourhood of the city, which has experienced a devaluation of prices in recent years as a result of urban violence. This neighbourhood was always associated with the large number of *favelas* that occupy the strands near the TNP. However, the neighbourhood has undergone a spectacular recovery after the occupation of the *favelas* by the police in the context of the preparations for the Olympics (Soares Gonçalves 2013). The Maracanã stadium is indeed adjacent to the neighbourhood. According to the website Zapimóveis.com, the price per square metre rose from 1,965 reais in January 2008 to 7,390 in July 2016, an increase of 276%. Online advertisements offer little mention of the park: the distance to *favelas* seems to be actually the most influential factor on the formulation and differentiation of real estate values. Thus, the proximity to the park and, as a consequence, to the adjacent *favelas*, contributes to the devaluation of real estate prices.

The neighbourhood of Jardim Botânico is located in the southern part of the city, near TNP, the Rodrigo de Freitas lagoon and is not particularly far from the beaches of Ipanema or Leblon. It used to be an industrial neighbourhood which evolved in the recent decades. Since the beginning of the nineteenth century, the Botanical Garden has been located in this neighbourhood, which was actually named after the garden. The advertisements available on the website Zapimóveis.com mention the existing green spaces in the neighbourhood, where a few houses owned by former employees of the Botanical Garden are still occupied by their heirs or were sold in recent years. Some other residential properties have been recently built. There is actually a land conflict between the Botanical Garden and the occupants of these houses: the Botanical Garden—with the support of the Neighbourhood Association of Jardim Botânico and with wide media coverage—started a judicial process to remove the residents, in the name of environmental protection. However, real estate speculation seems to be the main issue driving the conflict and the parties' interests: the price per square metre in the neighbourhood rose from 5,121 reais in January 2008 to reach 12,819 in July 2016, showing a 150% increase.

In relation to the neighbourhood of Alto da Boa Vista, property professionals utilise the presence of the TNP much more in their advertisements. This neighbourhood is close to the forest, and a few apartment buildings and other beautiful mansions are often used as reception halls for parties or events. The price per square metre also experienced strong inflation, rising from 1,755 reais in January 2008, to 5,932 reais in July 2016, an increase of 238% over the period. This area also benefited from the presence of the police in the slums of Tijuca and the south zone. The proximity to the forest and the mountains also provides a more clement atmosphere in contrast to the warm environment of the city of Rio de Janeiro, while extremely restrictive legislation on land zoning hinders the construction of new housing units. There may be some slums in this neighbourhood, but here the TNP has a greater power against the expansion of these *favelas*, precisely because of their proximity to the park's borders.

The mountains and the forest of the TNP are central elements in the production of the carioca landscape. But the importance of environmental externalities and green spaces needs to be evaluated within the wider dynamics of the real estate market, especially after the 2016 Olympic Games and the reshaping of the urban fabric.

6.3 Locating the Park within the Local Market Configurations and Dynamics

6.3.1 Attractive Land for Urban Dwellers and Developers: Looking for Housing, Looking for Profit

Diverging spatial trends actually characterise the property market when locating the park within the metropolitan contexts of Mumbai and Cape Town. Following its urban expansion, Mumbai is spreading towards the north (Shafizadeh Moghadam and Helbich 2013), in the direction of the Sanjay Gandhi National Park (SGNP). The scenario is the opposite in Cape Town: urban growth mostly occurred towards the northern and eastern parts of the city (Sinclair-Smith 2009), which, in terms of sales and new housing stock, concentrated market activities further away from the park (Fig. 6.2). However, the interplay of market forces resulted in intensifying the pressure on the park in both cities.

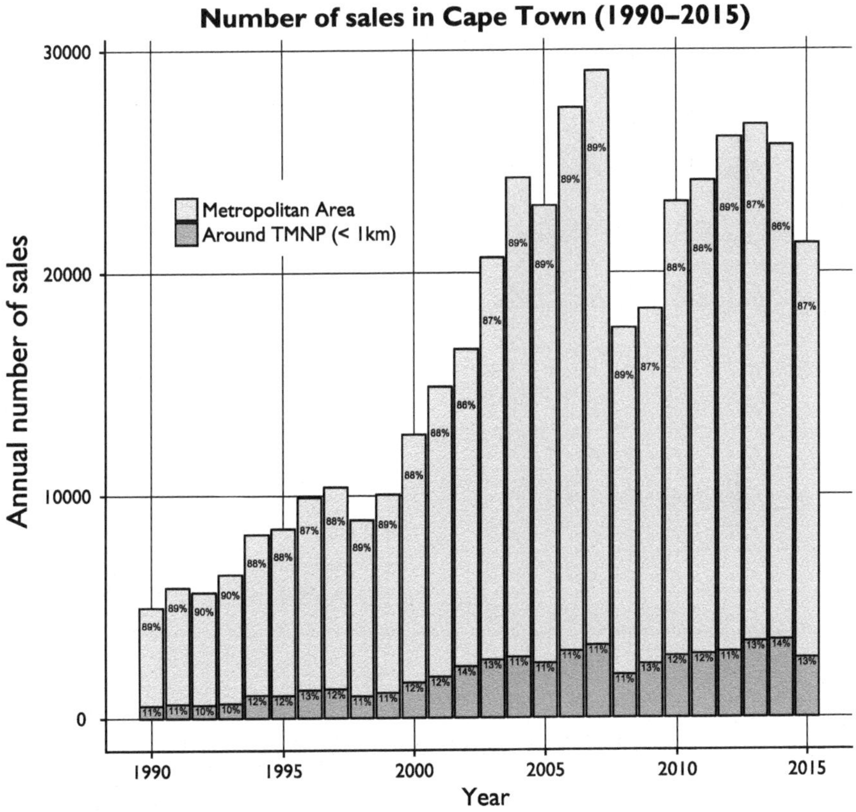

Author: J. Migozzi, 2017

Fig. 6.2 Number of sales in Cape Town (1990–2015)

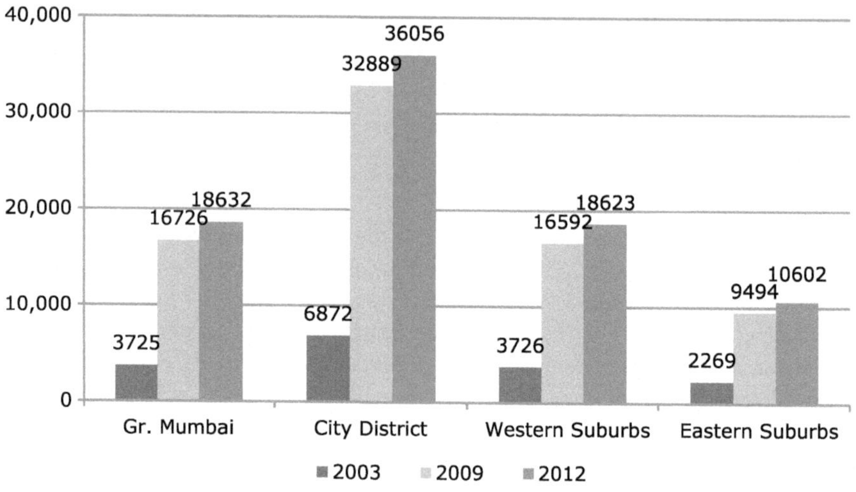

Fig. 6.3 Average prices (in Rs.) of flats (per square foot) in Mumbai. *Source Times of India* 2003, 2009, 2012. *Author* Abdul Shaban

Table 6.1 Comparison of factory workers' wages and prices of flats in Mumbai

Wages/Prices/Index		2003–04	2009–10
Average annual wages of factory workers	Rs. per annum	71,778	103,000
	Index no. of wages	100	143.0
Average price of flats in Gr. Mumbai (Rs)	Rs. per sq. ft	3,725	16,726
	Index number	100	449.0
Average price of flats in Mumbai District (Rs)	Rs. per sq. ft	6,872	32,889
	Index number	100	478.6
Average price of flats in the Western Suburbs (Rs)	Rs. per sq. ft	3,726	16,592
	Index number	100	445.3
Average price of flats in the Eastern Suburbs (Rs)	Rs. per sq. ft	2,269	9,494
	Index number	100	418.4
Flat price index–wage index ratio (2003 and 2010)	Gr. Mumbai	–	3.1
	Mumbai District	–	3.3
	Western Suburbs	–	3.1
	Eastern Suburbs	–	2.9

Author Abdul Shaban. *Sources* Maharashtra (2004, 2010) and *Times of India* (2003, 2009, 2012)

In the context of rising property prices and rental rates which reduced the affordability of housing (Fig. 6.3), the SGNP provided land for the urban dwellers unable to access housing through the formal channels of the residential property market. As shown in Table 6.1, the increase of flat prices was much superior to the increase of wages (Table 6.1), and the lower classes became increasingly priced out from the market: "only 5–6% of the population can afford a house in Mumbai"

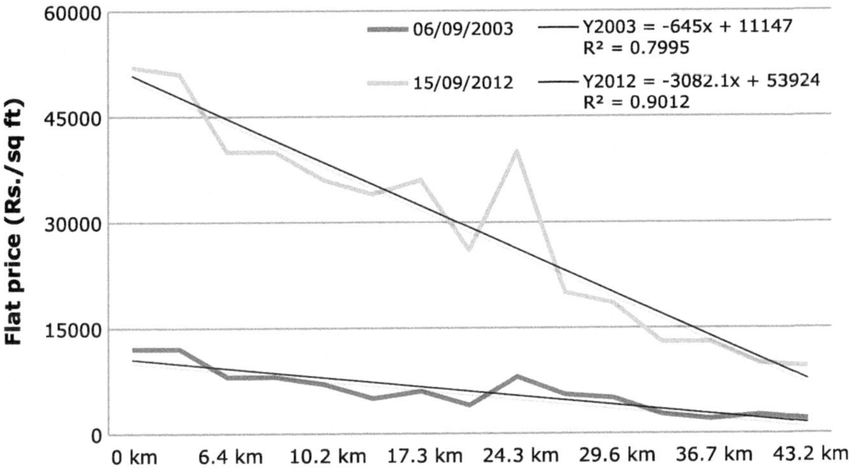

Fig. 6.4 Average price of residential flats along the Western Railway line and Western Expressway Corridor (2003 and 2012). *Source Times of India* (2003, 2009, 2012). *Author* Abdul Shaban

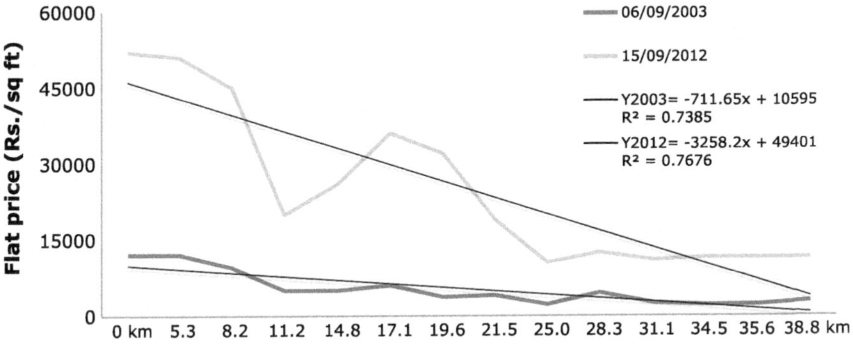

Fig. 6.5 Average price of residential flats along the Central Railway line and the Eastern Expressway corridor (2003 and 2012). *Source Times of India* (2003, 2012). *Author* Abdul Shaban

(Gandhi 2012, 226). The northward expansion of Mumbai and the rise in prices are also evident when comparing real estate prices recorded along the Western and Central Railways in Mumbai (Figs. 6.4 and 6.5), which respectively pass on the western and eastern side of the SGNP. Infrastructure megaprojects such as Mumbai Urban Transport Projet (I & II), Mumbai Urban Ifrastructure Project, Metro Rails or the Airport expansion further contributed to the forced displacement of urban dwellers, specifically the most vulnerable communities. A large segment of these displaced populations found shelter in areas where zoning prohibits development, such as the SGNP, but also on forest land or around mangroves.

From 1961 to 2001 the annual growth rate of the population around and in the park reached 9%, while that of Greater Mumbai grew by 4.2% in 1981–91 and 2.6% in 1991–2001.

Such a scenario of extensive land invasion did not unfold around the TMNP, where property developers targeted the remaining green land in areas historically appropriated through the market by the white, high-income inhabitants (Western 1996). Despite patterns of metropolitan growth shifting further away from the park, these privileged areas maintained their attractiveness on the housing market, especially during the 2003–2008 property boom. New housing schemes flourished all around the Peninsula, especially on the lower sections of the valleys (e.g., Sunnydale, Westlake, Silverdale, etc.). Market forces extended their reach to the upper sections of Noordhoek and Hout Bay, which recorded a notable increase in transactions as newly developed estates, mostly upmarket gated estates, entered the market (Guyot et al. 2015). Proportionally speaking, the number of transactions around the park increased over the last 25 years, but only minimally and during two different periods: in the initial years prior to the property boom (2000–2003) and from 2012 to 2015, when the market slowly recovered. Interestingly, our data sourced from the Deeds Office underline the fact that black townships such as Imizamo Yethu and Masiphumelele also registered a small number of transactions: formal market forces, despite the presence of informal settlements and the historic redlining of these neighbourhoods, extended their reach.

Unlike the SGNP, the TMNP itself was preserved from the urban expansion and the growing appetite of developers during the housing boom. In Mumbai, an important encroachment of the park, nurtured by the expansion of slums but also by economic activities (e.g., timber mafia, bootlegging, etc.) occurred indeed in eastern, western and southern directions, despite a May 1997 ruling by the Bombay High Court. Krishna Tiwari of City Forest Initiative counted more than 45 *padas* or slums in and around the SGNP boundary (Shalya 2012). In Cape Town, the park arguably benefited from a stronger and more restrictive zoning legislation. Influent and conservative residents' associations are still instrumental in the preservation of the "rural" atmosphere of their neighbourhoods, setting up exclusive architectural guidelines and building specifications (Guyot et al. 2014), which influence the type of housing products and the level of prices.

6.3.2 Parks and Housing Prices: Homogenisation in Cape Town, Differentiation in Mumbai

Looking at the spatial distribution of property values, Cape Town and Mumbai seem radically opposed in terms of how the park is located within the urban fabric. At the city scale, areas around SGNP are characterised by lower housing prices, while in Cape Town, properties situated on the edge of the TMNP remain the most expensive.

These spatial structures are time enduring in both cities. In 2012 and 2003, real estate prices in Mumbai declined along the Western and Central Railways, from

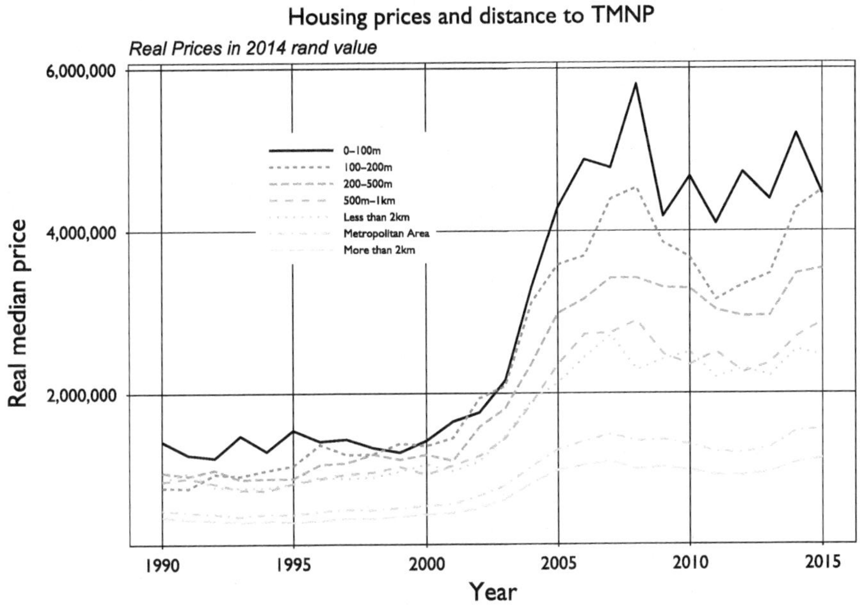

Fig. 6.6 Housing prices and distance to Table Mountain National Park

Colaba to the north. In Cape Town, the evolution of prices reveals the persistence of a strong distance gradient from 1990 to 2015: the closer to the park a property is, the great the value it is sold for (Fig. 6.6). This factor of distance strengthened during the housing boom from 2003 to 2007. Housing price indices, calculated for each distance gradient, highlight the disruptive impact of the property boom on the evolution of prices (Fig. 6.7). First evolving in parallel with the rest of the city, the values of properties closely located to the park experienced a dramatic increase from 2003. Inflation was much more pronounced around the park, until all areas started to display similar trends in 2010. In Mumbai, price growth peaked at 20% along the edge of the park, a level of inflation similar to the western seacoast or the gentrified areas of Dadar and Parel in central Mumbai.

Both cities experienced high growth rates, but for and with different spatial outcomes. In Mumbai, areas around the park experienced a process of social and spatial *differentiation* (Fig. 6.8), particularly visible in Kandivali East and Borivali East where upmarket, mixed-use condominiums mushroomed around the park's edge. First receiving migrants and low-income households, the northern part of Mumbai, historically cheaper, now also attracts higher-end projects, as land is being reclaimed from illegal settlements and recycled to the benefit of the upper and middle classes. A concentration of estate developments took place on both sides of the park, from Goreagon to Dahisar (in the Western Suburbs), and from Bhandup to Mulund (in the Eastern Suburbs), which created pockets of price inflation. In other words, the property market brought more heterogeneity around the park, which

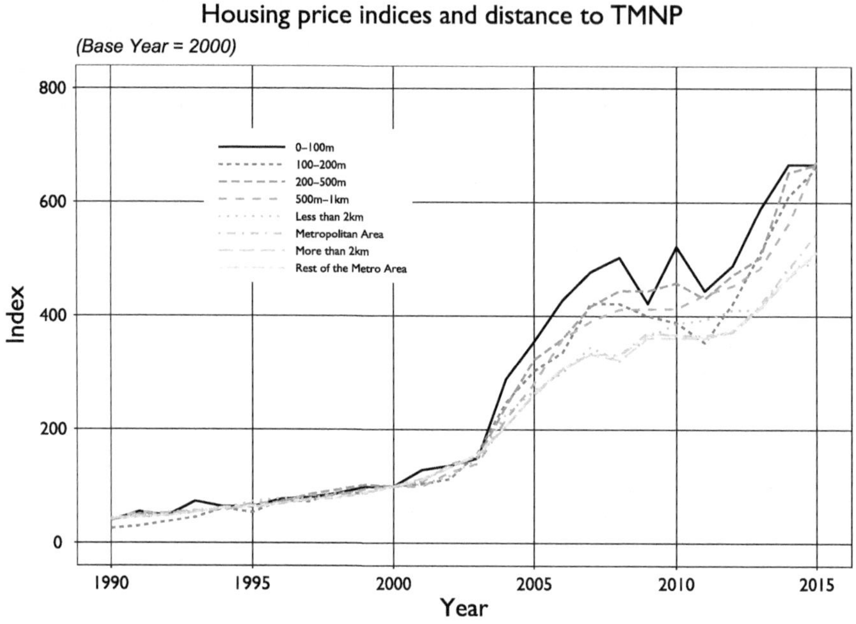

Fig. 6.7 Housing indexes and distance to Table Mountain National Park

materialised in a "cluster" structure through the stark juxtaposition of horizontal slums and vertical, upmarket estates. In Cape Town, the areas circling the park followed on the contrary a remarkable process of *convergence and homogenisation* in terms of housing prices and social profile (Figs. 6.9 and 6.10). Originally confined to the eastern side of the park (from Constantia to Newlands and on the Atlantic Seaboard), the highest prices remarkably extended to the eastern side of the Southern Suburbs especially during the 2003–2007 property boom. In the Peninsula, a similar evolution of prices underlines how most sections were increasingly established as upmarket areas, especially on the mountain's slopes. The lower sections of Hout Bay and Noordhoek valleys remain indeed the least expensive. The high values and the extreme volatility of housing prices located within a 100 m gradient confirm the consolidation of a luxury property market on the park's edges, especially along the Atlantic Seaboard, in Constantia and Noordhoek, where most buyers, relying on high incomes, assets and equity, buy with cash and don't need credit to finance their property acquisition. Overall, prices around the park tend to be 5 to 10 times more expensive than the median housing price in Cape Town (985,000 rands in 2014): the property boom and the evolution of prices reinforced the status of neighborhoods around the TMNP as exclusive areas.

Market dynamics remain nonetheless structured on the spatial hierarchy inherited from the colonial and apartheid planning regimes. All former townships are still associated with the lowest values: Ocean View, Masiphumelele or Imizamo Yethu, along with Hangberg, the south and predominantly coloured section of Hout

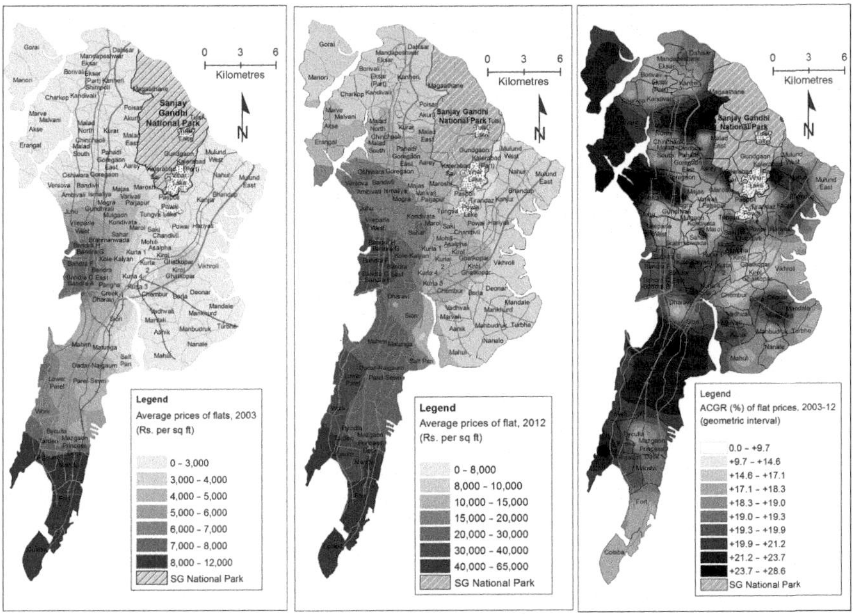

Fig. 6.8 Market value of residential flats in Mumbai (Rs./sq. ft)

Bay. These places share similar characteristics: smaller dwelling size, high rate of poverty and unemployment, building materials of a lower quality, presence of informal settlements, high density and limited market activity. In other words, despite the repeated claims of economic emergence, the spatial and social engineering of apartheid still heavily shapes property values and real estate dynamics in Cape Town: the maps underline that on the eastern side of the park, the road and the railways infrastructures, historically used as buffer zones for separating races, still produce visible barrier effects.

In Mumbai and Cape Town, market dynamics around the park, whether one is looking at the evolution of housing prices or at the type of developments, epitomize patterns of "*exclusionary urbanisation*". The stark contrast between Masiphumele and Lake Michelle, a secured and upmarket estate hosting a retirement village in the Peninsula, or between the slums and the newly developed, vertical compounds of Mumbai, epitomises how the park delineates a zone of housing opportunities for both ends of the social spectrum. Poor, and a land of for profit-oriented developers. Areas in and around urban national parks are attractive: in a context of urban growth and rising property prices, market forces turn them into a land of opportunity, both for private, profit-oriented developers and city dwellers aspiring to access housing, either through informal or formal market channels.

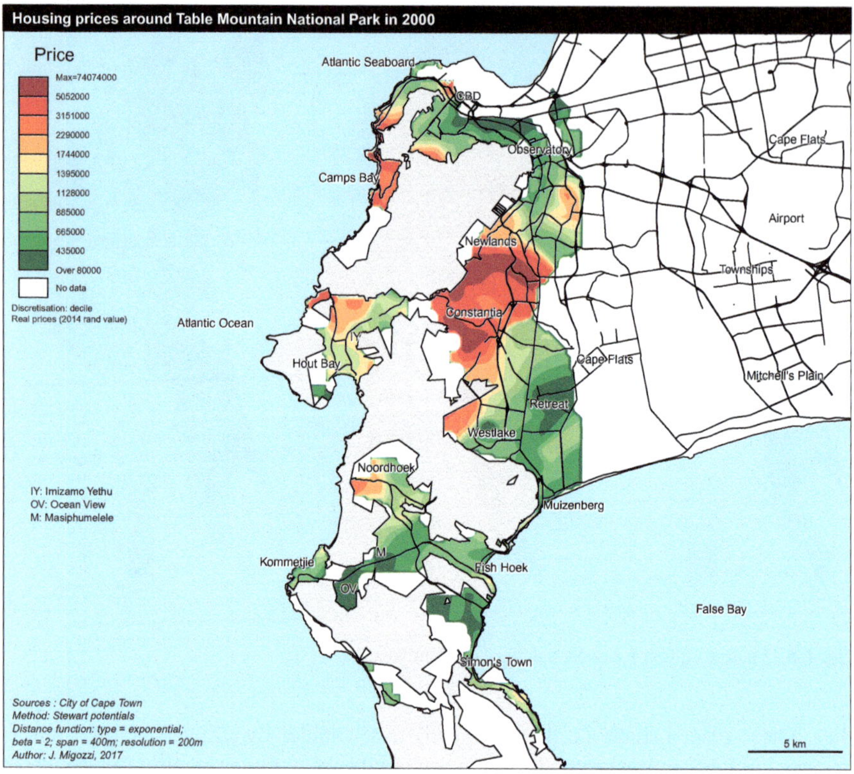

Fig. 6.9 Housing prices around Table Mountain National Park in 2000

6.4 "I Will Sell This Park Today": The Local Housing Market and Environmental Amenities

6.4.1 Marketing a Place, Selling a Property: The Park Incorporated into Market Devices

As previously demonstrated for urban markets of the Global North, pictures, narratives and storytelling patterns generated by the dominant actors of the property market (e.g., developers, real estate agents, etc.) form an essential component part of how housing markets are locally created and structured (Bourdieu 2005).

As the Internet became, on the buyer's side, the main source of information for real estate markets and a critical dimension of business and visibility for the property industry (Ba and Yang 2016), the presence of a national park is mainly used by market actors as a source of images. In Mumbai, on major websites such as Trovit.com, Propertywala.com or Thegooddeal.com, the SGNP is used for advertisement purposes, mainly to promote new developments located in the vicinity of

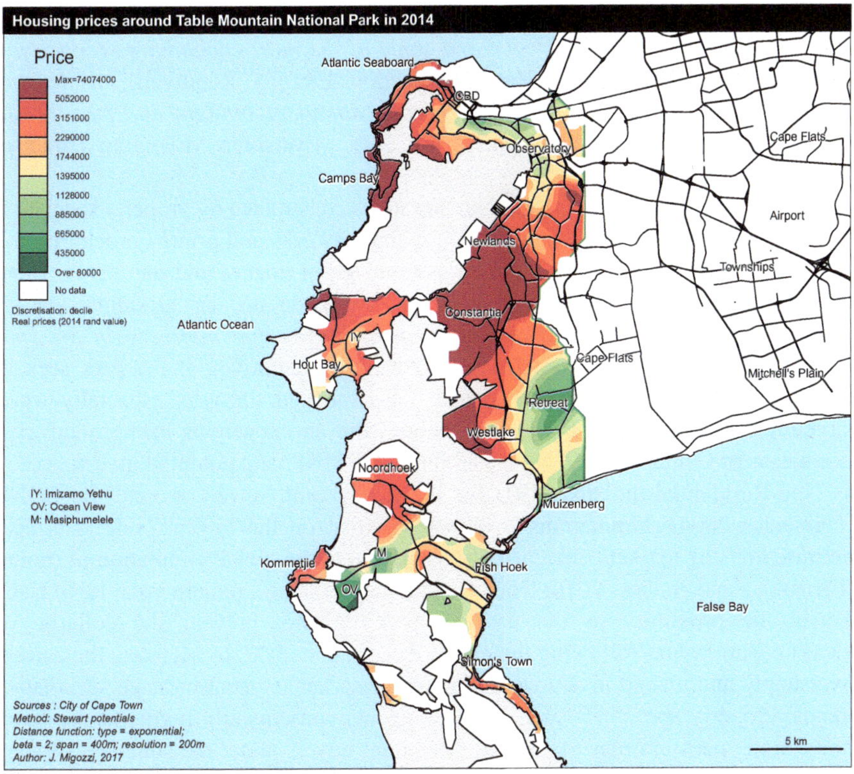

Fig. 6.10 Housing prices around Table Mountain National Park in 2014

the park. This is particularly striking for properties in the Western Suburbs. In Cape Town, sumptuous panoramic views of residential areas neighbouring the TMNP are widely used as visual marketing tools by the major estate agencies such as Pam Golding, Harcourts or Rawson Property. Most online advertisements, especially on local specialised websites such as Property24 or Gumtree, but also printed newspaper and electronic newsletters, feature appealing pictures of the property with the park in the background.

Unlike Cape Town where real estate agents mention the proximity of natural reserves, beaches and mountains for outdoor activities, very few advertisements directly mention the SGNP as an opportunity for such use of the park. The park is rather evoked through various linguistic proxies, whose similitude is striking in both cities. What matters, essentially, are the visual aesthetics offered by the park, which comes down in Cape Town to a "*green view*", "*a mountain view*", a "*scenic beauty*" or a "*mesmerizing view*".[3] Such descriptions strongly resonate with the

[3]See the following South African property websites: http://property.mg.co.za or https://www.privateproperty.co.za/.

Indian websites advertising properties around SGNP: "*own a home here and you will be spellbound by its opulence and some of the most enchanting scenery you have ever laid your eyes on*"[4] (Raheja Reflections Eternity, Kandivali); "*aside from a stunning garden-view, every home receives abundant sunlight and pure, fresh, unpolluted breeze from the neighbouring hills*" (Evershine Millenium Paradise, Kandivali East).

Through the discourses, images and narratives constructed by property specialists and circulating on websites, newsletters, social networks and printed brochures, the park is consequently absorbed into an assemblage of images and narratives which operate as "*market devices*" (Callon et al. 2007), defined as "*a simple way of referring to the material and discursive assemblages that intervene in the construction of markets*" (Boulay 2012). Pictures and narratives of the park seek to promote the property, the location and the associated lifestyle, especially when targeting high-income buyers and investors from an exogenous background: rich Europeans or Gauteng-based buyers around the TMNP; and potential investors such as Non-Residential-Indians (NRI) for the newly built towers around the SGNP (Fernandes 2004). Maintaining a climate of trust on the buyer's side is critical because housing markets, increasingly connected to the global circulation of capital (Fernandez and Aalbers 2016; Rouanet and Halbert 2016), remain volatile. In Cape Town, the proximity to the national park did not prevent the collapse of local housing values following the global recession in 2007. In Mumbai, the current oversupply and unsold inventory in the upper market segment indicate speculative trends. Property specialists such as estate agents consequently highlight the proximity to the park in order to convince potential buyers about the soundess of such investment opportunities. In the Cape Peninsula, where housing values have actually stagnated or even decreased in real terms since 2008, TMNP is depicted as a "*haven of tranquility*"[5] or "*serenity*", which makes the property a "*safe investment*" in areas characterised by "*exceptional growth*"[6] and amenities. Similar linguistic patterns target potential Indian investors. In a context of economic uncertainties and intense competition between real estate agents, asserting both the attractiveness of the area and the rationality of the investment is critical for property professionals in order to preserve the functioning of the market and secure an income.

[4]https://www.99acres.com/2-bhk-bedroom-apartment-flat-for-rent-in-raheja-reflections-eternity-kandivali-east-mumbai-andheri-dahisar-1200-sq-ft-r4-spid-L25416281?from_src=XIDPAGE_det&pos=SEARCH.

[5]http://www.property24.com/for-sale/noordhoek/noordhoek/western-cape/9089/104063437.

[6]http://propertywheel.co.za/2016/07/scarborough-property-sales-shows-exceptional-growth/.

6.4.2 Hedonic Modelling and Environmental Amenities: Evidences from Mumbai

The presence of the park clearly influences marketing strategies and discourses. Yet, the impact of the park itself on the determination of housing prices is not clear. Previous research inspired by hedonic modelling underlined that environmental factors are influential on property values (Nikolaos et al. 2011). Buyers in general would be willing to pay higher prices in the name of environmental quality in urban areas (Boyle and Kiel 2001). The main factors usually understood as determining property values are "*distances from place of employment and shopping, population density, the nature of public services (schools, police and fire protection, water supply, sewers,* etc.*), the amounts of traffic, noise, and air pollution, and so on*" (Grether and Mieszkowski 1974). Previous research on the impacts of environmental, structural and locational variables on housing prices in Mumbai concluded that the proximity to garden and water bodies are important factors in determining real estate prices. Despite some inherent limits of hedonic approaches in explaining the fluctuation of prices across time and space (Bowen et al. 2001), such a method was applied to Mumbai, in the hope to provide new materials and contribute to overcoming the lack of previous empirical research on Mumbai's real estate market from an environmental perspective (Gupta et al. 2009).

We tried to estimate the impact of various environmental amenities, including the SGNP, among other factors, on the real estate prices in the Municipal Corporation of Greater Mumbai. Table 6.2 presents the different factors selected. The analysis was conducted for the years 2003 and 2012, as the other independent variables were only available for these two dates.

It was found that in 2003 and 2010, the SGNP had not exercised any statistically significant influence on property prices. In other words, on average, the distance to SGNP does not influence the determination of housing prices when compared to other factors. As expected, the influence of the concentration of nitrogen oxides on real estate prices is negative, while distances to the CBD, to the western waterfront and to the western expressway positively impact on real estate prices (Pethe et al. 2014). The positive relationship between crime rate and real estate prices is surprising. This should be attributed to the large size and heterogeneity of the wards in Mumbai: the wards with higher real estate prices may have higher population density, resulting in a higher crime rate (Shaban 2010). While the impact of the SGNP on real estate prices might be statistically insignificant compared to other factors such as crime, pollution or transport, the park does polarise fast-growing areas of the local property market.

The method of Stewart potentials, which interpolates housing prices through a distance parameter, from georeferenced transactions, in order to map the evolution of prices around the park, confirms that other socioeconomic factors are instrumental in the determination of housing prices in Cape Town: former townships Ocean View or Imizamo Yethu might be extremely close to the park, yet they account for the lowest prices across the whole Peninsula. The time enduring and

Table 6.2 Determinant of flat prices in Mumbai

Determinants/*R*-Square/ *F*-statistics	Level variable					
	Flat prices 6 Sept 2003			Flat prices 15 December 2010		
	Coefficients	*T*-statistics	*p*-values	Coefficients	*T*-statistics	*p*-values
(Constant)	7935.56	5.547	0.000	36,370.76	5.92	0.00
Distance from CBD (km)	−168.14	−4.833	0.000	−739.20	−4.95	0.00
Distance from SGNP (km)	2.51	0.048	0.962	−75.12	−0.34	0.74
West waterfront (yes = 1; No = 0	459.53	1.133	0.260	3828.31	2.20	0.03
East waterfront (yes = 1; No = 0)	195.46	0.402	0.689	1123.37	0.54	0.59
Western Exp.way (yes = 1; No = 0)	460.12	1.165	0.246	3334.25	1.97	0.05
Western railway (yes = 1; No = 0)	306.52	0.749	0.455	1177.52	0.67	0.50
Eastern Exp. Way (yes = 1; No = 0)	−623.70	−1.102	0.273	−3795.00	−1.56	0.12
Central railway (yes = 1; No = 0)	−118.23	−0.260	0.795	−202.99	−0.10	0.92
Harbour railway (yes = 1; No = 0)	−914.69	−1.766	0.080	−247.79	−0.11	0.91
Crime rate (crimes per 1000 persons)	1.66	2.980	0.004	9.25	3.86	0.00
% of slum population	−1.89	−0.326	0.745	−40.81	−1.64	0.10
Density of population (persons/sq. km)	−0.01	−1.247	0.215	−0.01	−0.58	0.56
ACGR (%) of population 1961–01	−130.15	−1.455	0.148	−539.14	−1.40	0.16
PM10 load kg/d per 4 sq. km	0.31	1.061	0.291	−0.47	−0.37	0.71
NOx load kg/d per 4 sq. km	−0.16	−3.025	0.003	−0.54	−2.34	0.02
Adjusted R-Square	0.68	–	–	0.716	–	–
F-statistics	18.95	–	0.000	22.47	–	0.00

Author Abdul Shaban, 2016

uneven distribution of transactions and property values also underlines that the socio-spatial legacy of both the colonial and the apartheid regimes are more important than the presence of the park for explaining the spatial dynamics of the housing market. The social and racial profile of the neighbourhood, the specificity of the product (e.g., a gated community), and also the strategies and positioning of real estate actors on the market need to be investigated and taken into account when analysing the mechanisms of price determination, and identifying the specific role of the park:

> The creation of the Table Mountain National Park had a *huge impact*. I sold all these plots in Upper Noordhoek from 2000. People were rushing into buying these plots because they knew that with the park, the land was frozen, no further development would come. It was the last opportunity. And it definitely pushed the prices up.
>
> (Estate agent interviewed in San Michel, April 2016)

> We get a lot of foreigners visiting the Peninsula for the Cape Argus.[7] This year we had a Swedish gentleman who fell in love with the valley. I had something on the market for him. He didn't investigate the market much and agreed on the initial offer, which, to be honest, the seller overpriced a bit.... But this gentleman just wanted to buy immediately, so it was a done deal!
>
> (Estate agent interviewed in Noordhoek, June 2016)

6.5 Conclusion

Few urban national parks can be found inside emerging cities characterised by such an intense transformation of the urban fabric through the restructuring of market forces. Mumbai and Cape Town might share the existence of a park, but the local dynamics of the property market, both in terms of supply and demand, had contrasting impacts on the park itself.

In Cape Town, past and recent market forces, structured on decades of racial and racist zoning, gradually established the vicinity of the TMNP as a place for the "happy few": in the contemporary context of high interest rates and strong price inflation, the residential areas bordering the park became, socially speaking, more exclusive. Former black and coloured townships and informal settlements remain notable exceptions within this social landscape. These areas provide pockets of housing opportunities for the large part of the urban poor priced out from the local formal market by the interplay of price inflation, low wage regimes and mass unemployment. The situation in Mumbai is quite the opposite: rather than limiting the extension of affordable areas for the urban poor, market forces, associated with a strong urban growth, pushed vulnerable communities towards the park, leading to the extension of slums. As a result, the SGNP emerged as a place of housing opportunities for communities evicted and displaced by market dynamics (e.g., housing prices, large infrastructure projects) while property developers are now investing the land around the park.

The preservation of the park in Cape Town and the extensive mobilisation of the land in Mumbai underline the critical importance of the residential context in shaping the local patterns of land development and market structures. As affluent, predominantly white communities historically appropriated the park in Cape Town, the evolution of the housing market reinforced the status of the park as an upmarket area. Playing their trump cards of scenic views, picturesque landscape and

[7]The Cape Town Cycle Tour, also known as the "Cape Argus", is the largest cycle race in the world, and takes place every year in Cape Town.

privileged lifestyle, the property industry prioritised upmarket developments around the park, with a strong, strategic focus on gated and secured estates. Buying a property on the park's edge is certainly a marker of class and status in Cape Town, even more so behind high electric fences, safety cameras, and with a nice view. In Mumbai, patterns of urban development around the park actually highlight the disconnection between the historic, low-level and low-rise residential context (slums) and the recent market dynamics: the high-rise, speculative and vertical developments targeting high-income buyers and investors also underline the influence of property developers in seizing the land, orientating its use and extracting its value.

Engaging with the actor's discourses and strategies is therefore essential when investigating the making and the impact of environmental amenities through the lenses of the real estate market. Investigating the impact of the park through a simple factor of distance might not reveal, as the hedonic model concluded for Mumbai, the influence of the park on the determination of housing prices. Yet, the roll out of pictures and marketing narratives which directly mobilise the landscape or the presence of the park is particularly telling of how the park is actually incorporated into the making of local market configurations. Last but not least, the capacities of market actors to acquire land and invest (land recycling in Mumbai) or to restrict development (restrictive zoning policies in Cape Town) underline the instrumental role and differentiated impact of law implementation and policy instruments on the imbrication of the park within the local market structures and property dynamics.

References

Aalbers, M. (2016). *The financialization of housing: A political economy approach*. London: Taylor & Francis.

Abramo, P. (2011). A cidade COM-FUSA: A mão inoxidável do mercado e a produção da estrutura urbana nas grandes metrópoles latino-americanas. *Revista Brasileira de Estudos Urbanos e Regionais, 9,* 25–54.

Abramo, P., & Cravino, MaC. (2012). *Repensando la ciudad informal en América Latina.* Los Polvorines, Prov. de Buenos Aires: Universidad Nacional de General Sarmiento.

Abreu, Md A. (1992). *Natureza e Sociedade no Rio de Janeiro*. Rio de Janeiro: Prefeitura da Cidade do Rio de Janeiro.

Alegi, P., & Bolsmann, C. (2013). *Africa's World Cup: Critical reflections on play, patriotism, spectatorship, and space*. University of Michigan: University of Michigan Press.

Arfvidsson, H., Simon, D., & Oloko, M., et al. (2016). Engaging with and measuring informality in the proposed Urban Sustainable Development Goal. *African Geographical Review*, 1–15.

Ba, S., & Yang, X. (2016). *"Internet Plus" pathways to the transformation of China's property sector*. Springer Singapore.

Bénit-Gbaffou, C., Didier, S., & Morange, M. (2008). Communities, the private sector, and the state: Contested forms of security governance in Cape Town and Johannesburg. *Urban Affairs Review, 43,* 691–717.

Birch, E. L., Chattaraj, S., & Wachter, S. M. (2016). *Slums: How informal real estate markets work*. University of Pennsylvania Press, Incorporated.

Boulay, G. (2012). Assurer et rassurer le marché: géographie de l'incertitude économique et dispositifs de marché. *Rives méditerranéennes*, 113–133.

Bourdieu, P. (2005). *The social structures of the economy*. NY: Wiley.

Bowen, W. M., Mikelbank, B. A., & Prestegaard, D. M. (2001). Theoretical and empirical considerations regarding space in hedonic housing price model applications. *Growth and Change, 32,* 466–490.

Boyle, M., & Kiel, K. (2001). A survey of house price hedonic studies of the impact of environmental externalities. *Journal of Real Estate Literature, 9,* 117–144.

Brown-Luthango, M., Makanga, P., & Smit, J. (2013). Towards effective city planning—The Case of Cape Town in identifying potential housing land. *Urban Forum, 24,* 189–203.

Brum, M. (2013). Favelas e remocionismo ontem e hoje: da Ditadura de 1964 aos Grandes Eventos. *O Social em Questão, 29,* 179–208.

Callon, M., Millo, Y., & Muniesa, F. (2007). *Market devices*. Malden, MA: Blackwell Pub.

Cardoso, A. (2013). *O Programa Minha Casa Minha Vida e seus efeitos territoriais*. Rio de Janeiro: Letra Capital.

Chen, C., Zhao, N., Yue, T., et al. (2015). A generalization of inverse distance weighting method via kernel regression and its application to surface modeling. *Arabian Journal of Geosciences, 8,* 6623–6633.

Das, S., Gupta, R., & Kaya, P. (2010). Convergence of metropolitan house prices in South Africa: A re-examination using efficient unit root tests. *Applied Econometrics and International Development, 10*.

De Neve, G., & Donner, H. (2015). Revisiting urban property in India. *Journal of South Asian Development, 10,* 255–266.

de Soto, H. (2000). *The mystery of capital: Why capitalism triumphs in the west and fails everywhere else*. London: Basic Books.

Didier, S., Peyroux, E., & Morange, M. (2012). The spreading of the city improvement district model in Johannesburg and Cape Town: Urban regeneration and the neoliberal agenda in South Africa. *International Journal of Urban and Regional Research, 36,* 915–935.

Du Preez, M., & Sale, M. (2013). The impact of social housing developments on nearby property prices: A Nelson Mandela Bay Case Study. *South African Journal of Economics, 81,* 451–466.

Dupont, V., Jordhus-Lier, D., Sutherland, C., et al. (2016). *The politics of slums in the global south: Urban informality in Brazil*. India, South Africa and Peru: Taylor & Francis.

Fernandes, L. (2004). The politics of forgetting: Class politics, state power and the restructuring of urban space in India. *Urban Studies, 41,* 2415–2430.

Fernandez, R., & Aalbers, M. (2016). Financialization and housing: Between globalization and Varieties of Capitalism. *Competition & Change, 20,* 71–88.

Fotheringham, A. S., & Wong, D. W. S. (1991). The modifiable areal unit problem in multivariate statistical analysis. *Environment and Planning A, 23,* 1025–1044.

Gaffney, C. (2015). Gentrifications in pre-Olympic Rio de Janeiro. *Urban Geography*, 1–22.

Gandhi, S. (2012). Economics of affordable housing in Indian Cities: The case of Mumbai. *Environment and Urbanization Asia, 3,* 221–235.

GaWC. (2008). Classification of cities 2008. *The world According to GaWC 2008*.

GaWC. (2012) Classification of Cities 2012. *The World According to GaWC 2012*.

Giraud, T., & Commenges, H. (2016). SpatialPosition: Spatial position models. R package version 1.1.9000. https://github.com/Groupe-ElementR/SpatialPosition.

Grether, D. M., & Mieszkowski, P. (1974). Determinants of real estate values. *Journal of Urban Economics, 1,* 127–145.

Gunter, A., & Scheepers, L. (2011). "Crisylida Capital": Hatching informal township property markets to benefit low-income residents in Johannesburg, South Africa. *Urban Forum, 23,* 165–180.

Gupta, V., Mythili, G., Hegde, D., et al. (2009). Valuation of urban environmental amenities in Mumbai and its suburbs: A hedonic pricing approach. In *Indian Society for Ecological Economics 5th Biennial Conference*.

Guyot, S., Dellier, J., & Caillot, A. (2015). "Our rural sense of place" rurality and strategies of self-segregation in the Cape Peninsula (South Africa). *Justice spatiale | Spatial Justice, 7*, 1–36.
Guyot, S., Dellier, J., & Cerbelaud, F. (2014). L'environnement au profit des plus riches? Construction et hybridation d'un front écologique métropolitain dans la Péninsule du Cap (Afrique du Sud). *VertigO - la revue électronique en sciences de l'environnemen* 14.
Harris, R. (2015). International Policy for urban housing markets in the global south since 1945. In Miraftab, F., & Kudva, N. (Eds.), *Cities of the global South reader* (1 ed., xvi, 333 pages). London: Routledge.
India To. (2003). Property Times.
India To. (2009). Property Times.
India To. (2012). Property Times.
Kharas, H. (2010). The emerging middle class in developing countries. In: centre Od (Ed.), *Working Paper.* OECD Publishing.
Kopper, M. (2016). "My House, My Life": Experts, class meanings and the invention of a housing "market" in contemporary Brazil. *Horizontes Antropológicos, 22,* 185–215.
Le Goix, R. (2016). *Sur le front de la métropole. Une géographie suburbaine de Los Angeles.* Paris: Publications de la Sorbonne.
Leitão, G., & Araujo, H. (2013). Quando o futuro repete o passado: Uma análise dos conjuntos habitacionais produzidos pelo programa minha casa/minha vida na cidade do Rio de Janeiro. *Cuadernos de Vivienda y Urbanismo.*
López-Morales, E. (2015). Gentrification in the global South. *City, 19,* 564–573.
Magalhães, A. (2013). O" legado" dos megaeventos esportivos: a reatualização da remoção de favelas no Rio de Janeiro. *Horizontes Antropológicos, 19,* 89–118.
Maharashtra Go. (2004). *Economic Survey 2003–04.* Mumbai.
Maharashtra Go. (2010). *Economic Survey 2009–10.* Mumbai.
McFarlane, C. (2008). Postcolonial Bombay: Decline of a cosmopolitanism city? *Environment and Planning D: Society and Space, 26,* 480–499.
Michel, A., Denis, E., & Soares, Gonçalves R. (2011). Urban land trends in Southern countries. *Revue Tiers Monde, 2,* 240.
Nainan, N. (2008). Building boomers and fragmentation of space in Mumbai. *Economic and Political Weekly, 43,* 29–34.
Nijman, J. A. N. (2006). Mumbai's mysterious middle class. *International Journal of Urban and Regional Research, 30,* 758–775.
Nikolaos, K., Dimitra, V., & Agapi, X. (2011). Real estate values and environment: A case study on the effect of the environment on residential real estate values. *International Journal of Academic Research, 3,* 861–868.
Parnell, S., & Oldfield, S. (2014). *The Routledge handbook on cities of the Global South.*
Pethe, A., Gandhi, S., & Phatak, V. (2014). Real estate prices in Mumbai: Does the metro rail have an impact? *Economic and Political Weekly, 49,* 55–61.
Pirie, G. (2007). Reanimating a comatose goddess': Reconfiguring Central Cape Town. *Urban Forum, 18,* 125–151.
Robinson, J. (2006). *Ordinary cities: Between modernity and development.* London: Routledge.
Rouanet, H., & Halbert, L. (2016). Leveraging finance capital: Urban change and self-empowerment of real estate developers in India. *Urban Studies, 53,* 1401–1423.
Roy, A. (2015) Governing the Postcolonial Suburbs. In *Suburban Governance* (pp. 337–348). University of Toronto Press.
Samara, T. R. (2010). Playing Cape Town: Politics of stadium development for the 2010 World Cup. In *Urban spaces: Planning and struggles for land and community.* Lexington Books.
Sánchez, F., & Broudehoux, A.-M. (2013). Mega-events and urban regeneration in Rio de Janeiro: Planning in a state of emergency. *International Journal of Urban Sustainable Development, 5,* 132–153.
Searle, L. G. (2013). Constructing prestige and elaborating the 'professional': Elite residential complexes in the National Capital Region, India. *Contributions to Indian Sociology, 47,* 271–302.

Shaban, A. (2010). *Mumbai: Political economy of crime and space*. New Delhi: Orient BlackSwan.

Shafizadeh Moghadam, H., & Helbich, M. (2013). Spatiotemporal urbanization processes in the megacity of Mumbai, India: A Markov chains-cellular automata urban growth model. *Applied Geography, 40,* 140–149.

Shalya, C. (2012) First priority must be to remove encroachments, say experts. *Times of India.*

Simo-Kengne, B. D., Gupta, R., & Bittencourt, M. (2013). The impact of house prices on consumption in South Africa: Evidence from provincial-level panel VARs. *Housing Studies, 28,* 1133–1154.

Sinclair-Smith, K. (2009) The expansion of urban Cape Town. *Growth monitoring report.* Cape Town: City of Cape Town.

Soares Gonçalves, R. (2013). Une discipline olympique? Le retour des politiques d'éradication des favelas à Rio de Janeiro. *Mouvements, 74,* 24–32.

Steedley A. (2014) Understanding housing markets in South Africa. Expanding affordable housing market opportunities. South African Cities Network, Center for Affordable Housing Finance in Africa, 117.

Steinbrink, M. (2014). Festifavelisation: Mega-events, slums and strategic city-staging—The example of Rio de Janeiro. *DIE ERDE–Journal of the Geographical Society of Berlin, 144,* 129–145.

Tomlinson, R. (2003). *Emerging Johannesburg: Perspectives on the postapartheid city.* New York: Routledge.

Western, J. (1996). *Outcast Cape Town.* University of California Press.

Chapter 7
Why Did Leopards Kill Humans in Mumbai but not in Nairobi? Wildlife Management in and Around Urban National Parks

Frédéric Landy, Estienne Rodary and Bernard Calas

In 2004, leopards killed 19 people in Mumbai. Although fatal attacks have diminished since 2007 (Fig. 7.1), they show the scale of a problem that has received extensive local and even international media coverage. These attacks took place inside or on the edge of the Sanjay Gandhi National Park (SGNP), mainly in areas occupied by slums. They raised questions about the wisdom of siting a national park in a megacity of some 20 million inhabitants, the excessive density of the leopard population, and a way of managing space which in 1995 has seen more than 500,000 slum dwellers living in the park.

That at least is the view one might take on consideration of the Mumbai case alone. However, a comparative approach prompts a different way of thinking about things. In Nairobi, a city of 4 million people, the national park of the same name also borders slum areas and is home to a fairly dense population of leopards (as well as lions, hyenas and hippopotamuses), which represents a potential danger. Yet no wild animals have killed human beings in the park in living memory, other than a few incautious tourists who have left their cars to take close-up photographs of lion cubs jealously protected by their mothers.

How can this difference between Mumbai and Nairobi be explained? What is it that differs in the "humanimal" (Estebanez et al. 2013) relationship that sometimes makes the encounter between man and cat more dangerous for the man (and for the animal) in Mumbai than it is in the Kenyan capital? An initial hypothesis might be

F. Landy (✉)
University of Paris-Nanterre, Nanterre, France
e-mail: frederic.landy@ifpindia.org

F. Landy
French Institute of Pondicherry, Pondicherry, India

E. Rodary
French Research Institute for Development, Nouméa, New Caledonia

B. Calas
Department of Geography, University of Bordeaux-Montaigne, Pessac, France

F. Landy (ed.), *From Urban National Parks to Natured Cities in the Global South*,
https://doi.org/10.1007/978-981-10-8462-1_7

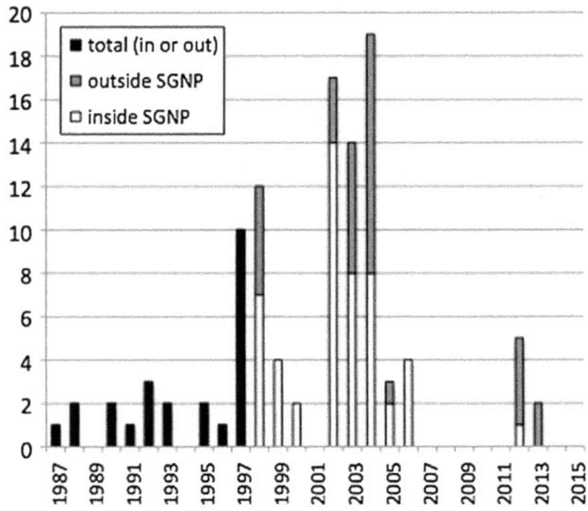

Fig. 7.1 Fatal leopard attacks in or around the park 1987–2015. *Source* SGNP

that a different way of managing the park, better coordinated with the management of the city, explains the absence of human victims in Nairobi. Perhaps the actors in the two spheres coordinate their efforts so that wild animals can be more effectively tracked in the city. And perhaps in Mumbai, by contrast, it is because the park and the city glare at each other, so to speak, across a barrier of mutual mistrust that the management of the leopard problem leads to such bad results. However, as we will see, this hypothesis proves to be largely erroneous. Sociocultural factors such as the representations of nature, or social disparities, which generate different vulnerabilities, also need to be considered. The management of wildlife and, more broadly, the way people live with it, should be seen not as neutral techniques or practices, but as signs of an institutional machinery and—beyond this—of an ontology of nature or at least a paradigm of protection. This leads us to a second hypothesis: much more than a simple question of the material management of wandering wildlife, there are more fundamental differences in the "ideological distances" between men and animals that may explain the contrast between the two cases studied.

7.1 Wildlife Conflicts: Situation Report

Mumbai is considered to be the world's most densely populated city. In terms of human population, this title is disputable, given that the estimates vary depending on the surface area considered. But in terms of the leopard population, there is no dispute. A count made early in 2015 recorded 35 leopards living in or around SGNP, making a density of 22 per 100 km^2, or less than 5 km^2 per leopard.[1] The

[1]Around the world, leopard densities in rainforests have been estimated to not exceed 1 per 10 km^2 (Bailey 1993).

attacks took place almost equally inside and outside the park (Fig. 7.1). The most recent deaths, in 2013, happened on the outskirts: a little girl urinating alongside her grandmother, after dark near her house; and a boy coming back in a group from school. The attacks thus resemble the spatial marker of a "buffer zone", an interface between the city and the park, extending in places more than 3 km beyond the external boundaries of SGNP (yellow squares on the map of Mumbai, see cartographic insert). At least five leopards have apparently established their territory outside the park, in Aarey Colony, an area on the outskirts of the park, situated right in the heart of Mumbai agglomeration but possessing "rural" characteristics, such as dairy cattle, forest and fields villages…

Human beings are far from the only victims. On the one hand, some leopards are killed (an average of two per year), sometimes in reprisal for attacks, usually hit by vehicles (section K on the map in the cartographic insert). In addition, the Adivasis, "indigenous" populations of Mumbai who live partly in the park, regularly lose goats (sometimes raised in the park despite prohibition of the activity) or dogs which, although semi-feral, are used as an early-warning system. A CCTV video popular on the Internet shows a leopard going into the entrance hall of a building in the Mulund suburb and coming out with a dog in its jaws.[2] So Mumbai's leopards are particularly intrusive, which prompts widespread fear of the animal, amounting at times to an urban psychosis.

Paradoxically, this psychosis affects all social groups (except to some extent the Adivasis, who are used to the presence of the animal), from those least exposed to risk to the most vulnerable: car-owning inhabitants of well-off households, with windows and doors in their houses, are much more strident in the media and social networks than slum-dwellers, who are actually the only victims,[3] but lack the means to publicise and express their concerns. What is often called bourgeois environmentalism (Baviskar 2002) has its limits, and there are plenty of people who are "non-environmentalist bourgeois" when it comes to leopards (Landy 2017).

The situation is compared with that in Nairobi, where leopards are rarely seen in the city, although one has perhaps established its territory in the Ngong Road Forest. Here, the cat, which has a maximum density in the park of around one individual per 6 km^2, kills "only" other animals: domestic animals (e.g. dogs, cats, etc.), and above all livestock belonging to the Maasai who live in the south of the park (164 pastoralists were compensated in 2012 for attacks by lions, leopards or hyenas, more than 40 km away from the national park, see Fig. 7.2).

A leopard that manages to enter a compound may kill all the goats and sheep inside it. So it is not that there are fewer conflicts in Nairobi than in Mumbai, it is their nature that is different. Local people sometimes call the Kenya Wildlife Service (KWS) because they have seen a carnivore in their garden, but the primary

[2]https://www.youtube.com/watch?v=75Yq0rOIsMY. Accessed on January 7, 2016.

[3]Though a lawyer from Mulund was killed in 2004, the only victim from the well-off social classes. He was in the habit of climbing into the park every morning before dawn… (Interview with members of the victim's family, January 2014).

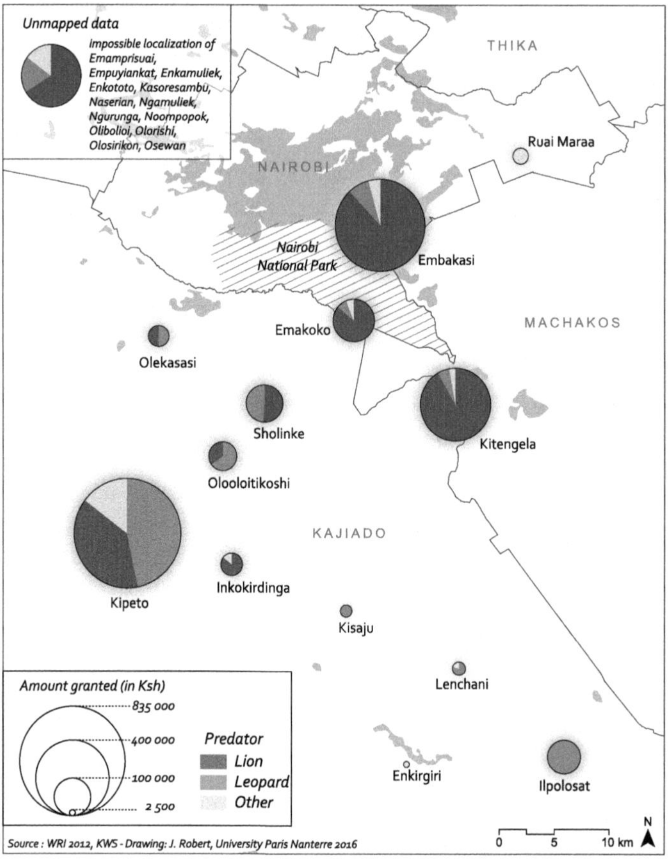

Fig. 7.2 Compensations ("consolations") granted by Kenya Wildlife Service to herders in 2012

victims of leopards or lions are livestock. That is why, despite the legal protection the big cats enjoy, the pastoralists often try to kill them in reprisal, which relocates and "humanises" the conflict.

In the city, the KWS intervenes in the whole conurbation, not only the low-density areas (e.g., villas with gardens, parks around institutions, etc.), which are the richest areas situated in the west of the city (Karen and Langata). For example, the largely abandoned industrial mining sites in the north-east of the park, where ponds can form during the rainy season, sometimes provide a habitat for hippopotamuses. And in the Kibera slum in the northern part of the park, which is too dense for many carnivores to wander, there is talk of warthogs and even hyenas digging up and eating corpses in the cemeteries.

It is time to explain why Nairobi's leopards, dangerous as these animals are to humans, are not deadly, unlike those of Mumbai. The hypotheses can be grouped into three broad categories: the first is ecosystem organisation, the second involves methods of managing the problem, and the third relates to sociocultural factors.

7.2 Dogs and "Naturban" Ecosystems

Another, highly contingent factor that lies outside these categories is chance. Given the statistical rarity of accidents, it cannot be ruled out. "We have been staying here for over 20 years and spot leopards at least three or four times a month, but to date no humans have been attacked", reported a woman from a slum in Mulund, a district of Mumbai, after the death of a little girl in 2012.[4] In this respect, wildlife-related accidents are entirely specific, given that, beyond structural determinants, they are always attacks by an individual. An animal suffering particular stress is very likely to show unusual behaviour. In the Indian countryside, villages may want a specifically dangerous elephant to be culled without necessarily wishing to see the species eradicated from their area (Sekar 2013). However, chance has to be ruled out as the only explanation, since the differences in mortality between the two cities are too large to be ascribed solely to (bad) luck. Beyond this, there is the question of the agency of the animals themselves, an issue recently rediscovered by certain authors (Midgley 1979; Haraway 2008), including geographers who—accustomed to considering environmental factors—had until recently neglected the animal dimension (Wolch and Emel 1998; Blanc and Cohen 2002; Buller 2013, 2014, 2015).

The possible role of taxonomy as a factor also does not seem to offer an adequate explanation. It is true that the Indian leopard (*Panthera pardus fusca*) differs from the African subspecies (*Panthera pardus pardus*). But the fact that fatal leopard attacks have almost ceased in Mumbai since 2007 obliges us to look elsewhere. Two explanations seem much more fundamental: predation by leopards on populations of domestic dogs, and the landscape configurations of the interfaces between park and city.

The leopard is an opportunistic predator, though its preference is for medium-sized ungulates. However, there is little documentation on the cat's real specialisations in terms of prey species (Hayward et al. 2006). A recent study on all the world's leopard subpopulations shows that the shortage of total prey biomass has a direct impact on livestock predation (e.g., cattle, sheep and goats). This impact is apparent regardless of the variations in body weight, spatial range and density of the different leopard populations. The authors therefore consider that the level of available wild prey biomass per square kilometre is a good predictor of conflicts between cat and human (Khorozyan et al. 2015).

---

[4]Ranjeet Jadhav, "Leopard devours 6-yr-old in Mulund", *Mid-Day*, July 17, 2012, http://www.mid-day.com/articles/leopard-devours-6-yr-old-in-mulund/172727. Accessed on June 7, 2016.

A study conducted in India's rural areas specifies the contribution of domestic animals to the leopard diet. Undertaken in Maharashtra, it shows that domestic animals represent 87% of the cat's diet, with dogs alone accounting for 39% of prey. Given the high density of goats and sheep in the countryside, dogs (together with cats) are overrepresented in the predator's diet (Athreya et al. 2014). This suggests a relationship between leopard and human that entails not a strategy of direct predation, but opportunism in a context of high predatory specialisation on domestic dogs. According to Butler et al. (2013), dog predation by large carnivores is low and infrequent around the world, but India's rural areas constitute an exception "where dogs clearly form a significant part of their diet, probably due to high dog densities and the relative scarcity of natural prey" (p. 135). This observation can be extended to urban India and in particular to Mumbai, where specialisation on dogs can once again be explained by the relatively low level of ungulate populations in SGNP (Edgaonkar and Chellam 1998; Tiwari 2008). The high density of dogs in Mumbai can cause declines in natural prey species for leopards (which in return leads to more predation on dogs). A study on the park has shown that wild animals account for 57% of leopard prey, as compared with 43% for domestic animals, including 24% for dogs alone (Surve 2015). Other, older studies gave an even higher percentage for dogs (47%, according to Tiwari 2009).

The situation is quite different in Nairobi, where mammal biomass and biological diversity are extremely high, in particular medium-sized mammals, both domestic (sheep…) and wild (impalas…), which constitute the preferred target for leopards (Foster and Coe 1968). This diversity has been on a downward trend for a century, but not to a significant degree (Western et al. 2009; Ogutu et al. 2013; Toth et al. 2014). Moreover, dogs are much less popular in Nairobi than in Mumbai, mainly for cultural reasons, but also because within the Indian metropolis and inside SGNP itself, the large number of slums, a habitat that offers little protection, encourages the use of dogs with varying degrees of domestication.

It is therefore very clearly the availability of prey that explains high dog predation by leopards in Mumbai and low levels in Nairobi. Human attacks in the Indian city would thus simply seem to be opportunities linked with this specialisation. Human beings are too big to constitute a preferred target for leopards, which will preferentially attack children or seated adults. The previously cited attack on a lawyer from Mulund in 2004 could only take place in steep terrain where the animal was able to spring on its victim as he was climbing among rocks.

However, these opportunities are also explained by landscape configurations. Here, two factors need to be taken into account: the dynamics of expansion in leopard populations, and their spatial hunting strategies. With regard to the dynamics of expansion, leopards are known to be solitary and territorial animals. Territories of a male and a female can overlap, but when young animals reach maturity they disperse until they find an area that is not already occupied by another individual of the same sex. Provided that prey is sufficiently abundant, these dynamics lead to a significant spatial expansion of the population. In the Mumbai case, the park—which is thought to be entirely occupied by the cat's territories—is surrounded by densely populated urban areas. The only area where the landscape connects to other wooded areas is located north of the protected zone, but is nevertheless crossed by an extremely busy road especially at night which is also the time when leopards use the human-used landscapes. By contrast, the interfaces between the park and the city are quite permeable: the boundary is only partially maintained by a wall, sometimes combined with barbed wire—and this often contains breaches through which animals and human beings can pass. In Nairobi, on the other hand, there is a contrast in the park boundary between its southern part, which is open to the 2500 km^2 of the Kitengela migratory corridor (Rodriguez et al. 2012), and its northern part which is enclosed by an almost impassable electric fence (although some animals do get through) that separates it from the city.

Leopard hunting strategies are therefore governed by the abundance of prey, by the availability of new hunting grounds and by the capacity of the landscape to offer the animals conditions for successful hunting. In this respect, the differences between the two urban ecosystems are significant, with Nairobi's being clearly less attractive than Mumbai's. The urban fabric is denser in Mumbai than in Nairobi and the interpenetrations between the park and areas of high population density are greater in India than in Kenya. More than 500,000 people were living in SGNP in 1995, and probably 120,000 still do so today. Aarey Colony is thought to be home to almost 20,000 Adivasis and 30,000 slum dwellers, fairly easy prey given the poor protection afforded by informal housing and the open areas around the settlements. Adivasis' use of natural resources is important: they compete for space with leopards. In Nairobi, by contrast, there are no houses in the park, and the Bangladesh slum, on the edge of the park, has only one or two hundred inhabitants, whereas Kibera, the city's largest slum, has a population of 250,000 but is relatively distant from the park. Nairobi's landfill sites contain almost no organic matter and are located far to the east of the city, whereas in Mumbai waste management is inadequate and garbage piles and bins are present in every neighbourhood, encouraging the proliferation of animals such as dogs, rats and pigs which are all attractants for leopards.

The degrees of interpenetration between urban areas and parks thus have a very clear impact on the structure and functionality of the ecosystems in which leopards operate. These ecosystems cannot be considered as natural environments, but should rather be seen as hybrid "naturban" milieus, where dogs occupy a position in the food chain that links the wildest and the most urban parts. The relation of humans to the leopard and the risks associated with it are therefore a secondary outcome of the big cat's predation on the canid, but this does not remove the specifically social dimension of the human link to the animal world and in particular the continuum between the domestic and the wild apparent in that link. Paradoxically, therefore, the canine presence invalidates a dichotomous vision of a conflict between a wild animal and human beings, and a vision of human beings moreover as separate from nature in their urban existence. Instead, it prompts us to identify the connections between those two spheres, in which the forms and varying effectiveness of spatial and species purification (porosity of interstices, presence/absence of domestic animals) is less a technical and administrative question than, as we will see, the reflection of different visions of the world.

7.3 Methods of Managing the Problem: The Law of Unintended Consequences

7.3.1 Recognition of Ecosystem Services in Nairobi

In both countries, the national park management services are known for their paramilitary character and their exalted vision of their role, which tends to be reflected in a certain lack of coordination with other actors and in often authoritarian decisions. Visitors to Nairobi National Park (NNP) are welcomed by a monument to the dead of the KWS, and by the statue of a heavily armed ranger. Many of these rangers are almost trained soldiers, who have notably followed a nine-month course in difficult conditions in Maniani. In Mumbai, the entrance to the national park is very different: no monument to the dead but a large fresco showing characters from Walt Disney's *Jungle Book*. However, appearances can be deceptive. Not only does India's Forest Department (FD) generally yield nothing to its Kenyan equivalent in terms of self-belief, not to say arrogance, but its management is even more rigorous, unwilling to leave anything to the municipality.

In Nairobi, the KWS organisation chart includes a Community Unit, responsible for relations with the outside world,[5] both rural and urban citizens, and the intervention unit against so-called problematic animals recruits Maasai, largely from Kitengela: the risks of collusion resulting from the social proximity between these recruits and the inhabitants seem less than the advantages of having rangers capable of negotiating with and understanding the needs of the pastoralists. Two decentralised sites in the heart of Maasai territory, in Kipeto and Kitengela, are also home to members of the Community Unit. There is no equivalently "social" approach in Mumbai, where it is only in recent years that the FD has begun recruiting Adivasis —but only for temporary contracts and menial tasks (see Chap. 5). There is no unit that specialises in managing the problems of relations with the city, and for the Territorial Division (which manages forestry issues outside SGNP), the problems are intractable: one Range Officer we spoke to even compared the situation to the war in Kashmir and the Line of Control between India and Pakistan.[6]

This is undoubtedly a matter of political will, but also a question of resources. The FD is under-resourced, because India, despite its ostensible interest in protected areas and iconic species such as the tiger, allocates proportionally fewer resources to protecting fauna than does Kenya. By contrast, fauna is a fundamental resource for Kenya, where tourism depends largely on wildlife, and where a significant percentage of the economy (17% of GDP) relies on this tourism industry. The nation and ethos of Kenya are partly constructed around the idea of Nature, both symbolically and economically the country's primary showcase—we will see that this also has consequences for the practices of citizens and the environmental

[5]Headed by the Community Warden, it also includes two corporals, two rangers, and a driver.
[6]Interview, February 13, 2013.

standards disseminated through education. Since the days of the first colonial conservation policies, nature was a fundamental imperial priority in the dominance of the British elite in this part of the world, subsequently transformed into a national priority adopted by Kenya's leaders and maintained by NGOs to the present day (MacKenzie 1988; Matheka 2005).

This explains why in 2005 the KWS accepted the program launched by a partly American NGO, The Wildlife Foundation, to finance two initiatives in NNP. First, the payment of compensation to pastoralists who had lost livestock to attacks by carnivores. True, the compensation was only partial, and the programme was ultimately suspended, but the idea was later implemented in the Maasai Mara reserve (Osano et al. 2013), and then in a more generous form at national level, by the Wildlife Act of 2013. In this respect, NNP has proven to be a laboratory—and its urban location, in the country's capital, has undoubtedly contributed to this role (see Chap. 14).

The second initiative was the payment of compensation to Maasai who undertook not to close their pastures to the south of the park and not to kill wild animals that had attacked their livestock. Here again, the initiative (2000–2012) fizzled out, but the idea was there: to pay for the ecosystem services provided by pastoralists who attracted wildlife by maintaining the pastures. NNP is a central space, politically dominant, but ecologically dominated by a whole hinterland through which wildlife travels: it depends on this hinterland, especially in the dry season, to be able to accommodate large numbers of animals.

There is nothing of this in Mumbai. With a few exceptions, the FD remains obstinately committed to a separation between park and city that no one may circumvent: neither people (it was out of the question that compensation should be paid to inhabitants of the park who have fallen victim to leopards), nor animals (leopards are supposed to stay in the park). This is a long way from the notion of ecosystem services, even though some foresters acknowledge that before its confiscation, the livestock held by the Adivasis in the current perimeter of the park maintained the pastures: the attempts at reforestation of almost the entire park were misconceived, in that they ran counter to the needs of the herbivores, potential prey for leopards. Fortunately, maintenance of pasture lands is now back on the agenda in the latest development plan.

7.3.2 Carnivores have never been modern

In Mumbai even more than in Nairobi, the carnivore represents a dual conceptual problem. First, it upsets the nature-culture dichotomy that underpins the delineation of protected areas and the separation between city and park. Indeed, following Latour (1993) and studies on the history of nature conservation (Adams 2004; Brockington et al. 2008), one may advance the idea that spatial arrangements such as national parks obey a twofold imperative of purification: by means of specific spatial arrangements, animals are rendered radically different from humans. Nonetheless, these arrangements produce hybrids that disrupt the grand categories through which "naturalism" operates (Descola 2013). The leopards that attack humans in Mumbai are a typical instance of these hybrids and show how hard it is for the naturalist and "modern" model to work.[7] It is a problematic issue for leopards and other potentially dangerous animals, which are seen not only as undomesticated, but also as non-urban. In this, the two cities reveal two different trajectories: the long-term resilience of a certain modern purification in Nairobi, and the historical weakness of that purification in Mumbai. In India, the issue is one of a history that remains resistant to spatial purification, understood as an effective and functional ("pure") segregation between nature and city. Which does not mean that tensions and diverging views do not exist in Mumbai today.

For most people and park managers, big cats are or should be living in protected areas only. Some of the leopards released in SGNP in the late 1990s had been caught in densely populated countryside and relocated to the park because it was imagined to be their "normal" habitat. Animals outside protected areas are supposed to be "stray" animals that must be "rescued", i.e. trapped and deported to protected areas. While few people in Mumbai dispute the fact that a protected area such as SGNP should contain leopards, as soon as the animals move slightly outside, it is considered a problematic exception that cannot be permanent.[8] In their survey of scientific literature on large felids in India, Ghosal et al. (2013) found that 90% of the published papers are based on research in and around protected areas, whereas about one-third of India's tigers live outside them—the proportion for leopards being probably higher. "The close association of felid research with PAs is also an outcome of these purification and translation processes discussed [by Latour and others]. Reports of large felids outside PAs and in multi-use landscapes are interpreted as the result of degradation of habitat and prey base within Pas….There is thus a 'correct place' for large felids, and their presence outside this moral geography is regarded as an abnormality" (p. 2678).

The result of these representations: in Mumbai, just as the slums need to be moved out of the park and the demolition of Adivasi villages remains on the official agenda, leopards that are deemed to have left the park must be returned to it. It is not only the political desire to limit the risks of conflict attributable to the FD, but also the belief in the non-urban character of these animals, which encourages the capture of any leopards seen outside the park. Such practices are counterproductive: not all such animals survive the trauma of capture, and competition with leopards

[7]See also Harrisson (1993): "*Forests mark the provincial edge of Western civilization, in the literal as well as imaginative domains .... We call it the loss of nature, or the loss of wildlife habitat, or the loss of biodiversity, but underlying the ecological concern is perhaps a much deeper apprehension about the disappearance of boundaries, without which the human abode loses its grounding.*" Leopard is to contribute to this transgression of boundaries….

[8]Although in charge of SGNP and the neighbouring Tungareshwar Wildlife Sanctuary, the FD does not appear to be very interested in enhancing the connectivity between these two forests. Yet, creating corridors and helping leopards move within a larger network of forests could help release the pressure on the national park by decreasing its population of leopards. Also, since mobility is critical to leopard survival, geographic isolation—like in the case of the SGNP—can lead to inbreeding, depression and increased extinction risk of the leopard population (Edelblutte 2016).

already present in the areas of relocation can prove dangerous to the introduced animal (Edgaonkar and Chellam 1998), and stress may lead to heightened aggression (Athreya et al. 2011). It was only after 2004 that the FD realised its error and stopped importing leopards—at the same time preventing further human fatalities, with the exception of a few rare tragedies.

In Nairobi, animal relocations also take place, but they are carried out with much greater circumspection. Above all, they apply primarily to herbivores, in the expectation that they will reproduce and attract big cats. Since the southern area is open, there is no clearly defined overpopulation threshold, in particular for leopards, since they are more mobile than lions (which tend to find themselves trapped in the national park in the rainy season when the herbivores have left).

7.4 Media Exposure and Perceptions: Social and Cultural Factors

7.4.1 Media Hijacking

Without wishing to underplay the human tragedies, there is no doubt that deaths in the slums of Mumbai are exploited by local political operators, in a very lively Indian democratic scene where the rate of voter participation is inversely proportional to income and qualifications. In this "political society", which does not really operate as a "civil society", so dominant is the role of political parties (Chatterjee 2004) that attacks in the slums, and also in the Adivasi settlements, generate big aftershocks. At the opposite pole of society, in the wealthy residences on the edge of the park, the attacks also make a lot of noise, because of the dual connections individuals enjoy: between each other, via mobile phones and "social networks"; and vertically, with the political community, senior local civil servants and the media. Although rarely the direct victims of attacks, wealthy Mumbai residents are the first to alert the authorities and the media.[9] Is not the "minor news item" the indicator of the emergence of a media society? Ultimately, Mumbai is experiencing a twofold process: the politicisation of attacks within poor populations, and media exploitation by the rich populations. The combination of these two processes explains the exceptional impact of the fatalities.

[9]They are also the first to call out on tribals undertaking illegal activities. During a field trip, E. Edelblutte visited a tribal village inside the park where a family was building a new house with better material than thatch since they had too many intrusions of leopards in their old house. A forest officer had to investigate after receiving a call from high-rise building residents living nearby who had reported "illegal activity from tribals". The officer, empathetic towards their situation, just gave them a warning and told them to be more discreet, so that he does not get other calls from urban residents. Also note that Adivasi sometimes "use" leopard attacks to expose their living conditions and ask for better amenities (especially electricity, waste management system and public lighting) (Edelblutte 2016).

In Nairobi, the situation is less tense, as much because of greater acceptance of wildlife, as because of the rarity of attacks on human beings. Nonetheless, as the commander of the KWS's Problematic Animal Control section explains, in the event of an attack, "the aim is to arrive before reporters, television, politicians, which is not always possible since we may have no vehicles or arrive too late. But the main thing is to listen to people, to provide reassurance. It is undoubtedly more difficult to deal with the Maasai than with the wealthy populations of Karengata: because they are armed, and in groups, and if they say that they are going to kill a lion, they will kill it, in ten minutes. You are talking to somebody who has lost animals! The definition of the conflict between human and wildlife is when you go to a situation where people are angry with you. But the situation has improved in the 12 years I have been here, there is more dialogue".[10] (We will see that this optimism has recently proved hollow).

In addition, the requests for intervention received by the KWS are socially and spatially very different: people living in Nairobi, particularly those in the wealthy western part, usually telephone *before* an attack, to report the presence of an animal, whereas the Maasai living around the city always call afterwards, to demand its capture, or perhaps compensation. The Community Warden claims that rural problems are the priority for the KWS, although calls from city dwellers represent half of reports: this is because the issue of conservation is less urgent in the city than in the countryside south of Nairobi. "It would make no sense to organise meetings in Langata on monkey problems! Those people contribute little to conservation, whereas the people in the south are key partners: they accept not to sell their land, they make the park unique, they don't fence or grow crops that much. We need to keep on engaging them".[11] This choice is not necessarily a choice in favour of the poorest: many of the Maasai are big landowners and economically powerful figures. However, it would seem that the situation is worse in Mumbai, where the powerful and the poor get very different treatment. In India, with the exception of attacks on people, when even the Adivasis alert the FD or the police, tribal people go unheard. Slum dwellers have no telephones or lack the courage to use them, whereas people from the wealthy residential areas will call even though they face less risk.

7.4.2 Perceptions

The crucial question is how the inhabitants behave in response to wild animals. Around Mumbai national park, but also in India's zoos and reserves, many people exhibit particularly inappropriate behavior, such as shouting or throwing stones at the animals. Some are absolutely determined to take photographs with their mobile phones, whatever the risk, while others succumb to noisy panic, instead of letting

[10]Interview, August 2, 1014.

[11]Interview, July 21, 2014.

the animal continue on its way. This causes stress to the animals, which may respond violently to human beings, either immediately or subsequently. It is clear that the simple "exposure to nature" during park visits is not enough to inculcate ecological values and behaviours (Mawdsley et al. 2009): most city dwellers have no notion of nature as heritage (to be respected) or as a resource (to be managed). It is a matter of perception of the environment, and more specifically of leopards: this is an essential difference between the two parks—in Nairobi, people have greater respect for wildlife and, while still afraid, have a general sense of how to behave if they encounter wild animals. "People in Kibera", a woman from the slum told us, "are aware. They had gone into the park with their school, they know that we must save lions outside the park if we want to see lions in the park".[12]

Nonetheless, this requires some qualification. Exceptionally, in March 2016, two lions were killed in two days in the south and east of the park, the second with a spear (suggesting Maasai reprisals), and the first by a KWS ranger who purportedly had no choice, since the anesthesia services were delayed and the crowd around the animal was making it aggressive. In the dry season, lions do not usually leave the park, but they were perhaps disturbed by work on the new bypass and railway line, which had apparently also caused the shutdown of electrification on the eastern fence. This also shows the limits of the self-control of Nairobians at the sight of carnivores in the city: as in Mumbai, they may respond by shouting, exciting the animal, or wanting to take photographs. The videos posted after the drama record the hooting of excited motorists. Nonetheless, accidents are less frequent in Nairobi, a difference partly explained by the different role played by wildlife in the national ethos of the two cities.

In Mumbai, two parallel initiatives are underway to try to reconnect the urban population with the animal world. The first was started by an activist previously affiliated to the Bombay National History Society, Krishna Tiwari, who is working with his group to run awareness-raising sessions on the danger of leopards in the slums and the Adivasi settlements. Signs posted on walls or trees explain how to stand tall to impress the animal in the event of an unexpected encounter, and encourage people to stick to certain times of day etc. (Photo 7.1). The other initiative is run by an association that is trying to work with the national park authorities, despite the reluctance they have shown until recently to collaborate with users or ecologists: Mumbaikars for SGNP (MfSGNP) works in schools to raise awareness amongst children and parents, as well as with police stations (often the first to be alerted in the event of an attack). MfSGNP also organises neighbourhood meetings, often in middle-class areas, to try to convince the inhabitants that leopards are not dangerous if one follows certain rules of behaviour: it is capturing them and releasing them in an environment that is not their territory that makes them aggressive.

[12]Interview, July 20, 2014.

Photo 7.1 "How to avoid leopard conflicts. Do's and don'ts in leopard areas" (Poster pasted in a tribal hamlet by the Forest and Wildlife Conservation Centre)

What we see here, therefore, is the emergence of an ambiguity linked to urban sociopolitical structures. In principle, the majority of the middle class or wealthy urban populations share a "naturalist" ontology, based on the modern dichotomy between nature and society—in this case, between park and city. For Mumbaikars, apart from a recreational area open to visitors, an urban national park should be less "urban" than "national", and above all "natural". So leopards have no legitimacy outside the boundaries of the park, for reasons of safety of course, but also because these animals have their territory in the sphere of nature, not of the city. In Mumbai, with the exception of a few activists mostly from the middle or wealthy classes, the majority of the population—both rich and poor—seems to have little attachment to urban (wild) fauna and more generally to the ecological project of the national park. In terms of ecological perspectives, the social contrasts are greater in Nairobi: whereas almost all the inhabitants of the Kibera shanty town seem to be frightened of wildlife, or at least not much interested in it,[13] those of the smart Karen district include plenty of households favourable to nature conservation and attracted to this very green area by the very proximity of the park and the prospect of seeing wildlife roaming free.

In both cities, this represents a sharp contrast with the Adivasi or Maasai minorities. For these two groups, there is no radical difference between human nature and animal nature, let alone a hierarchical order: they are practitioners of a blend of "animism" and "analogism", as classified by Descola (2013). Both the Adivasi and the Maasai consider themselves to be an integral part of an ecosystem, of a totality, which excludes neither humans nor animals. Sharing in this respect the perspective

[13]Animals can be associated with black magic. We were told about a hyena that had been sent to kill an enemy's livestock, but on the way had attacked the interviewee's female cousin. The latter agreed that "yes, we need national parks… because animals are destructive, and otherwise elephants could kill people". The park as protection for people, not wildlife….

of many Indian ecologists we met in Mumbai, they assert that God created an environment in which we cannot select certain elements and reject others (see Chap. 6). "The wildlife is our animals. If they kill one of our people we kill them. But we love them. We stay with them".[14] Wild animals are an integral part of the Maasai social system and of the mythology of the Adivasis: the Warli Adivasis of SGNP venerate Waghoba, the tiger-god, whose temple is in the park. The leopard is considered to be Waghoba's "little brother", and the two animals are often spoken of as one.[15] For some of the Adivasis interviewed, human fatalities are the consequence of a lack of respect for the leopard. The leopard's integration into the value system can also correspond to a spirit of resistance (Ghosal 2012, 120). Are not carnivores the ultimate "weapons of the weak"? (Scott 1985). In Nairobi, the KWS told us that some lions are killed less in reprisal than to send a message and attract attention: in this case, the conflict is not so much about dead livestock than about delays in compensation, or even about a well that the KWS had promised to help dig.

7.5 Carnivore Management by the Park, by the City, or by Both: A Matter of Postmodernity or A-Modernity?

Since wild carnivores are unaware of the official boundaries between city and park, there are in principle three ways of managing fauna outside the protected area: either the park is responsible for management everywhere, even outside its jurisdiction (type 1), or it is the business of the municipality (type 2), or the two institutions manage it together (type 3). In both countries, wildlife belongs to the state, and therefore falls under the authority and jurisdiction of the KWS and FD. However, it is only in Nairobi that the service invariably operates in the city (type 1). There is no dedicated hotline, but the KWS's telephone number is generally known, and people living in the areas most frequented by wildlife often have the private numbers of the Community Warden or rangers. Unlike in Mumbai, the police have little involvement with wildlife incidents. Ultimately, this type of management has proved quite effective—though the accidents of March 2016 seriously blotted the record—and the urban authorities certainly don't complain about being uninvolved in the issue.[16]

---

[14]Interview in Maasai Mara, March 24, 2012.

[15]On the contrary, for many non-tribal Hindus the leopard has no prestige since it is not the vehicle of any god—unlike the tiger who is attached to Durga, Parvati or Shiva deities. (Yet in Himalayan states the leopard is regarded as the vehicle of the Goddess).

[16]Locally, the KWS can be supported by other actors. For example, fauna in the Ngong Road Forest are partially managed by the Ngong Road Forest Sanctuary Trust, which sometimes also calls upon the Kenya Forest Service in the event of problems. Moreover, the recent decentralisation has given greater powers to the counties in the management of biodiversity (see Chap. 8), and the KWS could probably do more to gain support from elected officials.

Logically, since India's FD was not designed to operate in the urban environment, the second type of management—by the city—should be the mode that prevails in Mumbai. However, this is not the case and, in our opinion, here lies the heart of the problem. The city ignores the park, to the point that its inventories of open spaces completely elide any mention of the existence of this 104 km^2 area over which it has no oversight. As a result, until recently, when an inhabitant complained to the urban authorities about a leopard, he was referred to the FD on the grounds that the leopard belonged to the park, and that it was therefore up to them to handle the problem. However, when contacted, the FD would take the view that a leopard in the city lay outside its jurisdiction, and would refer the complainant back to the city authorities. It was only in 2012 that a joint telephone hotline was set up by the FD and the urban authorities, and that combined meetings of both institutions began to be held. Leopards have proved to be excellent go-betweens! (Landy 2017). However, these remain fragile and ad hoc structures, and there is no formalisation of relations with the city in the FD's organisation chart or activities.

Mumbai therefore has much to learn from Nairobi, whose approach suggests the benefits of granting compensation to limit reprisals against animals. It also shows—although the role of coercion in KWS's policy should not be underestimated—the need for mitigation, dialogue and coordination, and the advantage of a "proactive" prevention policy. For FD employees, this policy should be accompanied by the inclusion of a little social science in their training curriculum, to raise awareness of the human factor, of which they generally remain very ignorant.

As we can see, while both countries express the same concern for the protection of fauna and a wilderness often approached from a highly preservationist perspective, the political will in India is less strong, with the result that both financial and political resources are lacking. It is therefore for material as much as political reasons that the FD finds it hard to operate in the urban conurbation of Mumbai. Fortress conservation remains more or less dominant in official rhetoric in both Nairobi and in Mumbai, but in practice the KWS—as an instrument of that fortress—manages the urban boundary with an "efficiency" that is often authoritarian: this management of "purified" areas notably includes operations beyond the physical boundaries between the park and the city. In Mumbai, by contrast, actual practices contradict preservationist rhetoric. The FD's difficulties in intervening outside the park are not a sign of the strength of the fortress, but of its political weakness, since it has difficulty in managing the interface between humans and nature in a way that keeps them separate.

Baboons and Relocation in South Africa

In Tijuca park in Rio de Janeiro, agoutis and then howler monkeys have been reintroduced since 2010, but these animals remain scarce in a park that is otherwise devoid of large mammals. Table Mountain National Park offers a more interesting comparison with the Kenyan and Indian cases. It differs from the parks in Nairobi and Mumbai in that it does not form a single block, but several sections, each separated by road arteries or residential areas. It is also distinctive in being the country's only park that is not fenced, though paradoxically it directly borders the city, either via residential neighbourhoods or the edges of the city center itself.

This park contains neither leopards, nor any other predator that is really dangerous to human populations. It is home to a few caracals (*Caracal caracal*), but these cats are particularly discreet, not very wild, and too small to present a real danger to human beings. Large bovidaes (e.g., eland, *bubal*, etc.) and zebras populate the most southern part of the park, in the Cape of Good Hope section, but this zone is uninhabited and fenced. In the rest of the park, the most problematic and potentially dangerous species is the Cape Baboon (*Papio ursinus*), the largest of the baboon species, which can weigh up to 45 kg. While these monkeys do not present a direct threat to human beings, they can become very aggressive when attracted by food, not hesitating to enter cars or houses, or to make incursions into the Constantia vineyards which border the eastern edge of the park, causing very significant economic damage. These protected animals, now free of predators since the local disappearance of the leopard, have in fact become too numerous (around 400 on the peninsula): a degree of food pressure is therefore pushing them towards the city and resources that are fairly accessible, especially as the disappearance of pine trees has deprived them of a food source (pine nuts) and a habitat (they often lived in the trees). Tour operators and visitors want to have baboons close to roads, whereas at the same time, for safety reasons attempts are made to keep baboons out of the main visited areas: these animals are both an attraction for tourism and a threat to it.

Despite the means available to local authorities, this animal control problem is not being adequately managed. There is a baboon management plan that involves the city, the park and the regional Nature Conservation Department (CapeNature), but this focuses on problem individuals (which may be frightened off, relocated or euthanised, depending on the circumstances) and offers no strategic solution. In particular, there is no urban edge policy. None of the simple measures that exist to limit baboon incursions into inhabited areas is applied locally. The construction of fences along the houses at the edge of the park, and the use of baboon-proof waste bins, are neither funded nor compulsory. Apart from the case of the baboons, there are no policies for managing the park's ecological boundaries (see Chap. 8). Nothing, for example, prevents individuals keeping cats or invasive plants in their gardens, though these may have a crucial impact on environmental dynamics within the park.

It may be that this lack of coordination is a consequence of difficulties in the joint management between the administration of Table Mountain Park and Cape Town municipality in the last few years. However, these are genuine difficulties, partly the result of a specifically South African vision of the relation to nature, which is about more than a simple issue of local management. More specifically, they arise from a particular attitude to the boundaries that run across the country, designed both to separate social groups from each other and human beings from natural milieus. It is well known that the purpose of the establishment of apartheid in the mid-twentieth century was not only to keep social groups physically separate, but also to create an institutional segregation designed to maintain distinctions, even in cases where the black, Indian or colored populations might have practiced similar lifestyles (notably in urban and/or educated populations) and therefore claimed political equality. In this respect, the history of South Africa has been characterised both by highly controlled boundaries and by fear of their transgression. Against this general background, nature—in the form of parks primarily created to protect wildlife—has contributed to that spatial segregation by expelling marginal populations and then denying them access to those spaces. At the same time, protected wildlife itself constituted a potentially dangerous actor, capable of transgressing boundaries. Paradoxically, these concerns regarding transgression have persisted into the democratic transition, but shifted to the economic sphere. In this sense, intrusion, whether human or animal, constitutes a recurrent dimension in South Africa's political imagination. To employ the terminology used in this chapter, modern purification in South Africa has always been influenced by a threat of intrusion (Rodary 2016).

Apart from the construction of multiple barriers (picnic areas fitted with baboon-proof fencing), the material response to these fears in the sphere of wildlife conservation has primarily taken the form of relocation policies. Since the 1970s, this practice of wildlife relocation has become a highly lucrative business in the country. It affects all sectors of conservation, from private wildlife ranches to protected public spaces. In Table Mountain Park, this vision—which prioritises forced relocation over the negotiation of boundaries—is reflected, as we have seen, in a neglect of the issues relating to the boundary between park and city. It is also apparent in the direct management of the baboons, which are periodically transferred by truck to rebalance the populations in the south and north of the park. Rangers armed with firecrackers or paintball guns are tasked with intimidating the most intrusive individuals, and operate in the most touristic areas of the park and in isolated residential areas. And the problems of genetic impoverishment affecting the baboon population, which is divided into two groups by the residential area linking Noordhoek to Fish Hoek, are managed by regularly moving certain individuals from one group to the other. The creation of a corridor that would allow the animals to move from north to south is not on the agenda. This gives some idea of the problems—both institutional and symbolic—that may also arise in the event of a connection between the national park and the city's reserves, isolated in an urban environment that has plenty of connection issues to tackle other than those of wildlife.

Estienne Rodary

7.6 Conclusion

Our initial hypothesis was that a different form of park management, more in harmony with that of the city, might explain the absence of leopard attacks on humans in Nairobi. We speculated that the actors in the two spheres coordinate their efforts to ensure better oversight of wildlife. This hypothesis was not confirmed. Instead, we saw the importance of factors such as predation by leopards on populations of domestic dogs, the landscape configurations of the interfaces between park and city, and the diversity of representations of nature or social disparities, which generate differing vulnerabilities. This leads to a two-level conclusion regarding the role of the national trajectories in these countries of the Global South in respect of environmental concerns and their contribution to the new ways of understanding our relation to nature.

Leopards force us to consider in a new light the old controversy about environmentalism in developed countries versus the Global South. According to Guha and Martinez-Alier (1996), the theory of postmaterialism and the idea that only industrialisation and urbanisation lead to separation from nature, contributed to the belief that the deliberate environmentalist attempt to protect nature is only possible in developed countries. As Frykman and Löfgren (1987) wrote regarding Sweden: "One of the preconditions for a more Romantic and sentimental view was… the gradual withdrawal from an active or productive use of nature" (p. 78). Sentiment regarding animals, in particular farm animals, increased: "It was love at a distance." "Nature must first become exotic in order to become natural" (p. 83). However, Guha and Martinez-Alier disputed the view that the resistance to the creation of protected areas in developing countries is attributable to a lack of environmental awareness in their populations, arguing that the cause is simply a political conflict directly linked to the social marginalisation brought about by the establishment of protected areas. While Northern environmentalists tend to focus on nature conservation, Southern environmentalists address the interrelations between poverty and environmental degradation.

However, the terms of this controversy are being totally rewritten by studies such as this one, conducted in emerging economies where a growing elite is adopting a westernised *Weltanschauung*, while the majority of the population remains poor. We provide new answers to the controversy as to whether environmentalists are to be found only in developed countries or whether "empty-belly environmentalists" (Guha) exist in the Global South. At first sight, the former claim may seem to be validated by the SGNP case. On one side are middle-class environmentalists, whose way of life and culture are highly westernised; on the other, the slum dwellers who want to get rid of the leopards and are not interested in the national park. But where should the Adivasis or the Maasai be positioned on this "ladder of environmentalism", or bourgeois anti-environmentalists? It is clear that the question should not be posed in binary terms, since numerous types of "environmentalism" coexist. In Mumbai or Nairobi, as in developed countries, there are several types of middle-class environmentalism: the environmentalism of nature and wilderness has little to do with that of social ecology and environmental justice, nor with that of opposition to waste and industrial or electromagnetic pollution in cities (Blanchon et al. 2009). The first type of environmentalist would defend the leopards, the others not necessarily (Wolch 1998). Guha and Martinez-Alier are right to highlight the environmental mindsets of the Chipko activists and other peasant or indigenous movements; however, they should not set up the global North and South in opposition to each other, since many Indian environmentalists are quite close to the tenets of wilderness ecology, while often being "anti-poor". In contemporary India and Kenya, there are plenty of full-stomach environmentalists.

Ultimately, it is not because park and city are more integrated and harmonious in Nairobi than in Mumbai that there are fewer fatal attacks by carnivores. It is because landscape configurations and the availability of prey differ, but also because the big divisions of modernity between nature and culture have been tackled as a political issue in Kenya, whereas they have remained marginal in India's spatial representations. The paradox, therefore, is that though the two institutions responsible for the parks are formally organised on the basis of a very clear separation between the spheres of wild nature and the social world, the Indian trajectory has not really led to policies capable of establishing the boundaries that this separation entails.

This produces an apparent paradox, since current nature conservation policies around the world are moving towards greater integration, raising the question of whether Mumbai is not in fact better placed to pursue such integration than is Nairobi. The recent changes in nature conservation policies in Kenya, involving—as we have seen—a greater use of incentives, raise the issue of whether this is a genuine paradigm shift in conservation, or rather the empirical adaptation of an institution that seeks to perpetuate the national park model at any cost. In other words, far from making breaches in the "fortress", what we have is perhaps something of a preventative strategy conducted to maintain that fortress, not by extending its walls to annex further territory, but by building fences that are more social in kind and less visible. By incorporating conservation principles into the habitus of the population, the aim is to establish a new form of environmentalism as defined by Agrawal (2005): the creation of "environmentalist subjects" who will support conservation policies after adopting the imperatives promulgated by the authorities at all scales (e.g., national authorities, international NGOs with a strong presence in Kenya, etc.). In Mumbai, as we have seen, these policies were largely absent. It is not just a matter of education (there are as many school visits to the SGNP as to Nairobi Park, although ecological values are much better taught in

Kenya): in Nairobi, payment for ecosystem services, the creation in 2014 of a Management Committee including representatives of the "communities" and companies, the increase in visits by Kenyans to the park, where foreigners now represent a minority of visitors, the practices of companies that take their clients on trips to the park, have all helped to forge new forms of "mobilisation for nature" (Rodary 2009).

What we see in Nairobi, therefore, is an "adjusted modernity" rather than a genuine postmodernity that seeks to break down the dichotomies between nature and society, park and city. The fortress remains, and the recent "back to the barriers" movement, which promulgates a stricter conservation (Hutton et al. 2005), along with the revival of military anti-poaching policies in East Africa (Neumann 2004), are evidence that the tensions between integration and segregation remain extremely sharp within the conservationist world. In Mumbai, most city dwellers are without nostalgia for the countryside or forest from which many originate. They are fans of modernity, and the "concrete jungle" corresponds to an ideal largely founded on a progressive vision that places no value on the state of nature. Nonetheless, their spatial practices of proximity to animals, both domestic and wild, provide a glimpse of forms of connection and integration with the natural world which, even though they have no institutional correlative, may correspond—often inadvertently—to the objectives of integrated conservation, the "new ecology" and the incursion of nature into the city (Wolch 1998).

References

Adams, W. M. (2004). *Against extinction: The story of conservation*. London: Earthscan.

Agrawal, A. (2005). *Environmentality: Technologies of government and the making of subjects*. Durham; London: Duke University Press.

Athreya, V., Odden, M., Linnell, J. D. C., & Karanth, K. U. (2011). Translocation as a Tool for Mitigating Conflict with Leopards in Human-Dominated Landscapes of India. *Conservation Biology, 25*(1), 133–141.

Athreya, V., Odden, M., Linnell, J. D. C., Krishnaswamy, J., & Karanth, K. U. (2014). A cat among the dogs: leopard *Panthera pardus* diet in a human-dominated landscape in western Maharashtra. India. *Oryx, 50*(1), 156–162.

Bailey, T. N. (1993). *The African leopard. Ecology and behavior of a solitary felid*. New York: Columbia University Press.

Baviskar, A. (2002). The politics of the city. *Seminar, 516*, 40–42.

Blanc, N., & Cohen, M. (2002). L'animal. Une figure de la géographie contemporaine. *Espaces et Sociétés, 110–111*, 25–40.

Blanchon, D., Moreau, S., & Veyret, Y. (2009). Comprendre et construire la justice environnementale. *Annales de Géographie, 1*(665–666), 35–60.

Brockington, D., Duffy, R., & Igoe, J. (2008). *Nature unbound: Conservation, capitalism and the future of protected areas*. London: Earthscan.

Buller, H. (2013). Animal geographies I. *Progress in Human Geography, 38*(2), 308–318.

Buller, H. (2014). Animal geographies II: Methods. *Progress in Human Geography, 39*(3), 374–384.

Buller, H. (2015). Animal geographies III: Ethics. *Progress in Human Geography, 40*(3), 422–430.

Butler, J. R. A., Linnell, J. D. C., Morrant, D., Athreya, V., Lescureux, N., & Mckeown, A. (2013). Dog eat dog, cat eat dog: Social-ecological dimensions of dog predation by wild carnivores. In M. E. Gompper (Ed.), *Free-ranging dogs and wildlife conservation* (pp. 117–143). Oxford: Oxford University Press.

Chatterjee, P. (2004). *The politics of the governed: Reflections on popular politics in most of the world*. New York: Columbia University Press.

Descola, P. (2013). *Beyond nature and culture*. Chicago: University of Chicago Press.

Edelblutte, E. (2016). *Landscape and political ecology approach for long-term conservation of leopards (Panthera pardus) in a human-dominated landscape, India*. Report for UNPEC and EVS lab, University of Lyon 2. Unpublished.

Edgaonkar, A., & Chellam, R. (1998). *A preliminary study of the ecology of the leopard Panthera pardus fusca in the Sanjay Gandhi National Park, Maharashtra*. Dehra Dun: Wildlife Institute of India.

Estebanez, J., Gouabault, E., & Michalon, J. (2013). Où sont les animaux? Vers une géographie humanimale. *Carnets de géographes, 5* (online).

Foster, J., & COE, M. J. (1968). The biomass of game animals in Nairobi National Park, 1960–66. *Journal of Zoology, 155*(4), 413–425.

Frykman, J., & Löfgren, O. (1987). *Cultural builders: A historical anthropology of middle-class life*. Newark: Rutgers University Press.

Ghosal, S. (2012). *Cats in the city: Narrative analysis of the interactions between people and leopards in the Sanjay Gandhi National Park landscape, Mumbai*. Mumbai, A Mumbaikars for SGNP project report, http://www.mumbaikarsforsgnp.com/.

Ghosal, S., Athreya, V. R., Linnell, J. D. C., & Vedeld, P. O. (2013). An ontological crisis? A review of large felid conservation in India. *Biodiversity and Conservation, 22*(11), 2665–2681.

Guha, R., & Martinez-Alier, J. (1996). Introduction. In R. Guha, & J. Martinez-Alier (Eds.), *Varieties of Environmentalism*. Essays North and South. London: Earthscan.

Haraway, D. J. (2008). *When species meet*. Minneapolis; London: University of Minnesota Press.

Harrison, R. P. (1993). *Forests: The shadow of civilization*. Chicago: The University of Chicago Press.

Hayward, M. W., Henschel, P., O'Brien, J., Hofmeyr, M., Balme, B., & Kerley, G. I. H. (2006). Prey preferences of the leopard (*Panthera pardus*). *Journal of Zoology, 270*, 298–313.

Hutton, J., Adams, W. M., & Murombedzi, J. C. (2005). Back to the barriers? Changing narratives in biodiversity. *Forum for Development Studies, 2*, 341–370.

Khorozyan, I., Ghoddousi, A., Soofi, M., & Waltert, M. (2015). Big cats kill more livestock when wild prey reaches a minimum threshold. *Biological Conservation, 192*, 268–275.

Landy, F. (2017). Urban leopards are good cartographers: Human-nonhuman and spatial conflicts at Sanjay Gandhi National Park, Mumbai. In A. Rademacher & K. Sivaramakrishnan (Eds.), *Cities, towns, and the places of nature* (pp. 67–85). Hong Kong: Hong Kong University Press.

Latour, B. (1993). *We have never been modern*. Cambridge: Harvard University Press.

Mackenzie, J. M. (1988). *The empire of nature: Hunting, conservation and british imperialism*. Manchester: Manchester University Press.

Matheka, R. (2005). Antecedents to the community wildlife conservation programme in Kenya, 1946–1964. *Environment and History, 11*, 239–267.

Mawdsley, E., Mehra, D., & Beazley, K. (2009). Nature lovers, picnickers and bourgeois environmentalism. *Economic and Political Weekly, 44*(11), 49–59.

Midgley, M. (1979). *Beast and man: The roots of human nature*. London; New York: Routledge.

Neumann, R. P. (2004). Moral and discursive geographies in the war for biodiversity in Africa. *Political Geography, 23*(7), 813–837.

Ogutu, J. O., Owen-Smith, N., Piepho, H.-P., Said, M. Y., Kifugo, S. C., Reid, R. S., et al. (2013). Changing wildlife populations in Nairobi National Park and adjoining Athi-Kaputiei Plains: Collapse of the migratory wildebeest. *The Open Conservation Biology Journal, 7*, 11–26.

Osano, P. M., Said, M. Y., de Leeuw, J., Ndiwa, N., Kaelo, D., Schomers, S., et al. (2013). Why keep lions instead of livestock? Assessing wildlife tourism-based payment for ecosystem services involving herders in the Maasai Mara. *Kenya. Natural Resources Forum, 37*(4), 242–256.

Rodary, E. (2009). Mobilizing for nature in southern African community-based conservation policies, or the death of the local. *Biodiversity and Conservation, 18*(10), 2585–2600.

Rodary, E. (2016). *Politiques de connectivité. Conservation de la nature et fin de la modernité en Afrique australe*. Habilitation à Diriger des Recherches, University of Grenoble.

Rodriguez, L. C., Henson, D., Herrero, M., Nkedianye, D., & Reid, R. (2012). Private farmers' compensation and viability of protected areas: the case of Nairobi National Park and Kitengela dispersal corridor. *International Journal of Sustainable Development and World Ecology, 19*(1), 34–43.

Scott, J. C. (1985). *Weapons of the weak: Everyday forms of peasant resistance*. New Haven: Yale University Press.

Sekar, N. (2013). Tolerance for the charismatic marauders: Culture in wildlife conservation. *Economic and Political Weekly* June 15: 10–13.

Surve, N. (2015). *Ecology of leopard in Sanjay Gandhi National Park, Maharashtra with special reference to its abundance, prey selection and food habit*. Mumbai: Wildlife Institute of India; Maharashtra Forest Department, SGNP.

Tiwari, K. (2008). *City Forest Report 2006–07*. Mumbai: BNHS.

Tiwari, K. (2009). *Preliminary study on the diet composition of the leopard in Sanjay Gandhi National Park*. Mumbai: BNHS.

Toth, A. B., Lyons, S. K., & Behrensmeyer, A. K. (2014). A century of change in Kenya's mammal communities: increased richness and decreased uniqueness in six protected areas. *PLoS ONE, 9*(4), e93092.

Western, D., Russell, S., & Cuthill, I. (2009). The status of wildlife in protected areas compared to non-protected areas of Kenya. *PLoS ONE, 4*(7), e6140.

Wolch, J. (1998). Zoopolis. In J. Wolch & J. Emel (Eds.), *Animal geographies; place, politics and identity in the nature-culture borderlands* (pp. 381–404). New York: Verso.

Wolch, J., & Emel, J. (Eds.). (1998). *Animal geographies; place, politics and identity in the nature-culture borderlands*. New York: Verso.

Chapter 8
When Cities Host Parks: When Will Urban Frontiers Become Eco-Frontiers?

Sylvain Guyot and Estienne Rodary

8.1 Introduction

The issue of spatial, social or political encounters between cities and parks is central to the UNPEC research programme and more broadly to the future of biodiversity and the wellbeing of city dwellers in an increasingly urban world. The term "encounter" covers several forms, reflecting the diversity of encounters between urbanisation (urban space) and nature conservation (national park). Several meanings are given in an online dictionary (www.larousse.fr) for "meeting" (*rencontre*), including four which illustrate perfectly the relations between a city and a national park:

- "Meeting someone, finding oneself in someone's presence without having sought him/her out" refers to the autonomy of the urbanisation process compared to the conservation process, involving a sometimes unplanned co-presence.
- "A meeting [as a] concerted conversation between two or several people" refers, on the contrary, to dialogue strategies used between urban and park actors in order to improve the conditions of the encounter.
- "An engagement or unplanned battle between two enemy detachments on the move" highlights the sometimes extremely conflictual character of relations between cities and national parks, fairly close to the frontier logics described hereinafter.

S. Guyot (✉)
UMR 5319 Passages CNRS, Pessac, France
e-mail: sylvain.guyot@cnrs.fr

E. Rodary
Research Institute for Development (IRD), UMR GRED, Montpellier, France

F. Landy (ed.), *From Urban National Parks to Natured Cities in the Global South*,
https://doi.org/10.1007/978-981-10-8462-1_8

- Finally, "two things meeting, coming into contact or colliding" shows that contact arising from an encounter can lead to a new diminished or increased state.

Urbanisation and nature conservation are not set processes fixing borders within the metropolitan structure. On the contrary, they are dynamic processes, underlain by ecological, political, social, economic and territorial logics. In order to account for the mobility and intentionality linked to these processes, it appears useful and pertinent, in this case, to use the concepts of eco- and urban frontiers.

Eco-frontier refers to the appropriation of real or imagined spaces by ecological discourses or practices, where such spaces benefit from environmental and aesthetic amenities usually perceived as being highly significant (Arnauld de Sartre et al. 2012; Belaidi 2015; Guyot 2009, 2011; Guyot et al. 2014; Guyot and Richard 2009; Héritier et al. 2009). The eco-frontier corresponds to the spatial extension of different forms of nature conservation (e.g., institutional as with national park authorities, or more spontaneous as with residents' associations). These green appropriations tend to shape new spaces: nature enclaves for conservation, peri-urban environmental glacis, heritagisation of high places, etc.

The urban frontier (Legoix 2013) refers to the spatial extension of urbanisation, whether as urban sprawl or satellite cities, within areas considered as vacant in that they are not supposedly occupied, or in the form of recovery and renovation of already urbanised areas. The urban frontier seems to be determined by spatial, economic (real estate market), political (urban planning and development), cultural and social factors. In the context of urban national parks, eco-frontier inevitably refers to the urban frontier. This consideration is entirely in line with urban political ecology (Lawhon et al. 2013). Indeed, each one of these frontiers possesses a morphological substance, materialised by the advance of an urban artificialisation zone for the urban frontier (*urban edge*), and that of a suitable natural perimeter for the eco-frontier (*natural edge*, see Stephen 1998). This morphological substance is concomitant with a political substance: an urban development plan for the urban frontier, and nature conservation zoning for the eco-frontier.

In this chapter, we suggest the possibility that, in an emergent metropolitan context, eco-frontiers intertwine with urban frontiers. As a frontline between two processes that everything seems to oppose morphologically or, on the contrary, as a hybridisation space between naturalness and metropolitan-ness, as described in European or North American literature (Reymond 2007), eco-frontier and urban frontier cannot be dealt with separately. Hybridisation then becomes essential—as an explanatory notion—in trying to decipher the reality of the cooperation, reconstitution, domination or exploitation links between the two processes (Belaidi et al. 2011; Estabanez 2006; Kaufmann and Zimmer 1998; Zimmer 1998, 2004). Hybridisation can be construed differently at several levels, and spatialised within the park/city contact zone or at the upper level, particularly when the eco-frontier needs to legitimise its metropolitan membership, and conversely when the city needs to identify with its natural substratum.

What frontier dynamics are at play in the urban national park context in Rio, Cape Town, Nairobi and Mumbai? How is hybridisation produced? Who does it benefit?

8.2 From Frontier Dynamics to Hybridisation

The urban eco-frontier makes it possible to understand logics concerned with the spatial extension of nature conservation and the landscape, and entailing restrictions as well as a description of the urban frontier. In an emerging city endowed with a national park, the urban frontier is confronted with specific constraints. This meeting between emerging city and park entails divisions but also hybridisation principles that can lead to reconciling all or part of the urban populations with nature conservation, which is still sometimes perceived as being exclusive to certain residents.

8.2.1 Urban Eco-Frontier

Despite its late appearance (middle of the twentieth century, slightly earlier in Cape Town, see Chap. 2), the urban eco-frontier fits into wider generational logics (imperial, geopolitical and global) peculiar to the general eco-frontier working framework (Guyot 2015, 2011). Nairobi National Park (NNP), in Kenya, which was created in 1946 under British colonisation, is part of the imperial generation (conquest and nature conservation at the service of an imperialistic civilisation principle). The park exists because the British colonial power wanted to endow one of its colonial capitals with a protected "wilderness" area, leading to the creation of a buffer zone between the colonial city and the Maasai rural lands, and ensuring that the European colonial populations made the most of the wildlife nearby. Sanjay Gandhi National Park (SGNP) in Mumbai, which was created in 1950 after the country's independence from the British, is part of the geopolitical generation (conquest and nature conservation at the service of its nationalisation). It ensured the Indian national power had control over water resources and the forest ecosystem in a large city which, at the time, was barely decolonised and little controlled up until that time by the central State. Tijuca National Park (TNP), which was created in Rio in 1961, is also part of the geopolitical generation. Indeed, shortly after the transfer of the country's capital from Rio de Janeiro to Brasilia, the Federal State wanted to keep control of part of the city grounds (the *morros* in particular, i.e. the hills above the city) and the water resources. Lastly, Table Mountain National Park (TMNP) was inaugurated in 1998 in Cape Town, following the long and complex metropolitan eco-frontier progression. The national park is clearly part of the "global" generation of eco-frontiers (globalisation of conquest and nature conservation), because the park's creation was partly financed by international money lenders Global Environmental Fund (GEF) in the name of the international recognition of the fynbos ecosystem, but also meets a political objective by the African National Congress (ANC) in power, to control a substantial portion of the city's territory which is governed by the Democratic Alliance (DA), the opposition (Belaidi 2012). As such, during their creation, these four urban eco-frontiers met very different and sometimes interlinked objectives. Far from being set, these urban eco-frontiers subsequently experienced evolutionary cycles that made them (except for Mumbai) part of the global generation, in the image of the emerging global metropolises hosting them: highly developed international tourism, presence of the UN in Nairobi, major international sports events in Rio and Cape Town, etc.

These four metropolitan eco-frontiers embed two different logics. One is an "anti-urban" or "biodiversity logic" turned towards limiting the urbanisation of the most beautiful natural sites and endangered ecosystems. The other one is an "identification logic" from the city to the park, which can be named as "cultural logic". The biodiversity logic has more to do with park managers as well as local and international environmental NGOs, while the cultural logic is led by residents' associations or collectives which are sometimes supported by metropolitan authorities. In this light, the presence of an urban national park entails a disjunction of the eco-frontier in two forms, inevitably consubstantial in that they are both united by the same goal of nature conservation, but often competitive in that they are motivated by different groups of actors and divergent interests (Guyot et al. 2014). The "cultural logic" corresponds to an eco-frontier shift and is undoubtedly going to be underlain by hybridisation processes.

In Cape Town, Rio de Janeiro, Nairobi and Mumbai (see Table 8.1), the two urban biodiversity and cultural eco-frontier logics are present in different forms, sometimes highly unevenly. In Cape Town, the biodiversity logic is mainly promoted by South African National Parks (SANParks) which has been trying to expand the boundaries of the national park on former agricultural lands or forests, and to extract invasive plants. In Tijuca (Rio), this biodiversity logic is little represented, probably due to the "artificiality" of the natural carioca landscape recently being recreated (see Chap. 2). In Nairobi, the same logic concerns in particular the management of wildlife outside the southern boundaries of the park, in the Maasai pastoral area. In Mumbai, it corresponds to the current concern for maintaining ecological links between the park and Tungareshwar Wildlife Sanctuary in the north, and the rural area of Aarey Milk Colony in the south.

The biodiversity logic sometimes clashes with the "cultural logic", as in Cape Town, where certain residents' associations blame the national park for extending the park's boundaries without really restoring the damaged ecosystems, or for not properly managing the baboon issue or dealing with invasive plants. Likewise, in Mumbai, the recreation and urban cattle breeding area of Aarey Colony does not fit the park manager's objectives of fauna and flora protection. However, in Rio, with the project to reforest the buffer zone of TNP close to the *favelas*, and in Nairobi,

Table 8.1 Frontier dynamics in the four cities

	Cape Town	Rio de Janeiro	Nairobi	Mumbai
Urban frontier	Processes: – perimetropolitan gentrification on the Peninsula – Informal settlements and low-cost housing	Processes: – Occupation by the police of the *favelas* nearby the park – Land speculation in the *favelas* in the south zone (restoration/construction houses)	Processes: – Extension of peri-housing in the southern part of the park – Industrialisation in the northern part of the park – Encroachement of the park by railways and motorways	Processes: – Informal housing within the park: 120,000 inhabitants – Real estate projects near the park
Eco-frontier	1. Biodiversity eco-frontier extension of TMNP 2. Cultural eco-frontier: environmentalist + heritage activists 3. Global eco-frontier: World Heritage Site	1. Cultural eco-frontier: community reforestation in buffer zone *favelas* 2. Global eco-frontier: World Heritage Site, cultural landscape	1. Cultural eco-frontier: – Creation of conservancies – Compensation for wildlife attacks and financial encouragement to keep land unfenced	1. Biodiversity eco-frontier: Wildlife Sanctuary, northern part. 2. (possible) cultural eco-frontier: Aarey Colony
City/park frontline	Stabilised and pacified	Stabilised and relatively pacified	Stabilised and relatively pacified	Unstable and subject to infringements

Source Authors

with various attempts at creating private conservancies inside Maasai lands, the cultural logic strengthens the residents' commitment to the ecological values conceded by the urban national park.

8.2.2 Urban Frontier Close to the Park

The urban frontier, unlike the eco-frontier, is the very essence of the city, its progression and spatiotemporal reconstitution. In the case of emerging metropolises, the issues of urban sprawl, of increasing population and of urban reconstitution are crucial. It is often the suburbs that are the most representative of the global city which are being developed by urban policies. As such, the urban eco-frontier can impede the spatial expansion of the emerging city, especially at the expense of the poorest, but it can also lead to a reinforced selectivity of urbanisation and population types living close to the park. The relationship between eco-frontier and urban frontier in emerging metropolis is paradoxical. The four examples show that different situations exist, from the extreme case of Mumbai where the urban frontier is clearly expanding at the expense of the eco-frontier, to the case of Cape Town where the opposite is taking place. The urban frontier generates as many representations as it is multiform in the history of its advance. The urbanisation image in Rio is impossible to dissociate from the *favelas* located on the *morros*, near the boundaries of the TNP, just as it is inseparable from the seaside suburbs of Copacabana and Ipanema. The urbanisation image in Cape Town refers invariably to the racial segregation generated by the colonial and apartheid regimes, just as it is linked to the image of the luxury properties located in the Peninsula, near TMNP. In Nairobi, the image of the African city seems to predominate, with a strong opposition between former colonial suburbs (such as Karen) and informal settlements (such as Kibera Slum). In Mumbai, the presence of many informal settlements, especially near or even inside the national park, is reinforcing the image of extreme poverty which is often associated with this metropolis.

The current urban frontier of the four cities under study, near urban national parks, involves a great diversity of processes that are all representative of the emerging metropolis (see Table 8.1). In Cape Town, except for few slums located inside the park's boundaries (on the fringes of Hangberg, near Hout Bay), a very specific over-densification of poor populations has been taking place in four distinct areas near the park (Imizamo Yethu near Hout Bay, Masiphumelele and Ocean View near Kommetjie, and Red Hill in the south of the Peninsula). This process, whereby social, economic, environmental and health issues are being concentrated, is in contrast with the generalised enrichment of the Cape Peninsula and the construction of many luxury properties (Guyot et al. 2014). In Rio de Janeiro, the main characteristic of the urban frontier near the national park concerns rather a dynamic of urban reconstitution, with the pacification of the *favelas* and their partial gentrification. In Nairobi, the urban frontier near the park, corresponds to the advance of the rural and peri-urban habitat in the south of the park (through the privatisation and sale of former Maasai collective pastures), to industrial growth in the north and the east, and to railway and ring road projects that are denting the park. Lastly, in Mumbai, apart from the new luxury properties built near the park, informal settlements have been growing progressively inside the park since the 1980s–1990s.

Faced with these urbanisation dynamics, it is legitimate to question whether the boundary separating the urban edge and the natural edge is a conflictual, contested and often infringed frontier line, or whether it can lead to favouring hybridisation processes, of which the "cultural logic" of the eco-frontier seems to be a first variant.

8.2.3 Frontier Line: A Third-Zone of Hybridisation?

At first sight, the frontier-vs-frontier dynamics only seem to represent a problem in Mumbai (Chap. 4), while they are relatively pacified in Cape Town, Rio and Nairobi. In fact, in these three cities, recent current affairs have shown that the contact zone between the city and the park was the subject of recurrent conflicts, whether directly or indirectly. In Cape Town, a white ward councillor denounced the pollution of the wetland adjoining the park, by the poor residents of the township of Masiphumelele (interview dated 14/03/2013): "It's a huge problem. And we work on it every single day and it's just … it's a problem; I mean it's a huge problem. That wetland there is a nightmare. […] And it just … the next day you're back there and it's the same bloody thing, you know. It's all the night soil [that] goes into the canal and the nappies and … It's laziness I think. It's education and laziness. […] Because they're too lazy to go to the nearest dirt-bin, or to put it into a dirt-bin and tie it up and then … But they let their children play in it, that's the terrible thing, it's that those children go squish, squish, squish through that bloody stuff. That's how… I have a huge disaster risk programme going on in there." In Rio, *favela* residents carry out makeshift connections to retrieve the water resource which is found inside the park. In Nairobi, several urban development projects are threatening the very existence of the national park: a new ring road is to isolate the park entirely in the south, not to mention the railway line project which is to run through the middle of the park rather than cross the military zone of Langata Barracks and the suburb of Kibera Slum.

However, it seems that hybridisation dynamics prevail over "frontier-vs-frontier" conflicts, in that they enable the park to legitimate its presence within emerging metropolises. They also enable certain well-off residents to justify their control over land and real estate near the vast spaces of environmental amenities (greenification), and enable the metropolitan authorities to validate their integration into the club of world metropolises.

The hybridisation process (Fig. 8.1) is an immaterial (values, norms), material (landscape, territorial markers including neo-toponymy) and virtual (representations, web-based in particular) blend between the two a priori distinct or even contradictory processes of nature conservation and urbanisation. Hybridisation uses

WHAT? Blending between eco-frontier and urban frontier.

WHY? Acknowledgement by the stakeholders from either the urban fabric or the protected areas that "purity" is neither sustainable nor even desirable. Possible exploitation of this zone by stakeholders, in order to gain political power, and to use it as a territorial domination tool or as an economical resource.

WHERE? Proximity [frontline/contact between the urban frontier and the eco-frontier]; Projection [inside the protected area, or inside the city, or outside both]; Connection [connecting various protected areas, and/or specific urban areas using corridors]

HOW? Immaterial [discourses, philosophies, norms]; Material [landscape, territorial markers]; Virtual [web sites].

WHO? Stakeholders from various spheres. Interesting to see how personal and professional lives are hybridised.

WHEN? Genealogy of both urban and eco-frontiers.

Fig. 8.1 Hybridisation working framework in a context of encounter between the park and the city. *Source* Authors

a sort of "third-zone" as support that groups, selects and transforms elements included in the eco-frontier and urban frontier processes. Third-zone is a geographic neologism directly inspired by notions of third-space (Viard 1990; Vanier 2000) and third-landscape (Clément 2004). A third-zone, within a space dominated by a metropolis and a national park, refers to a space of encounter, a space of mixing and therefore of hybridisation of values borrowed from the two previous spaces, i.e. naturalness and urbanity. This "naturbanity" is consubstantial with the third-zone and comes with different effects that will be developed further on. As such, a third-zone is not equivalent to the buffer zone of the national park, or to a potential zone of adherence: it includes but also exceeds them (philosophically and geographically), in that a third-zone describes a more subtle and successful blend, for both residents and park managers, between naturalness and urbanity.

Cities such as Rio or Cape Town, where the national park seems to be well integrated into the metropolitan logic, result in levels of hybridisation that are more important than in Nairobi or Mumbai. Three different third-zones can be described: third-zones of contact, of projection and of connection. In contact between eco-frontier and urban frontier, third-zones can be located within the buffer zone of the park. In projection, third-zones can be situated inside the national park, inside the city, or outside of both. In connection, third-zones are found at the level of biodiversity corridors, linking the park to other protected wilderness areas of the city.

8.3 Who Benefits from Hybridisation?

In order to understand who benefits from hybridisation between eco- and urban frontier, we compare here the four case studies, and then propose a typology.

8.3.1 Comparing the Four Case Studies

A gradient of high/low hybridisation goes from Cape Town to Mumbai, via Rio and Nairobi (Table 8.2).

In Cape Town, six forms of hybridisation attributable to three groups of actors co-exist: the City of Cape Town, SANParks, and the residents (rich and poor). They each produce differentiated third-zones. The first hybridisation process, supported by SANParks, tends to produce third-zones of contact (picnic areas near the park's boundaries) and projection (recreational zoning inside the park). These third-zones sometimes have admission fees, which raises the issue of accessibility for all the residents, even though the motto of the park is "A park for all forever". The second hybridisation process organised by SANParks is linked to the establishment of environmental education, meant for all the children of the metropolis (see Chap. 12; Belaidi 2016). This third-zone is projected in the south of the Cape Peninsula, near Cape Point, one of the most touristic spots of the national park. Not only does this initiative make it possible for all the young residents of the city to know one of the most beautiful sections of the park, but it also aims, in particular, to hybridise the way children think and reason (immaterial hybridisation), so that they can subsequently apply these principles at home (connection hybridisation). In Cape Town, two hybridisation logics related to the metropolitan authority can be described: hybridisation through biodiversity network connection, which makes it possible to create third-zones in the form of corridors between the different protected

Table 8.2 Hybridisation between eco-frontier and urban frontier in the four cities

	Cape Town	Rio de Janeiro	Nairobi	Mumbai
Possible hybridisation	1. TMNP: park/city intertwined (leisure zones) 2. TMNP: environmental education (reconciliation ecology) 3. City: BIONET 4. City: Green Map 5. Rich residents: environmental "glacis" (rural sense of place, natural heritage, eco-estates) 6. Poor residents: urban gardens, green shacks	1. Park: religious use of the mountains/ forest/waters 2. Park: environmental education of *favela* inhabitants near the park 3. Poor residents: ecotourism projects in *favelas*	1. NGO + park: Agro-ecology in the southern part of the park (Maasai) 2. Park: Green line 3. Park: Leisure areas inside the park 4. Richer residents: environmental glacis in Karen	1. Park: Cultural hybridisation: tribal artwork inside the park/religious use of the park 2. City: Aarey colony/ jungle, zoo project 3. Poor residents: embryonic hybridisation in the informal settlements inside the park 4. NGO: Mumbaikars for SGNP

Source Authors

metropolitan natural spaces and the national park (e.g., Constantia Green Belt, etc.), and hybridisation through projection, from the Green Map, which indicates all the ecological initiatives within the City of Cape Town. Hybridisation advocated by the park and that advocated by the metropolitan authority are not well coordinated, but show that the encounter process between the two urban and nature conservation logics is maturing.

Another hybridisation logic in Cape Town is sometimes confused with the cultural eco-frontier, which is about the formation of a third-zone through contact hybridisation in the form of environmental glacis, from well-off residents (Guyot et al. 2014). Finally, from the point of view of the most destitute residents, several initiatives as regards slum greening and multiplying urban (and sometimes organic) vegetable gardens, can be attributed to projection hybridisation, often instrumentalised by environmental associations grouping together well-off city dwellers, desirous of making certain forms of socio-racio-spatial segregation long-lasting.

In Rio, the two main hybridisation processes concern *favelas* with environmental education (see Chap. 12) and the development of ecotourism. This is essentially contact hybridisation which creates third-zones recognisable within *favelas* and near the Corcovado, with the development of hiking trails.

In Nairobi, two hybridisation processes are similar to Cape Town's: a projection process fulfilled by the conservation authorities, with picnic areas inside the national park, and contact process with the existence of an environmental glacis in several well-off residential suburbs near the park's boundaries, including those of Karen and Langata. A survey conducted by UNPEC with members of the Friends of the Nairobi National Park (FoNNaP) association shows that most Western expatriates or Kenyans of European origin mobilised for the defence of the park and of wildlife, live in these suburbs. In these suburbs the landscape dimension (important vegetation), as well as the historical (colonial suburbs) and economic (many safari tour operators) identity, refer to proximity with the national park (Fig. 8.2).

Another contact hybridisation process in Nairobi seems to be active around the "greenline" project (http://nairobigreenline.org/), which consists of planting trees on

Fig. 8.2 Behind the gate, a villa with garden: is the Karen district in Nairobi a symbol of hybridisation? *Source* Photo by Landy (2014)

the border between the national park and the city. Finally, a last hybridisation process, is taken up through agro-ecological initiatives involving the Maasai populations in the south of the Park. This concerns a programme that, at some stage, remunerated "ecosystemic services", by giving money to the owners who undertook not to close their pastures. This is essentially projected hybridisation. Except for this last initiative, it seems that most hybridisation processes, in Nairobi, take place only by involving a fairly reduced fringe of the rather well-off and often white or expatriate population.

The few ecologisation processes mentioned in Table 8.2, concerning Mumbai, seem embryonic, except perhaps for MfSGNP (http://www.mumbaikarsforsgnp.com/). This association which emanated from civil society, managed to convince the Park Management of the importance of relying on city dwellers to help manage the park with regard to certain issues: animal census, renting bicycles, and working with the population towards better practices, with a view to protecting people from leopards—and vice versa (Landy, forthcoming). MfSGNP admits that it recruits from among the wealthy above all, a fact which is already presumed since the members are socially fairly powerful, are able to ensure that their actions are given a lot of media coverage, and are able to attract private funding. In fact, the policy of the Park Management goes in the same direction: the latter is thinking of reducing recreation areas for the general public (e.g., paddleboats, etc.) with more ambitious goals in mind in terms of environmental education, such as hiking with scientific guides, etc., with the same perhaps cynical but realistic principle: relying on the rich is simpler and more efficient in protecting the park.

However, other initiatives tend to integrate the poor, and "tribal" populations in particular: on the one hand, a group parallel to MfSGNP, founded by Krishna Tiwari, has been increasing the awareness of indigenous populations, and recognising the importance of their ecological know-how; on the other, a Warli painter living (illegally) in the park, has been commissioned to decorate the administrative buildings of the Forest Department (Chap. 5). Nevertheless, all this remains limited. Hybridisation appears particularly fragile as regards Aarey Colony, since the zoo project would undoubtedly have been more a victory of the urban frontier than an example of a "third-zone", since ring road projects are on the increase, in order to ease the east-west crossing of the built-up area. Would the more costly (although certainly necessary) alternative of a tunnel under the national park, manage to prevent the destruction of this fragile area, increasingly eaten away by more or less legal real estate projects?

8.3.2 The Political Uses of Hybridisation

We would like here to provide an analytical framework in order to expose the major processes of park/city hybridisation. We will then see whether they can be exemplified in the four case studies (Table 8.3).

Table 8.3 The different effects of the park/city contact hybridisation: generalisation

Hybridisation by: Third-zone dedicated to:	Contact	Projection	Connexion	Social dynamics
Recreation (spreading the park space towards the urban space)	**"Gateway"** effect: recreation areas located near the park's gates Examples: TMNP (Oudrekraal, Tokai Forest) NNP: Safari Walk SGNP: Tiger Safari, toy train TNP: picnic facilities, international tourism, ecotourism, sports	**"Centre"** effect: all areas accessible by city dwellers inside the park Examples: TMNP, TNP, NNP, SGNP. SGNP: trail with guide and prior permission	**"Corridor"** effect: Visitors spend more time in and around the park thanks to biodiversity corridors sometimes remote from the park's boundaries Examples: Green Belt, Constantia (Cape Town), Bionet. SGNP: Aarey Colony NNP: Bomas of Kenya TNP: establishment of green corridors (Mosaico Carioca)	Often fairly exclusive TNP: not necessarily exclusive. The park contains fairly popular areas
Education (spreading the park space towards the urban space)	**"All in the park"** effect: environmental education centres located inside the park Examples: Rio, TMNP (Cape Point)		**"Park for all"** effect: educating all young people wherever they live in the metropolis Example: Cape Town, bus dedicated to environmental education. Rio, initial workshops conducted in the *favelas* SGNP: awareness campaign on leopard hazards	Inclusive
Glacis (dwelling) (urban space)	**"Greentrification"** effect: enrichment near the contact line related to the real estate development of environmental and landscape value until the formation of an environmental glacis of whiteness (with potential ecotouristic function) Examples: Cape Town (Noordhoek, Constantia), Nairobi (Karen, Langata)		**Greenification** effect: greening different suburbs or facilities sometimes remote from the park but benefitting from the prevailing ecological logic Example: Cape Town: green shacks, green map	Exclusive
Production (spreading the urban space towards the park space)	**"Organic"** effect: development of organic agriculture, biodynamic agriculture and permaculture in urban agriculture Examples: Cape Town, Nairobi		**"Conservation easement"** effect: transfer of part of peri-urban land properties to nature conservation	Inclusive

Fig. 8.3 Silvermine Reservoir and Picnic Area (TMNP). *Source* Photo by Guyot and Dellier (2014)

Fig. 8.4 Hang-gliding from the national park above Barra de Tijuca. *Source* Photo by Landy (2012)

Four main functions of the third-zones can be brought out from the previous examples: recreation, education, housing-glacis and production. For most, they can be applied to contact, projection and connection spatial logics, denoting different forms of (inclusive and exclusive) social dynamics.

A dozen effects of park/city hybridisation can be brought out.

- the "**gateway**" effect indicates the importance of recreation areas, often the most used ones, located near the gates of the park in terms of the encounter between national park values and city dweller usage. Here hybridisation is high as it enables residents to identify with the park, and conservation actors to integrate sustained and diversified uses in their planning policy (Fig. 8.3).
- The "**centre**" effect refers to recreation areas that city dwellers can access inside the park, whether by car, bicycle or on foot. It transforms city dwellers into "naturban" dwellers who must adapt their equipment and habits to wilderness, and transforms the park that must host these populations by developing roads, paths or climbing routes (Fig. 8.4).

- The "**corridor**" effect relates to ecological corridors where connections allow inhabitants to experience naturban situations outside of the park. This effect sometimes makes it possible to connect several third-zones in the city together, and can lead to a high level of hybridisation beyond the park itself (Constantia Green Belt) (Fig. 8.5).
- The "**all in the park**" effect pertains to initiatives of environmental education centres inside the national park. It increases in situ hybridisation of urban populations that seldom visit the park (Fig. 8.6).
- The "**park for all**" effect offers environmental education to young city dwellers beyond the park limits, wherever they live on the metropolitan territory. As such, it projects hybridisation logics in the city, and connects in a functional and non-substantial manner several highly socially contrasted territories. This effect

Fig. 8.5 Green Belt Corridor (Constantia). *Source* Photo by Guyot and Dellier (2014)

Fig. 8.6 Environmental education, or simply recreation? School students in Mumbai park. *Source* Photo by Landy (2012)

Fig. 8.7 Baba in her garden created with the support of an NGO (Gugulethu Township, Cape Town). *Source* Photo by Landy (2014)

Fig. 8.8 Property towards Oudekraal (Camps Bay, Cape Town). *Source* Photo by Guyot (2013)

is borne by the national park institutions (and sometimes the metropolitan authorities) for environmental and social reasons (Fig. 8.7).

- The "**greentrification**" effect indicates the gentrification process of local residential areas in contact with the park, determined by the search for environmental amenities. Hybridisation takes place by transforming buildings in relation to the natural landscape (e.g., materials, rooms with a view, gardens with high biodiversity composition, toponymy, etc.) (Figure 8.8).
- The **greenification** effect follows a logic of environmental value projection inside the city, similar to greentrification but with different causes (Fig. 8.9).
- The "**organic**" effect refers to the multiplication of urban agricultural initiatives that have integrated environmental norms in their production. This effect generally concerns the same type of clientele as that visiting the national park.
- The "**conservation easements**" effect refers to the possibility given to some residents of the metropolitan outskirts to transform part of their property into a

Fig. 8.9 Hiranandani Gardens (Mumbai North): Despite the name, a derisory effort to greenify a middle class suburb. *Source* Photo by Landy (2009)

nature reserve. This effect leads to significant hybridisation through the naturalisation of residential areas.

Some effects concern most case studies (gateway, core), while others involve one example only (conservation easements). This typology (Table 8.3) offers a complete set of effects that can be used in other situations of park/city encounters, as in countries of the Global North.

At least four of these effects come mainly from the national park (gateway, core, endemic education and park for all), and concern third-zones situated inside and on the borders of the park, as well as in town. Four other effects are co-produced by city dwellers and the metropolitan authorities (corridor, greentrification, greenification, green edge), although rarely in collaboration with the park. The other effects relate to other economic dynamics at work in the city, such as organic production (although those dynamics may be more indirect effects related to the presence of environmentally friendly social groups living in the environmental glacis).

Who does park-city contact hybridisation benefit? Third-zones and the different effects of hybridisation are multiform and concern more or less all urban actors. Nonetheless, this is a fragmented hybridisation which operates through enclave systems. The hosting theory can in fact make it possible to go beyond this fragmentation to propose more successful intermixing between the different logics.

8.4 Conclusion: From Theory to Practice: Towards a Theory of Hosts

What can we make of those different dynamics that range from conflictual encounters between eco- and urban frontiers to hybridisation? In particular, what could the theoretical analysis of the four cities offer to practitioners and policy-makers? When moving from science to policy recommendation, the balance between scientific accuracy and governance relevancy is always hard to find. In such a context, the analytical framework that scientists sometimes want to promote may be at odds with the practical issues faced by park and city authorities, a partial misunderstanding that the programme UNPEC experienced and dealt with explicitly.

We suggest here that the understanding of nature/urban spaces relationships that underlies most institutions does not take into account all of the factors that actually make up a city with a park. This understanding of the interactions between the park and the city is largely informed by a narrow territorial comprehension of both the urban space and the ecological space. For parks, it ensures that conservation policies must be land-based, i.e. that large tracks of land have to be reserved for biodiversity conservation. For cities, it means that urban areas, because of their high population density living in a highly transformed environment, need more space for newcomers and new development. But this reading of city/park interactions is true as long as it concerns only two separate entities, the park on the one side and the city on the other side, which can live and function independently. Following on from this logic, the development of one side is seen as detrimental to the other side, e.g., the expansion of a national park can only happen at the expense of urban development. This "territorial trap" is partially a legacy of the fortress conservation ideology that has been predominant in rural biodiversity conservation for decades (Brockington et al. 2008), but is fuelled in the context of emerging cities by urban sprawl and real-estate speculation.

What the chapter shows, however, is a more complex picture of parks in cities. The concepts of eco-frontier, urban frontier and third-zone hybridisation illustrate the actual interwoven connections of nature and urban dynamics. If one sheds the territorial dimension of parks and cities and focuses on the different dynamics that shape their interrelationships, one will see networks and relational processes across various spaces (McCann and Ward 2010). Those networks are both material and ideal, ecological and social, as already detailed above. Because networks as a process are always more difficult to grasp than a material dimension of trunk of lands, science and practitioners alike have historically been prone to overlook networks. And when landscape ecology eventually reintroduced the structural and functional dimensions of habitat dynamics, it generally applied its results to conservation policies in rural areas (Merriam 1991; Bennett and Mulongoy 2006). In those spaces, ecological links are more obvious than in urban places. In the former the natural features of the connections are obvious while in the latter the naturalness is stuck among spaces of concrete and buildings. Hence the need to combine ecological and social networks. Because of the urban settings in which they exist, national parks cannot avoid the massive and diverse social thrusts that at the same time frame their management and justify their existence.

For these different reasons, the fortress mindset is not a rational answer to practical issues; it is rather a tapered view of city/park matters for the stake of management and administration. The bureaucratic division of authorities between the territorial park and the territorial city is thus a poor proxy of the actual connections that overflow and circumvent such divisions (a similar argument has been developed for climate change urban governance, see Bulkeley and Castán 2013). Besides, this division is largely a legacy of and an importation from other contexts in which antagonistic views of nature and society can lead to the marginalisation or even the erasing of one of the sides, for instance by degazetting a park or resettling residents. There are no such political solutions in big cities with national parks, where the non-exclusivity principle rules (Chap. 1). And since those connections and the social and ecological networks they create between the park and the city will endure, it is rational to take that into account in the management organisation and decision-making bodies.

Our analysis of eco- and urban frontiers and hybridisation dynamics calls for a renewed understanding of city/park governance. When the fortress is not able to integrate all the components and processes that make up the city/park connections and as such is confined by a defensive stance from both sides, we suggest viewing the city and the park as mutual hosts, that is, to understand city and park not as idle lands, but as dynamic networks that may mutually benefit from their interactions. It implies that both institutions should extend beyond their boundaries. Not only to the edges, but also to distant areas and to distinct institutions, in order to reach and jointly manage the different frontier and hybridisation zones.

"Host" is an interesting concept, because it bears a social meaning (receiving a guest) as well as a biological meaning. In the latter signification, host describes not a parasitic process but a commensalism, an association between two organisms in which only one benefits, or even mutualism, where the association is beneficial to both organisms involved.

In sum, moving towards city and park institutions as mutual hosts implies two steps. One is to institutionally follow the networks that social life and ecological dynamics create, as described and classified above in this chapter as eco-frontier, urban frontier and third-zone of hybridisation. The second step is to remove the fence and become active partners in each other's space. As the end of the day, one could imagine the national park managers becoming urban actors, and the city authorities becoming park managers. This may not be their respective mandate, but it seems the only way to fill the gap in institutional capacity (CBD 2012). Yet there is a last rational behind the host concept. As cities emerge as leading actors in politics, they manage to create their own political choices that are sometimes distinct from the neoliberal practices and discourses dominant at state and global levels and as such have a strong potential towards environment friendly policies, as experienced in the network of Agenda 21 European cities and the Aalborg Charter in Europe (Emelianoff 2007).

Forty years ago, the Stockholm Conference on the Human Environment brought together environment and development issues, but the conservationists missed this opportunity and remained marginal in the integrative policies. Conservation only started to articulate with development issues 10 years later with the promotion of the sustainable development agenda (IUCN et al. 1980). Today, as cities become "green", conservation cannot miss its chance again. Conservation must become a host of this change.

References

Arnauld De Sartre, X., Berdoulay, V., & López, R. S. (2012). Eco-frontier and place-making: The unexpected transformation of a sustainable settlement project in the Amazon. *Geopolitics, 17,* 578–606.

Belaidi, N. (2015). Théorie du droit et front écologique: Apport à la (re)définition de la justice environnementale. *Développement durable et territoires, 6,* 1.

Belaidi, N. (2012). Le Patrimoine mondial pour créer une identité commune dans l'Afrique du Sud post-apartheid? Exemple de la ville du Cap. In L. Bourdeau, M. Garavari-Barbas, & M. Robinson (Eds.), *Tourisme et patrimoine mondial* (pp. 269–284). Quebec: Presses de l'Université Laval.

Belaidi, N. (2016). Ecologie & Réconciliation. Etude à partir des initiatives d'éducation à l'environnement sur Table Mountain National Park. In R. Porteilla et al. (Eds.), *Afrique du Sud: 20 ans de démocratie contrastée?* Paris: L'Harmattan.

Belaidi, N., Berger, M., Houssay-Holzschuch, M., & Ninot, O. (2011). Cape Town, périphéries postapartheid, in Chaléard J.-L. (dir.). *Les métropoles des Suds vues de leurs périphéries,* Paris, PRODIG, coll. Grafigéo no. 34, 133: 79–92.

Bennett, G., & Mulongoy, K. J. (2006). *Review of experience with ecological networks, corridors and buffer zones.* Montreal: Secretariat of the Convention on Biological Diversity.

Brockington, D., Duffy, R., & Igoe, J. (2008). *Nature unbound: Conservation, capitalism and the future of protected areas.* London: Earthscan.

Bulkeley, H., & Castán, Broto V. (2013). Government by experiment? Global cities and the governing of climate change. *Transactions of the Institute of British Geographers, 38*(3), 361–375.

CBD. (2012). Cities and biodiversity outlook. Action and policy: A global assessment of the links between urbanization, biodiversity, and ecosystem services. Secretariat of the Convention on Biological Diversity/Stockholm University/ICLEI, Montreal.

Clément, G. (2004). *Manifeste du tiers paysage.* Paris: Sujet-Objet.

Emelianoff, C. (2007). La ville durable: L'hypothèse d'un tournant urbanistique en Europe. *L'Information géographique, 71*(3), 48–65.

Estebanez, J. (2006). Les jardins zoologiques et la ville: Quelle nature pour le Biodôme de Montréal? *Annales de géographie, 6,* 708–731.

Guyot, S. (2015). *Lignes de front: L'art et la manière de protéger la nature.* HDR dissertation, University of Limoges.

Guyot, S., Dellier, J., & Cerbelaud, F. (2014). L'environnement au profit des plus riches? Construction et hybridation d'un front écologique métropolitain dans la Péninsule du Cap, *VertigO - la revue électronique en sciences de l'environnement,* 14,1, online.

Guyot, S., & Richard, F. (2009). Les fronts écologiques - Une clef de lecture socio-territoriale des enjeux environnementaux ? *L'Espace Politique*, 3 (online).

Guyot, S. (2009). Fronts écologiques et éco-conquérants: Définitions et typologies. L'exemple des ONG environnementales en quête de Côte Sauvage (Afrique du Sud), *Cybergeo*, 471 (online).

Guyot, S. (2011). The eco-frontier paradigm: Rethinking the links between space. *Nature and Politics, Geopolitics, 16,* 675–706.

Héritier, S., Arnauld de Sartre, X., Laslaz, L., & Guyot, S. (2009). Fronts écologiques: Dynamiques spatio-temporelles et dominations multi-scalaires, *L'Espace Politique*, 9, 3 (online).

IUCN, UNEP & WWF. (1980). World conservation strategy: Living resource conservation for sustainable development. IUCN/UNEP/WWF, Gland/Nairobi.

Kaufmann, E., & Zimmer, O. (1998). In search of the authentic nation: Landscape and national identity in Canada and Switzerland. *Nations and Nationalism, 4*(4), 483–510.

Lawhon, M., Ernstson, H., & Silver, J. (2013). Provincializing urban political ecology: Towards a situated UPE through African urbanism. *Antipode.* https://doi.org/10.1111/anti.12051.

Legoix, R. (2013). *Sur le front: Pour une approche géographique du front d'urbanisation: Lotissements, voisinages, trajectoires*, HDR dissertation, University Paris 1 Panthéon-Sorbonne.
McCann, E., & Ward, K. (2010). Relationality/territoriality: Toward a conceptualization of cities in the world. *Geoforum, 41*(2), 175–184.
Merriam, G. (1991). Corridors and connectivity: Animal populations in heterogeneous environments. In D. A. Saunders & J. et Hobbs (Eds.), *Nature conservation 2: The role of corridors* (pp. 133–142). Chipping Norton: Surrey Beatty and Sons.
Reymond, H. (2007). Une évolution urbaine émergente: De l'ancienne symbiose ville-campagne à la future synergie des parcs urbains. In A. Da Cunha & L. Matthey (Eds.), *La ville et l'urbain: Des savoirs émergents* (pp. 207–220). Lausanne: Presses polytechniques et universitaires romandes.
Stephens, A. (1998). Co-managing the boundaries between urban and natural areas: A case study of scarborough (Cape Peninsula). *South African Geographical Journal, 80,* 101–107.
Vanier, M. (2000). Qu'est-ce que le tiers espace? Territorialités complexes et construction politique. *Revue de géographie alpine, 88,* 105–113. https://doi.org/10.3406/rga.2000.4626.
Viard, J. (1990). *Le tiers espace: Essai sur la nature*. Méridiens Klincksieck.
Zimmer, O. (1998). In search of natural identity: Alpine landscape and the reconstruction of the swiss nation. *Comparative Studies in Society and History, 40,* 637–665.
Zimmer, O. (2004). A unique fusion of the natural and the man-made: The Trajectory of Swiss nationalism, 1933–39. *Journal of Contemporary History, 39*(1), 5–24.

Chapter 9
National Parks in the Urban Arena: The Interplay of Actors for Incomplete Models of Governance

Véronique Fourault-Cauët and Jean-Fabien Steck

The question immediately raised by the study of urban national parks is that of scale, prompted by the telescoping between "national" and "urban". As a qualifying adjective, "national" refers here both to a form of conservation (IUCN (International Union for Conservation of Nature), category II), and to a status, and therefore to the ways in which said parks are governed and managed. For its part, the word "urban" relates to a category of space characterised by artificiality, and by forms of government and management that are increasingly devised and organised along decentralised lines. To work on urban national parks, especially in a metropolitan context, therefore means working simultaneously and concomitantly on two spatial entities whose nature and dynamics (conservation *versus* growth, in all the senses that may be ascribed to the latter, including its ideological dimensions) seem in principle contradictory, but are in fact combined, not to say interwoven. Beyond analyses that approach the subject by raising the question of the relations between nature and the city (see Chaps. 8 and 12), and beyond the close-grained study of the sectoral interplay and strategies relating to these parks (see Chaps. 6 and 7), there thus seems to be a need to tackle the question in political and territorial terms, by treating territory here as a political space that is a medium and a vehicle for projects. The relations between parks and cities thus need to be approached from a processual perspective that considers the strategies of actors within an interplay of policy and politics. This interplay is taking place in a context where prerogatives are being reconfigured, or even redefined, where power is being redistributed and management approaches renewed according to a principle of subsidiarity justified by its supposed effectiveness.

V. Fourault-Cauët (✉) · J.-F. Steck
University of Paris-Nanterre, Nanterre Cedex, France
e-mail: vfouraultcauet@parisnanterre.fr

V. Fourault-Cauët · J.-F. Steck
UMR LAVUE, Nanterre, France

F. Landy (ed.), *From Urban National Parks to Natured Cities in the Global South*,
https://doi.org/10.1007/978-981-10-8462-1_9

Through the problems raised by this distinctive proximity between urban fabric and protected spaces, we therefore propose to examine emerging forms of governance in relation both to national parks and to cities, in this case cities of the metropolitan type, by means of an approach understood as the "research field explored in order to explain, in particular, the contemporary forms and transformations of the state and the direction of societies, on the basis of actors and regulations beyond the state" (Le Galès in Lévy and Lussault 2013). Within this framework, the study of the local, empirical and often disordered interplay between the main actors offers a way to identify an emergent practice, or even a discourse, relating to the metropolitan connection between city and park. Our approach is therefore quite different from that of Chap. 10, which analyses the management and governance of the parks from the viewpoint of conservation, in other words from the perspective of the parks themselves.

Here, we will begin by identifying the distinguishing features of urban national parks, largely to do with population densities and the linked economic pressures. We will then demonstrate firstly the major role played by devolution (decentralisation policy), and the new distributions of prerogatives that it entails, and secondly the embeddedness of certain actors in an essentially global scale of analysis, in order to explain the timid and unequal ways in which these national parks are recognised as symbols and key instantiations of the integrated approach to protected spaces developed in the last few decades. Finally, we will see how the interplay of actors described is itself ambiguous, and how the integration of protected natural spaces into a model of metropolitan governance appears debatable. Overall, therefore, our argument will seek to show the distinctive nature of the interplay of actors around the particular geographical objects that are urban national parks.

9.1 Urban National Parks: New and Particular Configurations

9.1.1 The Initial Confrontation Between High Urban Densities and Protected Spaces

The emergence of the integrative paradigm in relation to protected spaces, which dates from the 1980s, considerably altered attitudes to national parks, and contributed to the change in their status to spaces that had previously received little protection of this kind. This sharp shift in approaches to conservation, less directly exclusive of previous populations and activities, considerably altered for a while the relations between parks and cities. In consequence, the proliferation of urban national parks contributed to the emergence of new sets of actors, either because conurbations expanded to become contiguous with formerly protected areas—as happened in Nairobi or in Rio—or because the creation of a national park directly adjacent to a conurbation now seemed possible, as was the case in Marseille.

Indeed, whereas many national traditions of protection prohibited permanent forms of housing within national parks or high population densities immediately adjacent to them, the growing proximity of certain national parks to large and expanding centres of population transformed these spaces into very popular recreational areas and leisure destinations for tourists.

Let us reiterate the key figures regarding access to the parks concerned: Tijuca National Park (TNP) receives 1.7 million visitors a year,[1] and Table Mountain National Park (TMNP) in Cape Town receives 2.67 million. The presence of tourists, facilitated by proximity to the city, thus clearly poses a major issue for these protected spaces. Moreover, certain national parks seem also to attract local visitors in addition to tourists: the majority of visitors to Nairobi National Park (NNP), for example, are Kenyan. In particular, the significant role assigned to education in the functions of the parks of Nairobi, Tijuca or Cape Town can lead,[2,3]—of course to varying degrees (Belaidi and Gonçalves Soares, Chap. 13) between places—to better understanding, and indeed greater participation, on the part of local actors, either through environmental education projects, or subsequently as part of local voluntary sector involvement partly driven by the proximity of the park and the (urban) usage arising from it. The recent bird count project staged in TNP is an example of local citizen initiatives of this kind, which draw on approaches widespread in global environmentalist practices.[4]

Finally, in practical terms, national parks in urban locations can provide a living space for people, whose status is variable, often uncertain and precarious. These populations therefore constitute a newly formed category of actors involved in the management of these protected spaces. Indeed, in several of the parks studied, recent years have seen the emergence of associations of users and local players, which structure the residential claims of the populations in the parks or in their immediate proximity: this is particularly the case in Rio and Mumbai (Bautès and Gonçalves Soares 2013; Edelblutte and Gunnell 2014). The emergence of policies of (Bautès and Gonçalves Soares 2013) "securisation" and forced eviction targeting populations in and around the parks has been widely documented, notably with regard to Rio (Ninnin 2014) and Mumbai (Saglio-Yatzimirsky and Landy ed. 2014), an example of how the environmental argument can be used to justify openly segregationist policies that to some extent mirror Yiftachel's analysis of the "dark side of urbanism" (Yiftachel, 1998). The project proposed in 2009 for the construction of a wall in Rio to separate the *favelas* from the forest areas is one illustration of this, where the perceived environmental threat of high urban growth is used to justify the construction of an "eco-boundary".

---

[1]http://www.ecobrasil.org.br/publique/media/Brazil%20National%20Parks%20mar%202010.pdf, accessed September 10, 2015.

[2]http://www.sanparks.org/assets/docs/general/annual-report-2015.pdf, accessed September 10, 2015 (data from March 2014 to March 2015).

[3]Source: KWS, http://conservationfinance.org/guide/WPC/WPC_documents/Reglcasestudies_Kenya_ppt_v1.pdf, accessed September 10, 2015.

[4]http://www.savebrasil.org.br/cidadao-cientista/, accessed June 3, 2016.

So while the question of the relations between national parks and populations is as old as such protected spaces themselves (Depraz 2008; Laslaz 2004), what distinguishes urban national parks from the perspective of the actors concerned is not only the high population density, whether permanent or transitory, around and within the park. It is also their territorial and institutional situation, which lies somewhere between the local and the national, or even, given the metropolitan nature of the cities and the conservation issues, the international.

While the "national" scale of the movers and shakers in the establishment of national parks has long been documented (Depraz 2008; Rodary 2007), the introduction of an additional metropolitan layer in the territories concerned, which is one of the major novelties of the national parks studied in this volume, has received less scholarly attention, although it is not without consequences. The Nairobi case is enlightening: the city is home to numerous international funding bodies such as UNEP (United Nations Environment Programme), UN-Habitat, or the World Bank, as well as the major bilateral development aid agencies, e.g., AFD (Agence Française de Développement), USAID (United States Agency for International Development), JICA (Japan International Cooperation Agency), etc., which together are responsible for the presence of significant numbers of highly educated people, often very familiar with environmental preoccupations and keen to participate in the management of these spaces in an institutional and/or personal capacity. FoNNaP—Friends of Nairobi National Park—would seem, for example, symptomatic of these new groups of metropolitan actors of local and international origin, who are firmly embedded in global environmentalist networks and committed to the protection of these nearby ecosystems. Its management clearly reflects the interest of a new metropolitan class in these protected spaces: FoNNaP's office is staffed of people who all belong to the wealthiest segment of Nairobi's population, with the largest membership of multiple networks and the greatest awareness of global ecological movements. Mirroring its leadership, FoNNaP thus contains large numbers of highly educated local "metropolitan" individuals who ally a global environmentalist perspective with their local experience, a combination extensively exploited by the organisation through sessions of waste collection or animal counting in collaboration with the Kenya Wildlife Service (KWS). Moving to Rio and the Friends of Tijuca Park, the President and Vice-President are respectively the head of a heritage management enterprise and director of sustainable development in a construction firm. They undoubtedly belong to an influential faction of Rio's metropolitan society.

From an institutional point of view, numerous international bodies may see measures in favour of the park as a way of raising their international profile, a source of additional interest linked to a specific metropolitan location, both in terms of development-related advocacy and conservation policies.

In addition, the distinctive location of these parks prompts the emergence of groups that may potentially draw on the image of the park to promote the metropolitan qualities of their city and its activities. Once again, this process can be seen in the different parks, but with varying intensity: the much hyped, largely industry-funded Greenline project underway in Nairobi since 2010, which entails the planting of a barrier of trees to protect the park, is typical of this trend, though it remains marginal in Kenya. In India, Mumbai First, a group that includes numerous businesses, but also the Indian government, the World Bank and the[5] Cities Alliance, whose central goal is to promote the metropolitan qualities of the Indian conurbation, give the environmental issue a high profile in its communication policy, but virtually without reference to Sanjay Gandhi National Park (SGNP), instead employing vague and non-territorialised arguments about urban sustainability.

All in all, therefore, the density and specificity of the actors concerned with the presence of urban national parks is something new. Some metropolitan economic actors seem to be heavily, if not directly, involved in the construction of these protected territories, at least in terms of promotion and advocacy. Local populations, because of their number and density, are also essential actors to consider, especially as they are to differing degrees sometime visitors to the parks. This raises questions about urban leisure activities and the role of the significant numbers of local people concerned with the future of these protected spaces.

9.1.2 Structural Discontinuities

While the IUCN definition of national parks (Category II protected area) states that they should be "*of sufficient size and ecological quality to maintain ecological functions and processes that will allow the native species and communities to persist for the long term with minimal management intervention*"[6] it also acknowledges the major social function of these protected spaces. The IUCN adds that the objective of national parks is also "*to manage visitor use for inspirational, educational, cultural, and recreational purposes, at a level which will not cause significant biological or ecological degradation to the national resources*". This somewhat paradoxical framework offered a difficult but tenable position while most national parks conceived from a strictly protective perspective were located in sparsely populated and largely unfrequented areas, and were themselves unpopulated. In this respect, the first national parks established in France (Laslaz 2009) were for a long time representative of protected spaces. Though sometimes violent, the conflict around them related essentially to the continuation of seasonal activities (e.g., breeding, hunting, tourist activities in the former central areas and their immediate proximity, etc.), and affected a small number of local actors.

The appearance of national parks near cities—with the exception of countries such as the United Kingdom in which national parks have traditionally consisted of peri-urban areas set aside for recreational use (and therefore outside IUCN Category II)—thus spotlights the difficulties inherent in the form of the park itself, conceived as the ordering of the world within a restricted and limited space, which in the cases

[5]http://www.mumbaifirst.org/mission_group.php, accessed April 13, 2017.

[6]https://www.iucn.org/theme/protected−areas/about/protected−areas−categories/category−ii−national−park, accessed April 13, 2017.

presented here introduce divisions between entities that are—to say the least—difficult to connect (André-Lamat et al., in Laslaz et al. 2014).

The proximity between protected space and urban fabric is a forceful example of the scale and complexity of the flows between the categories traditionally defined as "human" and "nonhuman", relating simultaneously to nature and to culture (Latour 2006; Swyngedouw and Heynen 2010). In concrete terms, this immediate proximity, not to say interpenetration, generates significant ecosystemic flows that are notably expressed by the different actors in terms of reciprocal dangers. So the risks associated with the proximity of wild animals are commonly raised within and around the parks of Cape Town (baboons), Nairobi (lions), or Mumbai (leopards), especially as they often generate extensive media coverage that triggers excited reactions and heated arguments (Chap. 7). The real dangers, but also the collective representations of wildlife, lead to a broadening of the spectrum of actors concerned with the proximity of the park. In return, problems linked with human presence in the protected space, whether occasional or quotidian, constitute both issues of practical management—for example, how to handle the spread of waste in the parks—but also of imagination, requiring the actors to accommodate to the proximity of a space that is perceived as both wild and everyday. The interactions or overflows of various kinds between the national park and the city lead to an expansion of the groups of actors concerned. In the outskirts of Nairobi, the wildlife migrations out of the park that occur in each rainy season cause depredations among sheep and cattle populations, and even among residents in the areas of ongoing urbanisation south of the city in the county of Kajiado. Once again, these phenomena are not specific to parks in urban locations, but they are distinctive in the sense that they affect increasingly large populations and, because of their atypical character, obtain a high media profile.

National parks are protected areas, but are also home to large human populations, both around and within the parks, as well as major tourist destinations. Today they very clearly face the problems caused by the juxtaposition, even the interpenetration, of city and protected space. In the areas covered in this chapter, we find a gradient of situations, from Mumbai's SGNP, whose boundaries contain temporarily tolerated slums, to the national parks of Table Mountain in Cape Town and of Nairobi, which have scarcely any permanent human presence, and between them TNP, where occasional encroachments are combatted through permanent surveillance. The problems of encroachment into protected space, observed in most national parks, are thus particularly acute in these urban national parks, partly because of population density, but also because of the fluidity of their land ownership status and the vagueness of their boundaries.

In Nairobi, the park—increasingly squeezed by a fast expanding urban fabric—has come to constitute one of the city's few land reserves. This explains the considerable pressure applied by the national authorities since 2012 to reclassify the protected area to make way for a bypass expressway and, more recently, a railway line. Here, the absence of any buffer zone has led to particularly violent conflicts between the authorities and environmental activists. The emerging importance of land ownership questions has led to the involvement of the National Land

Commission, a largely decentralised state agency that has come to play a central role in land regulation and the resolution of the conflicts arising around Kenya's national parks, notably in Nairobi.

More unusually, but in a way equally representative of the very powerful tensions caused by the contiguity between city and park, a reverse trend is emerging in Mumbai: SGNP has seen its boundaries extended over the decades, gradually absorbing areas inhabited by the Adivasis, "passive encroachers" (Zerah and Landy 2013) the absurdity of whose situation reflects the use of the status of protected space to undermine already precarious habitats (Landy et al. 2014 and cf. Chap. 5).

The question of the value of land situated near an environmental amenity of the scale of a national park (cf. Chap. 6) further deepens the tensions. The demand from a wealthy section of the populations of the cities concerned to live near a park in order to enjoy the benefits of the landscape, the absence of neighbours, or indeed the ecosystemic services attributed to it, is increasing land pressure on the border zones, areas often used up to now by other—economically marginal—actors. The eviction of the *favelados* of Rio, or the encroachments of the southern bypass into NNP, in order to avoid affecting expensive housing programmes nearby, are just two examples of the tensions caused by the presence of well-connected residents and wealthy neighbourhoods near the parks.

9.1.3 Divergent Rhetoric, Real Tensions?

The imperatives of protection in an urban environment are not the same as those of protection in general: in these locations, biodiversity objectives are sometimes less important than the wish to raise public awareness in order to encourage globally more environmentally friendly practices (Dearborn and Kark 2010). The specific question raised by the spatialised nature of protection here also contributes to the difference and specificity of the arguments.

In addition, the different national parks observed have evolved in ways that are partly unconnected with the conservation rhetoric associated with them, which ranges from calls for the sealing off of protected spaces, through to a form of protection that broadly includes the social uses of these areas. Amongst these diverging rhetorics, there are signs of a partial return to a so-called "fortress" approach to conservation, partly because of the qualified outcomes observed in policies to integrate communities into the management of protected spaces. South Africa is an example of this shift, sometimes described as a movement "back to the barriers" (Hutton and Adams 2005) after some 20 years of attempts to reconcile protection and development (Rodary et al. 2003). In this type of configuration, the urban national park seems destined for a state of structural conflict, so entrenched is the opposition between the factors present. Nevertheless, many national parks still maintain a rhetoric that emphasises the participation of the actors, or even the inclusion—though often partial in reality—of local populations in the management of protected space (Chap. 10). This is very superficially the case for NNP, which proposes to take more account of the positions of the mainly Maasai communities situated on the southern edge of the park. The rhetoric of participation developed around Tijuca takes a similar form, bearing little real relation to the actual practices observed, notably towards the local poor populations.

Urban national parks thus find themselves under the spotlight in these debates, presented as iconic battlegrounds where these tensions between different paradigms of biodiversity protection play out.

The contiguity between national parks and fast-growing cities is therefore leading to the emergence of new actors in this domain, often pursuing conflictual agendas. While conflicts are not confined to urban national parks, land pressure and media attention here are very high, giving greater visibility to the frequent participation of actors who are urban—or even in certain cases metropolitan—but have no strong local ties. Two particular rhetorical positions tend to emerge: on the one hand, a return to the barriers, the need for exclusion, often promulgated by the administrators (notably exemplified by the policies long predominant in Mumbai's SGNP); on the other hand, favouring a participation made particularly necessary by the presence of populations around, and indeed within, the parks.

9.2 New Players? Joint Advocacy of a New Urban and Environmental Governance

9.2.1 Favourable Conditions?—Devolution and New Territorialised Organisation of Powers and Prerogatives

While devolution was not simultaneous in the countries, parks and cities studied, in each case it has reshuffled the cards not just in environmental protection, but also in metropolitan urban policy and management as a whole. By sharply modifying the pre-existing balances of power, devolution has helped to challenge the image of national parks managed by all-powerful national administrations with little transparency in their practices. In 2010, for example, Kenya adopted a new Constitution, culminating in the creation of new administrative entities such as counties. Article 62 of the Constitution maintains the public status of the national parks, which remain under the jurisdiction of central government, with landownership questions administered by the National Land Commission. With respect to the planning and management of these protected spaces, however, the situation is more complex. The Wildlife Conservation and Management Act of December 24, 2013 assigns shared prerogatives for the planning and management of these parks. It rules that: "Every national park, marine protected area, wildlife conservancy and sanctuary shall be managed in accordance with a management plan that complies with the requirements prescribed by the Fifth Schedule. In preparing and adopting a management plan, the Service shall consult with the county wildlife conservation committee.

In the case of protected areas, the formulation and implementation of management plans shall involve the participation of neighboring communities" (Art. 44).

In the case of NNP, therefore, three counties (Nairobi City, which contains the entire park, Kajiado, which shares a long southern border with it, and Machakos which borders its south-eastern edge) now have a theoretical right of oversight over the planning and management of the national park, which nevertheless remains under the control of the KWS, an arm of central government. In addition, the Act recognises new entities, the "communities" neighbouring the NNP, giving them a degree of involvement in management plans—in concert with the KWS—via community wardens, who existed before the promulgation of the Act, but whose relations with the communities are likely to evolve. We are thus seeing a proliferation of actors, some of whom are achieving certain forms of recognition, making the question of their connections, coordination, and regulation, even more imperative. This is particularly noteworthy in that a migratory corridor—whose legal status is under discussion—has just been superimposed on the territory of the park itself, passing through the county of Kajiado.

Devolution therefore brings more or less direct and clear obligations to consult over the future of the national park, between authorities which previously inhabited largely separate domains. While the national park itself still remains partly outside local and metropolitan control, the practical environmental priorities on its margins (e.g., wild animals in residential areas, pollution, land ownership, etc.) prompt numerous debates that are bringing new, partly local, actors into play.

By creating a proliferation of local layers of power, and by giving more prerogatives to the actors involved, devolution thus seems to encourage new ways of managing territories, in particular urban national parks, by placing them structurally under the eye of metropolitan actors, and even in certain cases under their partial control. This raises questions about new ways of organising park management services: these institutions, originally paramilitary in India or in Africa, and emanating from ministerial experts in the case of Brazil, are undergoing administrative devolution at the same time as political devolution. This administrative devolution raises the question of the capacity of the local actors to respond to the new forms of shared management and governance to be put in place with the local authorities. The adaptability needed by people with little experience of any form of power-sharing is not without difficulties, given the loss of legitimacy that can result from administrative reorganisation into sub-regions, at a time when the central structures still expect—as they have always done—to exercise a great deal of power in management practices. This point is particularly sensitive in the case of NNP, where the local KWS officials still report to a deeply hierarchical administration, whose offices are located in the park, while at the same time being required to begin unprecedented discussions with actors who have been invited to participate in debates on the future of the park, such as the Maasai communities in the Kitengela triangle.

9.2.2 Urban National Parks, Emerging Areas of Specific Governance Methods

Numerous insightful studies have been written on the types of governance prompted by national parks, whether relating to differences in the acceptability of these protected spaces (Depraz 2008), or the transition from conflict to compromise (Laslaz 2004).

Does the metropolitan character of urban national parks produce a specific type of governance? Is there a flow of practices between metropolitan urban governance (practices that vary greatly in their nature and effectiveness from one case to another) and models of park governance? Are the same people involved, in particular activists, and do they employ the same tools to advocate the same participatory requirements? This is both a general question about the systemic practices of the actors involved in park management and a specific question about the identity of the actors and their metropolitan character.

Despite the new importance ascribed to the environment in international "sustainable development" goals, and the shift in the scales at which decisions are made, in cases where this shift applies to conservation areas (Cohen and McCarthy 2015), it is questionable whether a metropolitan model of governance for these protected spaces is truly emerging.

The urban national parks studied here can seem like improvised laboratories for different approaches to protected space, with a spectrum ranging between integration and sanctuarisation. In this respect, the WCPA (World Commission on Protected Areas) Urban Specialist Group set up by the IUCN is indicative of this major international player's new interest in these types of protected space and the specific problems affecting them. The programme sought to measure whether these entities—a very small minority amongst the large number of usually more remote areas in the IUCN's stable of Category II protected spaces—could serve as an example. As mentioned above, it was thought that these protected spaces, because of the participation and integration demanded by their urban character, could constitute—if not models—at least key testing grounds for this new approach. Conversely, the sharp tensions associated with the contiguity of city and park might favour the return to a more exclusionary approach.

In terms of actors, professional groups have established themselves even in national parks regarding which the arguments are least developed and the political intentions least clearly outlined. These international experts progressively develop global environmental arguments and aspirations based on a protected space from which they quite quickly withdraw. For example, Paula Kahumbu, the high-profile former chairwoman of FoNNaP (until 2014), has distinguished herself by her international academic career (e.g., Universities of Bristol, Florida, Princeton, etc.), her membership of a large number of international associations and bodies (Wildlife Direct), and a very large presence on the web and social networks. The many figures of this type, who correspond in part to the definition of "nomad experts" (Verdeil 2005), who simultaneously propagate models, pursue higher scale agendas (global environment) and back demands for local management that are sometimes postcolonial in tone (where the park constitutes a key component of the quality of life of a wealthy segment of the metropolitan population), constitute a distinctive feature of these parks. Partly due to their involvement and promotion, the parks are becoming arenas for new forms of environmental protection. Their emphasis on a few iconic species and on purportedly innovative practices (e.g., species counts and knowledge networks based on citizen participation, massive use of social networks, etc.) illustrates how these parks are gradually being structured into symbolic spaces for a new kind of appropriation and—at a time when the territories of government and administration are being redefined—of a new kind of governance of natural environments. This governance presents a threat to the potentially important role that—through the actions of certain leaders—urban and local populations have begun to take in developing knowledge about the parks and, thus legitimised, in decision-making processes relating to them.

A final factor is the substantial mobility of these groups of environmental players. Most of the people interviewed in Nairobi reported having been members of several bodies involved in the management of the park, either simultaneously or successively: for example, a former KWS official responsible for community management in the south of the park had joined an environmental association, and another was involved in several voluntary structures. The case of the elected governor of Kajiado, an activist and specialist in animal biology, seems typical of the multiple status of the environmental players, a factor that cannot but considerably change the relations between administrations and non-government actors. This multiplicity is epitomised by the strategies observed during the conflict over a bypass that would encroach into the park. The environmental organisations, in particular FoNNaP, took legal action against the permits granted to the builders by NEMA (National Environment Management Authority), initially with the discreet support of the KWS, or at least some of its agents, that was unable to enter into conflict with another national administration. Symptomatically, during this period, FoNNaP was housed in the premises of NNP, a situation that ended in January 2015. Same paradox in Mumbai when, in 1995, an environmental organisation brought a complaint against the Forest Department because of the encroachment of slums into the park, while at the same time being supported in its action by part of that same Forest Department. Overall, therefore, it is hard not to subscribe to Rodary's observation: "Professional conservation is a collaborative rather than conflictual world, characterized by long-standing global integration, highly connected networks of actors and a flourishing industry of 'ecological crisis'" (Rodary 2011).

Ultimately, the relation to the urban national park seems ambiguous in the case of many conservationist players. The big metropolitan actors, in the full sense of the term metropolitan, priorities global causes (e.g., animal survival in the case of Nairobi, biodiversity in the case of Rio, etc.), enabling them to bypass the imbroglio of prerogatives in which most of the parks mentioned are currently entangled. Moreover, this also makes it easier for them to hold a discourse on participatory governance in environmental issues than if it is linked specifically with the spatial and territorial problems of urban national parks, which are often treated as no more than local versions of those big causes. Nevertheless, other networks can emerge, more focused on the territorial conflicts associated with the parks. In Cape Town, for example an environmental discourse emerged through the mobilisation of citizens from relatively well-off social categories, initially around issues relating to the safety of their neighbourhoods. In Mumbai, the conflict about the slums took the form of a legal battle, lot number by lot number, over whether one piece of land or another belonged to the Forest Department. In the case of Nairobi, finally, while the adherents of "local" conservation are much less audible than the big global organisations (e.g., Wildlife Direct, African Wildlife Foundation, etc.), they are the ones who are now combatting the administrators of the park over encroachments by the road or railway. Amidst these dynamic events, the city appears only rarely as one of the conservation actors, and even less as a category that requires specific protection.

Changes in governance, therefore, both in the national parks and with regard to environmental issues in metropolitan areas, seem attributable not so much to the construction of an argument around these hybrid objects—protected spaces within the city—as to the presence of metropolitan actors closely involved with the global arguments and practices of conservation, without direct connection to the specificities of the urban context. The most audible conservation arguments revolve around global categories, and on the occasions where local territorial problems are considered, little attention is paid to the specific urban status of the territories concerned.

9.3 From the Myth of the Green City to the Reality: A Fantasy Model of Governance?

9.3.1 Weak Writings, Weak Contacts: Mutual Ignorance Between Actors

While the recent reconfigurations of power and additional layers of authority relating to urban national parks seem theoretically to have the potential to generate new forms of governance between actors, the reality observed is more qualified, raising questions regarding not just vertical, but also horizontal connections and integrations. In the case of Nairobi, an analysis of the planning and management documents for both NNP and the city reveals a great deal of ignorance about the territory and the actors present beyond the boundary separating the two spaces. The Management Plan for NNP dates back more than 20 years, and has still not been updated. In particular, it scarcely mentions the metropolis, except in critical terms, treating it as no more than a negative externality (waste, occupancy described as excessive). The stated objectives for the park are the conservation of Kenya's natural environments as world heritage, the sustainable use of Kenya's ecosystemic resources for the economic development of the nation, and the protection of people and property from injuries or damage caused by wild animals. There is thus no question of the park's role vis-a-vis the city, and even less of a local scale that is fundamental in the construction of the management plan. The scale of the perspective remains national, even international, despite a highly particular local context. The layout and design of Safari Walk, a part of the park accessible on foot, equally emphasises the fact that NNP encompasses almost all the ecosystems of Kenya in a sort of national zoo, while totally ignoring its proximity to the city.

Conversely, the initial work for the future city masterplan pays very little attention to the park: the feedback on the preliminary participatory processes established in 2013 refers to questions of water and waste as the city's essential environmental priorities, not the national park. In the old plan for the metropolitan region (Nairobi Metro 2030), the national park received only marginal mention in relation to the potential development of tourist activities or as a green space with the same status as all the city's other open spaces or parks.

The interviews conducted in 2013, both within the KWS and with officials of Nairobi City Council (in the environment and the spatial planning departments) all highlight the very low level of familiarity between the actors, both in the respective structures of each of the institutions and—the two things are linked—at the interpersonal level, which we have seen to be so important within conservationist networks. True, this lack of interpersonal relations between authorities whose prerogatives over the national park are shared, is attributable in part to the recentness of devolution in Kenya. Nonetheless, we argue that this lack of contact also reveals the absence of a connection between city and national park in most of the areas studied. This is probably because the transfer of prerogatives to local levels remains very limited, as well as asymmetrical, as long as the management of the parks remains under the authority of a state administration which, though deconcentrated, continues to be deeply hierarchical.

Opposing political adherences can also contribute to the poverty of exchanges between the authorities responsible for managing the national parks and the urban authorities, and may even be the source of new forms of conflict: at TMNP, for example, one of the reasons for the lack of dialogue between the park management authorities South African National Parks (SanParks) and the urban authorities resides in the fact that the former fall within the ambit of the African National Congress (ANC), whereas the city and the province are dominated by the Democratic Alliance (DA). As a result, the only points of discussion between these politically adversarial and partly competing institutions, correspond to situations of crisis where coordinated action is essential, for example in the handling of the difficulties associated with baboon attacks on tourists. Conversely, competing environmental education policies have steadily been introduced between the national park, which is developing its own programme, and the city, which primarily promotes indigenous knowledge around Kirstenbosch Botanical Garden.

9.3.2 The Reduced Role of Urban National Parks in Metropolitan Development Models

The cities of Rio and Cape Town, in particular on the occasion of big international events, have constructed a highly structured political discourse focusing on both spaces and, as a result, the two institutional entities of the park and the city. Conversely, the two other areas studied reveal an almost total absence of integration between metropolitan challenges and the issues of protected space.

The parallel flows of experts and expertise observed in big cities and urban national parks are contributing to a "rearrangement of the structures of economic and political domination" (Verdeil 2005). They are also contributing to the emergence of metropolitan models of development understood—usually from a processual perspective—as "simplified and symbolic schemas that account for some reality" (Moriconi-Ebrard in Lévy-Lussault 2003), and assumed to be reproducible. Nonetheless, can these models be considered today to give an explicit role to the national parks that are contiguous with a growing number of global cities?

Urban national parks, because of the "expert" actors whose interest they arouse, seem at first sight to be capable of integration into a more global perspective on the development of the cities that contain them. However, the different terrains chosen here show that this model is ultimately highly uneven in construction. Paradoxically, it is in fact the absence of thought about local issues prompted by the presence of the national park that seems to correlate very closely with the weakness of the metropolitan model encompassing the park. In Cape Town, for example, the city shines a highly political spotlight on the integrated nature of the park's environmental management and its de facto contribution to the metropolitan status of the city. While the emergence of an "expert" population helps to highlight the environmental value of the park, the result is often an immediate shift to the global scale, without consideration for the metropolitan. In Cape Town, the focus of the city's political spotlight falls on a few high-profile factors, such as fauna management and the safety issues associated with the park, but ultimately reveals a limited interest in the national park in the metropolitan model it promulgates.

It might be argued that national parks today remain peripheral in the construction of metropolitan environmental arguments, consigned to the margins of more general environmental issues. The first international networks, prompted by commitments such as those of the Aalborg Charter on the "sustainable city", formed since the 1990s without consideration for the national parks and biodiversity protection in general. The primary aim of sustainable urban planning is to design cities with the smallest possible environmental footprint, not to respond to the imperative of protecting local ecosystems. As Emelianoff (2007) notes, "*its novelty lies in a linking of scales that distinguishes it from urban ecology approaches: the intergenerational, but above all the problems of the global environment*". The real spatial problems caused by the local character of protection in a context of systematic pressures and tensions between cities and parks, preclude any simplistic—and utopian—reading of the environmental quality of the metropolis in question.

In other words, the presence of a park—national though it may be—cannot be taken as a serious argument that a metropolis can be defined as a green, or even sustainable, at least outside its own strategy of communication on the subject, if it happens to have one.

So, while urban national parks might seem in theory to constitute complex territorial configurations that are favourable to the emergence of new sets of actors, in practice the image of them as laboratories for an emergent metropolitan governance around environmental issues, seems highly debatable. The confiscation of the participatory processes identified, but also the poverty of the exchanges between institutional actors on the part of both city and park, prompt scepticism as to the emergence of a form of joint management that might be able to smooth out extreme territorial tensions. Moreover, these areas of friction give a glimpse of the difficulty of building a local scale of reflection and participation around these parks which, despite the immense local challenges they face, remain largely invisible to the urban institutional actors.

References

Bautès, N., & Goncalves Soares, R. (2013). De la qualification juridique à l'usage social du droit: les favelas en prise avec l'intervention publique. *Géocarrefour, 88*(3), 217–226. www.cairn.info/revue-geocarrefour-2013-3-page-217.htm.

Cohen, A., & McCarthy, J. (2015). Reviewing rescaling: Strengthening the case for environmental considerations. *Progress in Human Geography, 39,* 3–25.

Dearborn, D., & Kark, S. (2010). Motivations for conserving urban biodiversity. *Conservation Biology [serial on the Internet], 24*(2), 432–440.

Depraz, S. (2008). *Géographie des espaces protégés* (p. 320). Paris: Armand Colin.

Emelianoff, C. (2007). La ville durable: l'hypothèse d'un tournant urbanistique en Europe. *L'Information géographique, 71*(3), 48–65. www.cairn.info/revue-l-information-geographique-2007-3-page-48.htm.

Hutton, J., & Adams, W. M. (2005) Back to the barriers? Changing narrative in biodiversity conservation. *Forum for Development Studies*, 341–370.

Landy, F., Bon, E., & Zerah, M. H. (2014). Entre ville et forêt: conflits d'acteurs dans le Parc national Sanjay Gandhi de Mumbai (Bombay, Inde). In L. Laslaz et al. (Eds.), *Espaces protégés et territoires* (pp. 228–247). Belin: Conflits et acceptation.

Laslaz, L. (2004). *Vanoise: 40 ans de Parc National, Bilan et Perspectives, 2004* (p. 428). Paris: L'Harmattan.

Laslaz, L., Gauchon, C., Duval, M., & Héritier, S. (dir). (2014). *Les espaces protégés, entre conflits et acceptation* (p. 431) Paris: Belin.

Latour, B. (2006). *Nous n'avons jamais été modernes. Essai d'anthropologie symétrique*. Paris: La Découverte. «Poche/Sciences humaines et sociales», (éd. originale, 1991).

Ninnin, J. (2014). Le rêve carioca: entre planification urbaine et déplacements forcés de population. *L'Espace Politique.* http://espacepolitique.revues.org/2950.

Rodary, E. (2007). Mobiliser pour la nature, ou la construction et la disparition du local. *L'Espace géographique, 1*/2007 (Tome 36), 65–78. www.cairn.info/revue-espace-geographique-2007-1-page-65.htm.

Rodary, E. (2011). Crises et résistants: les écologies politiques en Afrique. *Ecologie & politique 2*/ 2011 (N° 42), 19–32. www.cairn.info/revue-ecologie-et-politique1-2011-2-page-19.htm. https://doi.org/10.3917/ecopo.042.0019.

Rodary, E., Castellanet, C., & Rossi, G. (2003). *Conservation de la nature et développement, l'intégration impossible?* (p. 308). Paris: Gret/Kathala.

Saglio-Yatzimirsky, M.-C., & Landy F. (Ed). (2014). *Megacity slums. Social exclusion, space and urban policies in Brazil and India*, London, Imperial College Press.

Swyngedouw, E., & Heynen, N. (2010). Urban political ecology, justice and the politics of scales. In G. Bridge & S. Watson (Eds.), *The Blackwell City reader* (pp. 79–86). Chichester: Blackwell Publishing.

Verdeil, E. (2005). Expertises nomades au Sud. Eclairages sur la circulation des modèles urbains. *Géocarrefour, 80*(3). http://geocarrefour.revues.org/1143.

Yiftachel, O. (1998). Planning and social control: Exploring the dark side. *Journal of Planning Literature, 12*(2), 395–406.

Zerah, M.-C., Landy, F., (2013). Nature and urban citizenship redefined: The case of the National Park in Mumbai, *Geoforum, 46*, 25–33.

Part III
Circulation of Practices, Discourses and Models

Chapter 10
A Conservation View: Towards New Urban Protected Area Governance?

Louise Bruno-Lézy

> *The acceptance of territorial heritage, considered as a common good, requires ... that we go beyond the dichotomy between natural heritage and cultural heritage, which underlies the principles of conservation based on the inventory of specific assets, and that we go towards a definition of territorial heritage which reintegrates the nature/culture pair, where a territory is the result of co-evolutionary processes between the environment and human civilisations*
>
> Magnaghi (2014, p. 20)

> *There are many conservations, and it is time to stop pretending otherwise*
>
> Sandbrook (2015, p. 566)

In his opening speech at the 5th IUCN World Parks Congress in 2003, South African President Nelson Mandela put the issue of governance at the heart of the discussions held in Durban around the theme of "Benefits Beyond Boundaries": "*We know that the key to a sustainable future for protected areas lies in the development of partnerships. It is only through alliances and partnerships that protected areas can be made relevant to the needs of society .... A sustainable future for humankind depends on a caring partnership with nature as much as anything else*" (IUCN and WCPA 2004).

This chapter offers a discussion on the evolution of the governance of urban protected areas, based on three major notions: the importance of partnerships for conservation, the need to develop a global approach concerning the territory as a common good reintegrating nature and culture, and the recognition that we need a wide variety of conservation methods to address global ecological, economic, social and political challenges.

The traditional approach to Conservation Biology, known as Reservation Ecology, has been translated spatially into the territorial exclusion of natural and cultural spaces to be protected (from development). Since the creation of the first

L. Bruno-Lézy (✉)
Urban Conservation Strategies Specialist Group, IUCN-WCPA, Gland, Switzerland
e-mail: louise.bruno@yahoo.fr

F. Landy (ed.), *From Urban National Parks to Natured Cities in the Global South*,
https://doi.org/10.1007/978-981-10-8462-1_10

national parks at the end of the nineteenth century, this has resulted in a dual system of land use,[1] leading most surface areas to be used to economic ends and extracting from them a few specific spaces, considered as exceptional, as a form of compensation for the damages caused by development (Magnaghi 2014).

Putting part of the territory under a bell jar comes from a colonial logic that brought colonisers to appropriate territories at the expense of local indigenous populations,[2] who depended on these for their survival (Crosby 2004). This dichotomous system, which opposes nature and culture, has resulted in a paradox which is the cause of today's ecological crisis. This enabled the over-exploitation of the planet's natural resources, in favour of an economic development believed to be infinite.

If this territorial exclusion system that protects what is considered as natural or cultural heritage, has shown its limits, what should come next in building a new approach to the territory and the way to manage it? This issue concerns two levels. At the spatial level, it affects the conception of the territory as a whole and as a common good as defined by Magnaghi (2014). At the social level, it involves the way the management and governance of natural areas are integrated into this holistic logic.

The multiplicity of meanings linked to the word "conservation", at the beginning of the twentieth century, makes it possible to envisage approaches to biodiversity conservation other than that which consists of putting some territories under a bell jar (Sandbrook 2015).

In order to tackle these issues, we will first try to apprehend the evolutions that can be observed in the governance of protected areas around the world, particularly those identified by the International Union for Conservation of Nature (IUCN) following the fifth World Parks Congress.

We will then analyse the governance systems of four urban national parks in the metropolises of emerging countries: Tijuca in Rio (Brazil), Table Mountain in Cape Town (South Africa), Sanjay Ghandi in Mumbai (India) and Nairobi (Kenya), with respect to the international typology defined by the IUCN and the Convention on Biological Diversity (CBD). This will make it possible to further analyse and compare the mechanisms and requirements orienting the management of these protected areas facing urban dynamics characterised by high inequality.

From these examples, we will propose a few scenarii concerning the evolution of the governance of protected areas, abandoning the nature-culture dichotomy and reintegrating protected areas into the territory defined as a common good.

[1]The notion of territory, as used here, goes beyond the geographic area or spatial entity. It is understood as a dynamic system which is the result of interaction processes between people and the environment.

[2]The expression "indigenous", as used here, does not refer only to populations characterised by common genetic and cultural heritage. It also refers to an autocatalysis capacity between the society concerned and its living environment (Bruno et al. 2016).

10.1 Changes in Conservation Governance

A growing consensus among scholars and development practitioners points to the importance of improving the governance of man-environment systems, which is an essential condition for achieving sustainability, reducing inequalities and creating conditions for socio-environmental justice …

Sowman and Wynberg (2014, p. xiii)

Although the management and governance of protected areas are objectively different, they are intrinsically linked. Management concerns the way the resources of protected areas are managed, i.e. the conservation method as well as the human, technical, financial, legal and communication means. Management touches on what is being done to reach the objectives given by governance authorities.

Governance, according to Graham et al. (2003, pp. 2–3), refers to the processes through which management is carried out and "the interactions among structures, processes and traditions that determine how power and responsibilities are exercised, how decisions are taken, and how citizens or other stakeholders have their say". Governance concerns power relations between interested parties; the decision-making processes (Who is responsible for making decisions?); the conditions for their efficient implementation and for assessing actions. These mechanisms have evolved since the first national parks.

The modern conservation movement goes back to European imperialism (Grove 1995; Jepson and Whittaker 2002). The "national park" is above all a colonial geopolitical and cultural construction, with a strategic role related to identity: USA (Yosemite Grant: 1864; Yellowstone: 1872), Australia (Royal: 1879), Canada (Rocky Mountain, today Banff: 1885). This construction is based on a portion of the territory being controlled by a given authority and a certain social category (the urban elite), with its representations of nature usage (Depraz 2008, p. 131). The influence of this model, attributed to the United States and conveyed internationally, has been highlighted by Everhart (1972) in what he calls the family of the "Yellowstone children".

The London Convention relative to the preservation of the fauna and the flora in their natural state (1933) only concerned Africa, which at the time was for the most part colonised. This convention provided for the creation of national parks and the strict protection of certain animal species. The Convention on Nature Protection and Wild Life Preservation in the Western Hemisphere (Washington Convention 1940), called on American countries to create protected areas based on the North American national park model. Partly inspired by this convention, the African Convention on the Conservation of Nature and Natural Resources (1968) encouraged parties to create protected areas where local populations were excluded and touristic activities, such as sport fishing, were accepted. While the USA and Canada, as early as 1965, organised training sessions intended for the leaders of conservation worldwide, the International Office of the US National Park Service helped many countries to establish their national parks. International organisations such as the IUCN, the FAO (UN Food and Agriculture Organisation) and later the UNEP (UN Environmental Program), also played an important role in propagating the concept of "national park as *primus inter pares* among the other types of protected areas" (Phillips 2003, pp. 10–11).

This model spread throughout the world, promoting the dichotomy between "parks" and "people". Its effect was devastating for traditional populations, whose relationship with nature was different from that proposed by John Muir and the first ideologists of American "national parks" (Cronon 1995) that were nonetheless created on sacred Native American lands (e.g., Yellowstone, Yosemite, Grand Canyon, Grand Teton, etc.). Transposing the resident-free park model, in the countries of the South in particular, in places where traditional populations live, has been at the root of many conflicts which reflect the fact that the initial concept was inadequate. This combined with other factors, such as land conflicts, inadequate monitoring, the corporatism of park agencies, urban sprawl, poverty and economic crisis, are behind the fact that this conservation model has been questioned (Diegues 1994).

Tendencies began to change during the 1970s. In order to relate changes in the way protected areas were managed and governed, Phillips (2003) analysed the themes of the recommendations stemming from four IUCN World Parks Congresses: Seattle in 1962, Yellowstone/Grand Teton in 1972, Bali in 1982 and Caracas in 1992. The themes dealt with, which practitioners and researchers forwarded to the IUCN, reflect the preoccupations and practices found on the ground. It was only from 1982 onwards, during the Bali Congress, which is considered as a turning point in the way conservation was being conceived, that the first questions on the links between protected areas and development began to appear, including the recognition of the key role played by local populations in supporting conservation. These tendencies were transcribed in the World Conservation Strategy (IUCN, UNEP and WWF, 1981) and confirmed in 1991 in Caring of the Earth (IUCN, UNEP and WWF). During the Caracas Congress which took place a few months before the Earth Summit held in Rio, ideas relating to regional conservation strategies arose, with networks of protected areas linked by corridors (McNeely 1993; Holdgate and Phillips 1999). However, the links between conservation and development were widely criticised, particularly as regards the balance between stakeholders, power relations and the relationship between development and ecological destruction (Anderson and Grove 1987; Redclift 1989; Adams 1990; Brandon and Wells 1992; Noss 1997).

In order to support the discussions on governance of protected areas at the fifth World Parks Congress, a world survey was conducted with 100 agencies managing protected areas, with a view to assessing changes in the governance systems between 1992 and 2002 (IUCN and WCPA 2004). The results showed increased participation from various actors and a greater use of the mechanisms of social accountability and participative techniques. Close to 90% of respondents deemed that protected areas governance had improved during the decade, and 67% that this had also improved management efficiency (Dearden et al. 2005).

Despite an apparent consensus in Durban, conservation biologist Kent Redford, who had shocked his audience by asserting that even indigenous populations could contribute to biodiversity loss (Redford 1991), had to avoid an open conflict between them and the conservationists. According to Christensen, the Indigenous people had come to the Congress ready to tell the entire world that "parks and protected areas are fundamentally incompatible with the rights and aspirations of impoverished local communities" (Christensen 2004, p. 34). This brought to light the contradictions of integrated conservation and development projects (ICDPs), a notion that was established by Katrina Brandon and Michael Wells the same year when the Earth Summit was held in Rio (1992), and where the union between conservation and development had been sealed (Wells and Brandon 1992).

For Redford, there was no win-win situation as far as the conservation-development coupling was concerned. That is why, together with the World Conservation Society's (WCS) John Robinson, they proposed to rephrase "integrated conservation and development projects" as either "conservation projects with development" or "development projects with conservation" (Christensen 2004, p. 39). This rewording clarifies the need for stakeholders to clear up, at the outset of the project, their interests and expectations, and draw attention to negotiations and compromises essential for the project. As such, a project aimed at reaching this double objective but ignoring the necessary compromises to this end, would be destined for failure, to the detriment of the disadvantaged people and biodiversity.

The model of centralised governance of protected areas (controlled by a government irrespective of its level) was being questioned, as were ICDPs. This called for developing a new governance paradigm (Phillips 2003), because areas of strict protection, belonging to States and managed by governments, were no longer (and are still not) considered as sufficient to ensure the objectives of biodiversity conservation, or those of economic and social development (Chape et al. 2005). Elaborating a new paradigm for the management of protected areas would require a new governance direction, towards that where the agency in charge of conservation would work in partnership with other levels of government, with local authorities, indigenous people, the private sector and NGOs.

The new paradigm envisaged at the time, advocated a wider vision of protected areas at three levels. First of all, it proposed *a widening of the range of stakeholders interacting with protected areas*, beyond those behind their creation and management, including indigenous people and local communities. Second, it was proposed that the work should be carried out on a *much wider scale than the protected area stricto sensu*, as was done in the past, through the bioregional planning of ecological networks and buffer zones. Thirdly, these two advances would open one's *understanding of the definition of protected area*, making it possible to include parts of the inhabited landscape (for example, in Category V of the IUCN's protected areas).

However, despite huge conceptual advances in the discussion on the types of protected area management during the last 40 years, and despite increasing experiences in participative governance, results have remained fairly mixed (Ghimire and Pimbert 1997; Utting 1994; West and Brechin 1991). While, in theory, the conservation movement seemed to have assimilated the notion of integrated governance of protected areas, the challenge, as always, is to apply it in the field.

Analysing the evolution of governance types and their related policies and practices, has led the IUCN and the CDB to distinguish four main types of governance for protected areas, according to the actors taking part in the fundamental decisions (concerning creation, main objectives and management in particular).

- Governance by government (Type A): at different levels and also by combining various institutions;
- Shared governance (Type B): between various right holders and stakeholders;
- Private governance (Type C): individuals and private organisations (generally the land owners);
- Governance by indigenous people and/or local communities (Type D).

This typology of governance of protected areas recognises the need to open conservation systems as well as dialogue with society, particularly with local populations impacted by the creation and the management of protected areas. However, when trying to apply this typology to the governance of urban protected areas, through UNPEC case studies, these categories seem much vagues than they appear at first (Table 10.1).

10.2 Very Different Sociocultural and Political Contexts…

The governance of a protected area, or of a system of protected areas, is shaped by history, culture and the interplay among local, sub-national, national and international actors and institutions. It is the result of processes of developing and exercising authority and responsibility over time.

Borrini-Feyerabend et al. (2014a, p. 26)

While the four national parks share the same management objective (which is why they are classified in the IUCN Category II), the path each one followed to fulfil this objective differed according to the context. Other management issues came up over time and governance methods changed in accordance with new parameters (see Table 10.2).

The drivers of park governance are still national or federal agencies in charge of managing protected areas. However, the governance methods observed are situated between Types A and B, with a gradation revealing great complexity (Type A transiting towards Type B; a hybrid type "AB"). The protected areas under private governance (Type C) can be integrated into shared governance (Type B). This is the case in Cape Town. Our case studies do not (yet) contain protected areas managed by local communities (Type D).

The management of Sanjay Gandhi National Park (SGNP), by the Maharashtra Forest Department, is highly centralised (Type A). The management of Nairobi

Table 10.1 Analysis of the governance of the four parks studied by UNPEC in the light of IUCN typology

Governance type	National Park	Type A			Type B			Type C			Type D	
		Government-managed protected areas			Co-managed protected areas			Private protected areas			Community conserved areas (CCAs)	
	IUCN Category II National parks	Federal or national ministry or agency n charge	Regional/ Local/ municipal ministry or agency in charge	Government delegated management (e.g., to an NGO)	Transboundary management (multiple scales)	Collaborative management (various forms of pluralist influence)	Joint management (pluralist management board)	Declared and run by individual landowner	Declared and run by non-profit organisations (e.g., NGOs, universities, etc.)	Declared and run by for-profit organisations (such as individual or corporate landowners)	Declared and run by indigenous people	Declared and run by local communities
Type A	Sanjay Gandhi	Maharashtra Forest Department										
Type A▸B	Nairobi	KWS				County Wildlife Conservation and Compensation Committee					In the future, in the south of Nairobi National Park?	
Type AB	Table Mountain	SANParks	Cape Nature		BioNet	CCT +TMNP		private landowners	x (e.g., Western Cape University)	x (e.g., Eskom)		
Type B	Tijuca	ICMBio	Pedra Branca (Rio State)/ locals & municipal PAs		Carioca Mosaic	PNT Advisory board	Co-management PN + Rio City + Rio State				CCAs as part of the Carioca Mosaic	

Source Adapted from Borrini-Feyerabend et al. (2004b)

Table 10.2 Evolution of parks in context compared to a national park traditional conservation model

	As it was	*As it is becoming*			
	Traditional model	SGNP	NNP	TMNP	TNP
Objectives	– *Set aside for conservation* – *Established mainly for spectacular wildlife and scenic protection* – *Managed mainly for visitors & tourists* – *Valued as wilderness*	– Set aside & established for conservation (forest & water supply for the city) and education to a lesser extent – Valued as wilderness	– Ecosystem conservation (Savannah) & spectacular wildlife; wildlife migration – Managed for visitors & tourists – Valued as wilderness	– Ecosystem conservation (Fynbos) & scenic protection – Run also with social and economic objectives – Managed for visitors & tourists, with local people in mind – Valued for cultural importance of so-called wilderness; natural monument (Table Mountain; Cape Point & Cape of Good Hope)	– Atlantic rainforest conservation & scenic protection – Run also with social and economic objectives – Managed for visitors & tourists, with local people in mind – Valued for cultural importance of the landscape (Corcovado, Pedra Branca, Pedra da Gavea, etc.)
Governance	– *Run by central government*	– Run by central and State government (Forest Department of Maharashtra State)	– Run by central government (KWS) – Emergence of a partnership via the CWCCC	– Run by central government (SANParks) and other institutional partners (City of Cape Town, CapeNature, etc.)	– Run by central government (ICMBio) and other partners (Rio City, RJ State, institutions, civil society)
Local People	– *Planned and managed against people* – *Managed without regard to local opinions*	– Managed without regard to local opinions	– Managed up to now without regard for local opinions; future dialogue with the CWCCC?	– Managed with regard to local people	– Run with, and in some cases for, local people

(continued)

Table 10.2 (continued)

	As it was	*As it is becoming*			
	Traditional model	SGNP	NNP	TMNP	TNP
Spatial Context	– *Developed separately* – *Managed as "islands"*	– Developed separately from the city – Managed as an "island"	– Developed separately of the city – Managed in coordination with the wildlife migration area south of the park	– Planned as part of national, regional, and international conservation & heritage systems (e.g. UNESCO Natural Heritage Site) – Developed considering the urban context – Developed as "networks"(strictly protected areas, buffered and linked by green corridors: Biodiversity Strategy—BioNet)	– Planned as part of national, regional, and international conservation & heritage systems – Developed within the city context (e.g., UNESCO Cultural Heritage Site) – Developed as "networks"(strictly protected areas, buffered and linked by green corridors: Carioca Mosaic)
Perceptions	– *Viewed primarily as a national asset* – *Viewed only as a national Concern*	– Viewed primarily as a national asset – Viewed only as a national concern	– Viewed as a national asset (conservation, tourism) with a local concern.	– Viewed as a national and local asset as well as an international concern	

(continued)

Table 10.2 (continued)

	As it was	*As it is becoming*			
	Traditional model	SGNP	NNP	TMNP	TNP
Management Techniques	– *Managed reactively within a short timescale* – *Managed in a technocratic way*	– Managed in a technocratic way – About protection		– Managed adaptively in long term – Perspective with social & economic considerations – About protection, restoration and rehabilitation	
Finance	– *Paid for by taxpayer*	– Paid for by taxpayer – Income from tourism (perceived by the central government, not directly by the park)		– Paid for by taxpayer with private partnerships for projects – Income from tourism (perceived by the central government, not directly by the park)	
Management Skills	– *Managed by scientists and natural resource experts* – *Expert-led*	– Managed by scientists and natural resource experts – Expert-led		– Managed by multi-skilled individuals with local knowledge also	
Governance Evolution Period	– 19th → 20th	→1960s	→1980s	→2000s	→2010s

Source Adapted from Philips (2003, p. 20)

National Park (NNP), which up until then was highly centralised, is currently undergoing a major transition following the publication of the Wildlife Conservation and Management Act (No. 47, 2013). This law is leading the Kenya Wildlife Service (KWS) to create an entity promoting dialogue between all actors concerned and/or impacted by its management: the County Wildlife Conservation and Compensation Committee (CWCCC, Type A transiting towards Type B?). Table Mountain National Park (TMNP, hybrid type "AB") in Cape Town, and Tijuca National Park (TNP, Type B) in Rio, are already more advanced as far as dialogue and collaboration with other local stakeholders are concerned. Both have created partnerships, with different types of commitments of course, with the municipal and provincial or regional levels. They have been integrated into conservation projects beyond the parks' boundaries, on a bioregional scale: in Cape Town, TMNP is part of the Cape Town Bioregional Plan (BioNet); and in Rio, TNP is part of the Carioca Ecological Mosaic. These ecological networks operate according to the principles of Connectivity Conservation Management (Pulsford et al. 2015) with different protected area managers (namely city, province, state or private), a global strategy, a steering committee and federative-type governance. This does not mean in any way that relations between stakeholders are perfectly fair and well-balanced, but that despite conflicts, compromises, divergent positions and interests, and even sometimes "paternalistic" authorities, conservation initiatives involving local populations are established and can be of benefit to both sides (cf. the reforestation project in Rio or the programme to fight against invasive species in Cape Town).

In Mumbai and Nairobi, parks were created in what were then rural areas, and which at the time were considered as wilderness (if one ignored the cultivated lands in the case of Mumbai and the pastures in the case of Nairobi). They were progressively brought into contact with the city, through uncontrolled urbanisation which is characteristic of the countries of the South. In Rio and Cape Town, the parks were born in the cities. The areas that became national parks had been protected for a long time already,[3] making it possible to extract them from urbanisation even if, since then, the urban fabric has been developing and densifying all around. This is why relationships between cities and parks are so different in these cases.

In Mumbai in 1950, the national park was created to protect the semi-deciduous forest and the two dams in the north of the city, which played an important role in supplying drinking water to the city, and which have since become residual. The Forest Department manages this space of 104 km^2 (which contains the highest density of leopards in the world). The fact that it keeps expanding (fivefold between 1967 and 1982) remains a source of conflicts. Despite its location in the centre of an urban area of 21 million residents (UN 2014), the management of the national park

[3]Preserving Tijuca, as the national forest, goes back to 1861. In Cape Town, the Cape of Good Hope Nature Reserve was created in 1939, Table Mountain Nature Reserve in 1963 and Silvermine Nature Reserve in 1965.

follows the old fortress conservation model, as do most rural national parks in India. Even if, recently, the more open management has tended to modify its position, managers are still oscillating between conservation and urbanity: policy of openness to the public (e.g., promenade, contemplation, yoga and visiting Buddhist caves, etc.), environmental education intended for the city's schoolchildren, and *manu militari* expulsion of encroachers. Their management method is very much turned towards conservation, and dialogue with urban institutions, city dwellers and other actors in the city remain secondary (Bruno et al. 2016).

The KWS manages NNP, which covers 117 km^2. The park is situated at the extreme north of a semi-arid savannah ecosystem of more than 2,000 km^2, based on the ungulates' seasonal migration between the north of Tanzania and the south of Kenya. These large scale migration movements are important for the functioning of the entire ecosystem, and supply ecosystemic services and economic benefits to the whole region, well beyond the city of Nairobi.[4]

Animals take refuge in the national park during the dry season from June to November, and in December, when the rains return, they scatter towards the south, on the plains of Athi-Kaputiei. These are part of the traditional territory of the Maasai pastoralists, who bring their herds there to graze and cohabit with the wildlife, constituting what Deep Ecology calls an anthropo-ecosystem.[5] As such, the dispersal areas are situated outside the protected area, on community and private lands, where animals must face increasing anthropic pressure (Fynn and Bonyongo 2011).

The City of Cape Town is situated in an area of great geobiological wealth, known as the "Cape Floral Kingdom", which in 2004 became a global biodiversity hotspot and part of the World Heritage list.[6] The national park (which covers 250 km^2 on land and 1,000 km^2 at sea) was established in 1998 in the high zones of the Cape Peninsula, where protected areas managed by different organisations existed already. The idea of protected areas being managed by a unique authority appeared in 1974 and again in 1977, but was strongly contested.[7]

The solution was then to adopt a form of management that was based on federative governance, with all the authorities subscribing to common conservation objectives. This is how the Cape Peninsula Protected Natural Environment

[4]In 2012, income from tourism in Kenya was estimated at USD 1.2 billion (KNBS 2013).

[5]Deep Ecology integrates the human species in the biosphere as a whole. The notion of anthropo-ecosystem developed by Deep ecology concerns a co-evolutionary process that takes into account the needs of the fauna and flora, as well as those of human beings (Næss 1973; Foreman 1991).

[6]The Cape Floral Kingdom contains 9,600 indigenous plant species, of which 70% are endemic and 1,406 are on the IUCN's Red List of endangered species (Katzschner et al. 2005).

[7]A report published in 1974 by two botanists from the University of Cape Town, Eugène Moll and Bruce Campbell, pointed out the flaws of managing public natural area properties, which at the time were managed by 14 different organisations, and put forward for the first time the idea of management carried out by only one authority. In 1977, a new report concerning the "Future Control Management of the TM and the Southern Peninsula Mountain Chain" concluded in favour of a form of management controlled by the South African National Parks Board (SANParks 2004).

(CPPNE) Management Advisory Committee was born, covering 30,000 ha of land in the peninsula. The first 14,000 ha of the then Cape Peninsula National Park were validated in 1998. Currently, TMNP consists of 25,000 ha of the CPPNE and the objective is to integrate the remaining 5,000 ha into the park (SANParks 2008).

As such, TMNP is not made up of only one park but of different sites, eight in total, managed by four authorities, i.e. the South African National Parks (SANParks), CapeNature (which is the provincial nature conservation authority), the Eastern Cape Nature Conservation Board (ECNCB) and the South African Biodiversity Institute (SANBI).[8] As it is, this situation requires a model of federative governance in order to reach conservation objectives in a city of 3.75 million residents (UN 2014).

Tijuca National Park, which consists of a reconstituted tropical forest, is the smallest and at the same time the most visited of the national parks in Brazil (more than 2 million visitors per year). The park's 40 km^2 are managed by the Chico Mendes Institute for Biodiversity Conservation (ICMBio), the national organisation in charge of protected federal areas. Despite strong anthropic pressure, the park is part of the city's identity and contains Rio's major symbol: the statue of Christ on the Corcovado Mountain.

However, along the park's boundaries, the expansion of informal settlements and wealthy communities is creating considerable pressure on the ecological quality of the forest, while persisting urban violence poses a threat to the city's recreation potential. Considering these deep interactions, the park and the city seek to reduce urban impacts on the forest and to promote a socio-environmental balance along their borders through shared management.

10.3 Where Issues and Conflicts Contribute to Forging Governance Methods

The reflexive self-organisation of independent actors involved in complex relations of reciprocal interdependence, with such self-organisation being based on continuing dialogue and resource-sharing to develop mutually beneficial joint projects and to manage the contradictions and dilemmas inevitably involved in such situations. Governance organised in this basis need not entail a complete symmetry in power relations or complete equality in the distributions of benefits …

Jessop (2003, p. 1)

[8]The eight sectors of TMNP and their managers: Table Mountain National Park–SANParks; Kirstenbosch National Botanical Gardens–SANBI; Cedarburg Wilderness Area–CapeNature; Groot Winterhoek Wilderness Area–CapeNature; Boland Mountain Complex–CapeNature; De Hoop Nature Reserve–CapeNature; Boosmansbos Wilderness Area–CapeNature; Swartberg Complex–CapeNature; and Baviaanskloof Protected Area–ECNCB.

Anthropic pressure linked to urban expansion is a trait shared by the four parks. In Mumbai, the forest has been progressively caught between two urban growth axes, and the park has become an enclave in the northern part of the city where population growth is the highest. The particularity, in this case, is linked to the scale of the urbanisation phenomenon, and to the progressive extension of the park towards the city. In 1995, more than 500,000 people were living in slums inside the actual park, making the presence of the 8,000 Adivasis indigenous population almost anecdotal. The latter harvest firewood as well as medicinal plants and sometimes farm illegally. In addition to slums, many plush residences have also been eating away at the borders of the park, which are fairly loose. The current borders are the result of the park's extension into agricultural lands and forests. Generally, in Mumbai, the city has been advancing on the park and the park's borders have been widening the limits of the city.

This is why the park, seen from the city, always represents an important land reserve. In the 1990s, following the invasion of the park's fringes, ecologists and slum defenders confronted one another, going as far as the Supreme Court which, in the end, ordered the destruction of the slums and the rehousing of people who could prove that they resided in the park before 1995. Spaces liberated in this way have since then become the object of desire of property developers.

Moreover, in Mumbai, the park does not represent a common denominator between the different actors; it does not embody the symbol of the city, as is the case in Rio or Cape Town, and is not linked either to its identity. The most powerful urban actors put as much pressure on the park as do the poorest populations in search of space to settle in. The Adivasis are opposed to the conservation policies supported by the ecologists, as are they to the residents of the slums who invade their forest but of whom they also take advantage (by subletting living spaces and selling products of the forest to them). The Forest Department is opposed to the builders' lobby, while the municipal corporation of Greater Mumbai is fighting against the claims of the Forest Department concerning disputed plots (Zérah and Landy 2013; Landy and Bautès 2013). The complexity of the situation is such that it compels everyone to put things into perspective before applying the principle of shared management of the protected areas.[9]

In Nairobi, urban developments in the traditional Maasai territory south of the park, and the increasing number of property fences (more than 20% of the ecosystem is fenced), are blocking an important number of migratory corridors linking the national park to the adjacent plains (MEMR 2012). The migration of the wildebeests between the park and the plains has reduced by more than 90% (30,000 in 1978 to fewer than 2,000 in 2011). The growing urbanisation, the roads and

[9]Joint participation in managing Indian national park resources was not legally possible before the adoption of the Scheduled Tribes and Other Traditional Forest Dwellers (Recognition of Forest Rights) Act, 2006, aiming at recognising rights for certain rural minorities on the protected areas where they live. The law was notified close to one year after its adoption and has been in force since the January 1, 2008. However, only the northern part of the park of Mumbai is considered as "rural" and is therefore affected by this law.

highways (the Athi River-Namanga Road and the Greater Southern Bypass Road project along the southern border of the national park), the railways and other industrial developments, including mining, have been aggravating the situation (Ogutu et al. 2009, 2013; FoNNAP 2011). The urban concentration and usage conflicts have also been increasing human-animal conflicts.

Dispersal areas and migratory corridors can only be kept open by developing wildlife-friendly uses. To this end, cooperation with the local communities and private land owners is necessary.[10] In other words, the conservation of migration paths, and therefore the preservation of the dynamic ecosystem of Nairobi's National park, calls for the implementation of conservation (and social development) initiatives beyond the borders of the protected area, in areas that are urbanised or coveted by urbanisation, in a regional planning-integrated approach. As a necessary although unprecedented experience in Kenyan administration for the park as much as for the city, a new administrative level was introduced: the county, with that of Kajiado being of particular interest here.

County Wildlife Conservation and Compensation Committees (CWCCCs), entities that in principle remain independent from the KWS, supposedly have a role to play within this framework. Indeed, among their functions, these committees are responsible for linking the KWS and the main actors of each county as regards the conservation and management of the territory's faunistic resources. They have the right to examine the elaboration and implementation of management plans produced by the KWS on the one hand, and must take part in planning land use on the other, so as to ensure the conservation of habitats, corridors and dispersal areas. The CWCCCs must also accompany the ecological compensation system[11] and implement human-animal conflict attenuation/compensation mechanisms, in collaboration with the community associations for the protection of wildlife. The committees are also responsible for setting up environmental education and public awareness initiatives (Wildlife Conservation and Management Act, No. 47, 2013). The first calls for candidates with a view to constituting CWCCCs were sent out in 2015. Their role in the field still needs to be elaborated.

In South Africa, except for national parks and botanical gardens as well as marine resources which are managed at the national level, the spheres of competence of environmental management and nature conservation are shared between the three levels of government. This, in principle, opens up the possibility of cooperation between the different levels, but also leaves a certain vagueness in the definition of the role of the city vis-à-vis provincial and national governments.

The value of the experience in Cape Town is the development of a biodiversity strategy with conservation objectives a priori shared by the three levels of

[10]In order to discourage the construction of fences in the scattering area south of the park, NGO Wildlife Foundation launched the Lease Programme (no longer running today) which paid land owners an indemnity in exchange for their accepting not to subdivide their plots or prevent the wildlife from passing during migration (Trzyna et al. 2014).

[11]Compensation measures make up for the impact of a planning exercise or project execution on the environment.

government, as well as other public institutions and civil society. This strategy, despite all the operational difficulties relating to its effective establishment, sets a goal. The way to achieve this is tortuous and fraught with pitfalls, whence the necessity to elaborate a model of governance involving the different parties concerned.

To this end, a City Biodiversity Task Team, made up of representatives from the departments of the city and their main partners, was created to supervise and implement the strategy. In addition to the working groups responsible for elaborating action plans, a Biodiversity Forum was established to ensure communication and participation by all city biodiversity stakeholders. The first biodiversity strategy of Cape Town, defining the Biodiversity Network, was adopted in 2003 and has been revised periodically since.

The strategy includes, in particular, the most critical biodiversity protection issues found in the Cape Flats, these highly urbanised suburbs where the poorest population live. In addition to 22 protected areas managed by the municipality, other public and semi-public organisations and private owners are also managing natural areas in this sector.[12] Some private properties are managed through an agreement with CapeNature (at the provincial level), which has been working with the city on the basis of an agreement. SANParks, at the national level, manages the Table Mountain and Cape Peninsula sectors, which are classified as national park. The links between city departments and SANParks have been formalised via a bilateral forum and working groups (CCT Biodiversity Report, 2008, p. 51). However, the ecological links between the national park and the city protected area network (BioNet) still have to be developed.[13]

Tijuca, among the four parks, can be considered as the most assiduous as far as shared management practices and participative governance are concerned. While the first shared management initiative between the park and the city of Rio dates back to 1999, the establishment of the participative governance system is in fact the local translation of the national conservation policy (Act no. 9,985 of June 18, 2000).

This law introduced two innovative elements into the Brazilian conservation system. On the one hand, it introduced protected areas with sustainable uses, which aim at ensuring compatibility between nature conservation and sustainable natural resource uses. On the other hand, it introduced a two-level integrated conservation approach: first, from the spatial point of view, where it brought in a bioregional landscape scale through the establishment of ecological corridors and mosaics of protected areas; and second, from the point of view of governance, where it adopted

[12]The provincial nature reserve of Driftsands is managed by CapeNature; the Cape Flats Nature Reserve is managed by the University of the Western Cape; the Plattekloof Natural Heritage Site and the Koeberg Nature Reserve are managed by Eskom (the country's electricity supplier).

[13]A first draft was realised in 2014 during the international conference on Biodiversities, organised by the UNPEC programme in partnership with SANParks and the city of Cape Town. On that occasion, a field trip had been proposed in order to follow two rivers linking the park and the city (www.upa-network.org).

the obligation to consult the public and to create committees (consultative for high protection categories and deliberative for the others), in order to ensure that the population will take part in the management of protected areas. This approach corresponds to what has been identified as an emerging paradigm during the IUCN Congress in Durban, in 2003.

The first consultative committee of TNP was created as early as 2002, by the park's authorities, with a view to rapidly meeting legal requirements (IBAMA 2002). Subsequently, thanks to the park's programme on environmental education, which was focused on training and developing projects that involve the local population, the issue of water resource protection became a central element of dialogue (Camphora 2005). From 2005 onwards, as part of the "Water in the Protected Area" programme, a participative diagnostic was carried out with the double objective of identifying social actors considered as essential partners for managing the park, and building another consultative committee (IBASE 2006; Lézy-Bruno 2014).

The participative process for establishing the new committee lasted over a year, from November 2004 to December 2006. This work, which was run by the Brazilian Institute for Social and Economic Analyses (IBASE), took into account the differences in the actors' capacity for social representation. It sought to reinforce democracy by ensuring that people and strategic groups interacting on the park became qualified, including disadvantaged populations living in the *favelas* and in the deprived neighbourhoods. The new 24-member committee, finalised in 2006, observed the principle of equality between the representatives of civil society organisations and public as well as private institutions. Other stakeholders considered as important, were included into the "guest observer" category, with no voting right (even though they can ensure their lobbying work).[14] Subsequently, an action plan was elaborated (2007–2008) and two technical chambers were established: one dedicated to conservation and the other to tourism (Irving et al. 2013).

At the end of the project, the initial mobilisation ran out of steam and the biannual renewal of the committee did not take place. In addition, the issue of representativeness of the many disadvantaged communities neighbouring the park was being raised (only 4 communities out of more than 46 identified). A new process for rebuilding the committee was only finalised in 2012. In this new formation, priority was given to the representativeness of associations from adjoining suburbs, as well as from the tourism industry since it is highly active in the park. This, it seems, led to greater interaction between stakeholders, more efficient participation from civil society representatives in the management choices, and greater acceptance of the proposed actions (Lézy-Bruno 2014). In the course of 2015, participative workshops were conducted, in anticipation of a new formation of the

[14]The following were part of this category of governmental and non-governmental institutions identified as important partners for the park: the Brazilian Fund for Biodiversity (FUNBIO); the Institute of Urbanism Pereira Passos (Town Hall of Rio); the Environmental Commission of the Chamber of the Deputies of the Federated State of Rio; the Research Institute of the Botanical Garden of Rio de Janeiro; the Nature Conservancy and the School of Visual Arts of Lage Park.

Committee, seeking to inform stakeholders about park issues and offer them training in this regard, and to perpetuate the participative process in Tijuca.[15]

This experience, initiated more than 10 years ago, does not in any way exclude the conflicts and actions involving public institutions, which could be perceived as "paternalistic". The renovation of the Paineiras-Corcovado complex is testimony to this: an architectural and environmental project in the form of public–private partnership estimated at more than USD 1 million, co-run by the Association of the Friends of the Park.[16] Managing the benefits of ecological compensation linked to the project and intended for environmental education actions, remains a subject of discord. Local associations defend their capacity to implement pragmatic projects meeting in the best way possible the needs of local communities, while the authorities insist on the centralised management of actions having sufficiently vague practical effects for those concerned.

10.4 The Future in the Mirror: Identified Tendencies

> *In ecology there is now a growing understanding that human processes and cultures are fundamental for sustainable management of ecosystems, and in urban planning it is becoming more and more evident that urban management needs to operate at an ecosystem scale that is beyond the traditional boundaries of the city, both with respect to the bio-physical and social*
>
> Elmqvist et al. (2008, p. 3665)

Do opening up a park's governance and changing the scale (from protected areas to bioregions) provoke a loss of efficiency at the level of conservation? Chris Sandbrook speaks of plural conservation in the twenty-first century, and proposes a fairly wide definition which encompasses the diversity of conceptions of what conservation can be today: "*actions that are intended to establish, improve or maintain good relations with nature*" (Sandbrook 2015, p. 565). While conservation is from now on multiple, each park, in its cultural, socioeconomic and political context must find where to position the cursor between a traditional form of conservation (the biological diversity-based model), and opening up to the territory as well as its stakeholders and inhabitants.

"*Fundamental to success is the recognition of the significant trade-offs that occur between conservation and development goals.*" (Campbell et al. 2010). In the case of South Africa, Dahlberg and Burlando (2009) point out that Integrated Conservation and Development Projects (ICDPs) have limited success, partly because the role played by negotiations and compromises lacks recognition.

---

[15]Oficina de renovaçao do conselho consultivo. http://www.parquedatijuca.com.br/#noticia?id=238 Consulted on March 6, 2016. Accessed on June 11, 2018.

[16]http://www.paineirascorcovado.com.br/proj.html.

Magnaghi (2014, p. 18) goes further and calls for breaking with this dual system of territory utilisation, where one part is occupied by human establishments and submitted to "economic rules" with no heritage value, and the other is made up of spaces related to the natural and cultural heritage to be protected from development. He compares this dualism with the notion of "common-good territory", which is the "product of co-evolution processes between human labour and nature", producing what he calls "neo-living ecosystems" (p. 10). The idea is to integrate city and nature, to conceive the territory as a set and not extract from it spaces of natural heritage to be protected. This calls for going beyond park borders and reintegrating them at the level of the regional, and urban as well as suburban landscape, but also at the level of natural and agricultural spaces (integrated landscape management[17]). What is essential, therefore, is to change the civilisational approach, in order to forge links between human and natural spaces, physically in space, but also in the systems of governance and participation of residents and institutions in a collective project. From this point of view, what would be the main issues linked to changes in the governance of the parks under study?

In Mumbai, despite, rather than because of, the fact that it tried to expand, the fortress park has been besieged by the city. However, while it seems that making a real attempt at participative management is unimaginable due to the scope of the urban phenomenon as well as the socioeconomic and cultural characteristics of India, the park's conservation cannot be imagined without an alliance with the metropolis, which would entail the mutual recognition of services rendered by the park to the city and vice versa. A first level of collaboration between the different levels of public authorities, as was developed in Cape Town, could contribute to inventing an urban forest model fully integrated into the urban area. Could not the deteriorated zones—following the encroachments of the 1990s—be the subject of a reforestation programme, in partnership with the municipality, supplying jobs and training to the poorest population (as in Rio and Cape Town), thereby beginning to forge links between the park and the city?

In Nairobi, the recent legislative evolution, reminiscent of the Brazilian law requiring that a permanent and close dialogue with urban actors in managing protected areas be established, requires the construction of a negotiating arena, beyond the CWCCC dialogue forum, applicable to the cultural territory of the Maasai, south of the metropolis. This opens up the possibility to complete actions that have already been set up, such as the land use control community plan, concerning the sustainable management of wildlife dispersal areas, in relation to pastoral herding, settlements and other land uses (Nkedianye et al. 2009). Developing in this sector an indigenous and local community conservation area, would enable the Maasai to take an active part in managing their territory and reaping the benefits, while contributing to the conservation objectives of the national park.

Faced with major challenges linked to poverty and the need for basic infrastructure and services, biodiversity conservation in Cape Town is not usually

[17]http://peoplefoodandnature.org/. Accessed on June 11, 2018

perceived as a priority, but as a luxury. The main challenge then consists in changing the perception of the various sectors of society on biodiversity (Katzschner et al. 2005). In this context, urban policies seek coherence between socioeconomic development and biodiversity conservation, based on a synergetic approach highly influenced by ICDPs. However, this takes place in a very institutional manner and the populations, while being associated with the organisation of operational projects, such as those run by CAPE,[18] are not associated with decision-making processes. Even though governance is collaborative between the bodies of the different levels of authority, it remains highly institutional and top-down. The actual participation of civil society remains to be established in the decision-making processes, so as to build a "Park for All Forever".[19] Indeed this could perhaps contribute to transforming society and changing completely old power relations (Cock and Fig 2002).

In Rio, the hopes raised by their experience are based on transparency in the objectives defined by the stakeholders and their capacity to negotiate compromises needed to speed up, on the one hand, the conservation of existing protected areas within the city, among which is the national park, and on the other, the requirements of social development in a metropolis confronted with the impacts of socioeconomic dynamics based on exclusion. Currently, the national park is undergoing a globalisation process, where the urban-natural landscape of Rio has been classified as World Heritage Site (UNESCO 2012) and the Paineiras-Corcovado complex has been renovated. These international projects place the national park in a global logic of competition between the world's greater metropolises, which also corresponds to choosing a city as a host for major international events (as with the 2014 World Cup of Football and 2016 Olympic Games). The park then becomes a showcase of the city and the highly urban conception of conservation. However, does not the park's internationalisation run the risk of leaving aside the most disadvantaged local populations that are being excluded from an excessively regulated project confined to liberal reasoning?

The survival of these four national parks inevitably requires them to dialogue with the city and its stakeholders, and to build a new model of protected area which is no longer that of the first national parks, and which no longer strictly corresponds to the definition of Category II, still in force today. These parks, in addition to their primary function of biodiversity and natural landscape conservation, are (or could be) at the same time: a conservation showcase (as practiced in rural areas but also increasingly closer to urban dwellers, through the restoration and rehabilitation of damaged environments); a tool for the education and awareness of urban populations; a major element of the cities' green infrastructure; an instrument of economic and social development, through tourism in particular; and a means of social

[18]"Cape Action for People and the Environment (CAPE) is a 20-year partnership of government and civil society aimed at conserving and restoring the biodiversity of the Cape Floristic Region and the adjacent marine environment, while delivering significant benefits to the people of the region." http://www.capeaction.org.za/.

[19]Motto of Table Mountain National Park.

inclusion, via projects that seek to integrate disadvantaged populations into the activities of the park and in relation to it. Prompted by these dynamics, we have identified what we call "emerging urban parks", like those in Cape Town and Rio, because they seek to associate economic development, the increase in value of ecosystemic goods and services, and the quality of life of populations to nature conservation, while showcasing conservation and the city (Bruno et al. 2016). As such, the park is breaking with the nature/culture dichotomy and is becoming the representative of a resolutely urban nature.

In fact, the real challenge being observed is to break with the park model where humans are excluded, and to build a new model where they are integrated into nature conservation. In the metropolises of the South, a further challenge is to succeed in reconciling nature conservation with social and economic development. This is where more or less successful sustainable development experiments are being carried out. Through the evolution of the system of governance of parks and the changes in the relationships of the urban arena, stakeholders are building the necessary compromises for stability in the conservation-development marriage, considered by some as incompatible.

Bibliography

Adams, W. M. (1990). *Green development, environment and sustainability in the third world*. London: Routledge.

African Convention on the Conservation of Nature and Natural Resources. (1968). https://treaties.un.org/doc/Publication/UNTS/Volume%201001/volume-1001-I-14689-English.pdf.

Anderson, D., & Grove, R. (1987). *Conservation in Africa: People, policies and practice*. Cambridge: Cambridge University Press.

Borrini-Feyerabend, G., Bueno, P., Hay-Edie, T., Lang, B., Rastogi, A., & Sandwith, T. (2014a). *Lexique sur la gouvernance des aires protégées et conservées, Courant Renforcer la diversité et la qualité de la gouvernance, Congrès Mondial des Parcs 2014 de l'UICN*. Gland: UICN.

Borrini-Feyerabend, G., Dudley, N., Jaeger, T., Lassen, B., Broome, N. P., Phillips, A., et al. (2014b). *Gouvernance des aires protégées: de la compréhension à l'action. Collection des lignes directrices sur les meilleures pratiques pour les aires protégées No. 20*. Gland: UICN.

Brandon, K. E., & Wells, M. (1992). Planning for people and parks: Design dilemmas. *World Development, 20*(4), 557–570.

Bruno, L. (2005). Chez soi en ville: un regard sur Rio. *Travaux de l'Institut de Géographie de Reims, 115–118*, 9–30.

Bruno, L. (2008). La forêt au cœur de la ville. Le Parc National de Tijuca, Rio de Janeiro. *Géographie et Cultures, 62*, 27–42.

Bruno, L. (2012). Y a-t-il une place pour la nature en ville ? In L. Bruno, E. Lézy (dirs.), *Biodiversities. Les aires protégées urbaines, des laboratoires grandeur nature* (pp. 251–283). Paris: Editions Le Manuscrit.

Bruno, L., & Lézy, E. (Eds). (2012). *BiodiverCities. Les aires protégées urbaines, des laboratoires grandeur nature*. Paris: Editions Le Manuscrit.

Bruno, L., Landy, F., & Lézy, E. (2016). Des parcs nationaux et leurs métropoles: L'évolution des rapports à Rio de Janeiro, Mumbai, Nairobi et au Cap. In S. Barles, & N. Blanc (dirs.), *Ecologies Urbaines* (pp. 79–116).

Campbell, B. M., Sayer, J. A., & Walker, B. (2010). Navigating trade-offs: Working for conservation and development outcomes. *Ecology and Society, 15*(2), 16. [online] http://www.ecologyandsociety.org/vol15/iss2/art16/.

Camphora, A.-L. (2005). *Análise socioambiental dos condicionantes da gestão de recursos hídricos no Parque Nacional de Tijuca.* Rio de Janeiro: instituto Terrazul, relatório, programa Petrobrás ambiental – Projeto água em unidade de conservação.

Chape, S., Harrison, J., Spalding, M., & Lysenko, I. (2005). Measuring the extent and effectiveness of protected areas as an indicator for meeting global biodiversity targets. *Philosophical Transactions of the Royal Society B: Biological Sciences, 360*(1454), 443–455. https://doi.org/10.1098/rstb.2004.1592.

Christensen, J. (2004). Win-win illusions. In *Parks* (Vol. 14, No. 2, pp. 34–41). Durban World Parks Congress, Gland. Produced by the Nature Bureau, UK.

Cock, J., & Fig, D. (2002). From colonial to community-based conservation: Environmental justice discourse. In D. A. Mc Donald (Ed.), *Environmental justice in South-Africa* (pp. 131–135). Cape Town: UCT Press.

Colloque International Biodiversities Le Cap (2014). http://www.upa-network.org/.

Colloque International Biodiversities Paris (2010). http://biodivercities.net/.

Colloque International Biodiversities Rio (2012). http://upa-network.org/.

Cronon, W. (1995). The trouble with wilderness; or, getting back to the wrong nature. *Uncommon ground: Rethinking the human place in nature* (pp. 69–90). New York: W. W. Norton & Co.

Crosby, A. W. (2004). *Ecological imperialism: The biological expansion of Europe* (2nd ed., pp. 900–1900). Cambridge, UK: Cambridge University Press.

Dahlberg, A. C., & Burlando, C. (2009). Addressing trade-offs: Experiences from conservation and development initiatives in the Mkuze wetlands, South Africa. *Ecology and Society, 14*(2), 37. [online] http://www.ecologyandsociety.org/vol14/iss2/art37/.

Dearden, P., Bennett, M., & Johnson, J. (2005). Trends in global protected area governance, 1992–2002. *Journal of Environmental Management, 36*(1), 89–100.

Depraz, S. (2008). *Géographie des espaces naturels protégés: genèse, principes et enjeux territoriaux.* Paris: Armand Colin.

Derose, A. M. (2004). Overview of community participation at the Vth IUCN World Parks Congress. In *Parks* (Vol. 14, No. 2, pp. 18–29). Gland.

Diegues, A. C. (1994). *O mito moderno da natureza intocada*, 4e éd., Hucitec. São Paulo: USP.

Elmqvist, T., Alfsen, C., & Colding, J. (2008). Urban systems. In S. E. Jørgensen, & B. D. Fath (Eds.), *Ecosystems. Encyclopedia of ecology* (Vol. 5, pp. 3665–3672).

Everhart, W. C. (1972). *The national park service*. New York and London: Praeger.

FoNNAP (2011). How the proposed greater southern bypass will affect Nairobi Park. Friends of Nairobi National Park. http://fonnap.wordpress.com/2011/08/23/how-the-proposed-greater-southern-bypass-will-affect-nairobi-park/. Retrieved from March 11, 2016.

Foreman, D. (1991). *Confessions of an eco-warrior*. New York: Crown Publishing.

Fynn, R., & Bonyongo, M. (2011). Functional conservation areas and the future of Africa's wildlife. *African Journal of Ecology, 49*(2), 175–188.

Ghimire, K. B., & Pimbert, M. P. (Eds.). (1997). *Social change and conservation: Environmental politics and impacts of national parks and protected areas*. London: Earthscan.

Graeme, L. W., Lockwood, M., Kothari, A., Feary, S., & Pulsford, I. (2015). *Protected area governance and management*. Canberra: Australian National University Press.

Graham, J., Amos, B., & Plumptree, T. (2003). *Governance principles for protected areas in the 21st century*. Ottawa: Institute on Governance in collaboration with Parks Canada and CIDA.

Grove, R. H. (1995). *Green imperialism: Colonial expansion, tropical island Edens, and origins of environmentalism 1600–1860*. Cambridge: Cambridge University Press.

Holdgate, M., & Phillips, A. (1999). Protected areas in context. In M. Walkey, I. Swingland, & S. Russell (Eds.), *Integrated protected areas management*. Boston: Kluwer Academic.

IBAMA (2002). *Como o Ibama exerce a educação ambiental*. Brasília.

IBASE (2006). *Educação ambiental em unidades de conservação*. Rio de Janeiro.

Irving, M. A., Corrêa, F. V., & Zarattini, A. C. (Eds.). (2013). *Parques nacionais no Rio de Janeiro. Desafios para uma gestão social da biodiversidade.* Rio de Janeiro: Editora Letra e Imagem.

IUCN, & WCPA (2004). 2003 Durban World Parks Congress. In *Parks* (Vol. 14, No. 2). Gland: IUCN.

Jepson, B., & Whittaker, R. J. (2002). Histories of the protected areas: Internationalization of conservation values and their adoption in the Netherlands Indies (Indonesia). *Environment and History, 8,* 129–172.

Jessop, B. (2003) *Governance and metagovernance: On reflexivity, requisite variety, and requisite irony.* Lancaster University. http://www.lancaster.ac.uk/fass/resources/sociology-online-papers/papers/jessop-governance-and-metagovernance.pdf.

Katzschner, T., Oelofse, G., Wiseman, K., Jackson, J., & Ferreira, D. (2005). The city of Cape Town biodiversity strategy. In T. Trzyna (Ed.), *The urban imperative* (pp. 91–104). Sacramento: California Institute of Public Affairs.

KNBS (2013). *Economic survey 2013 highlights*. Kenya National Bureau of Statistics. http://www.knbs.or.ke/Economic%20Surveys/Cabinet_Secretary_Presentation_on_Economic_Survey_May_2013.pdf. Retrieved from March 11, 2016.

Landy, F., & Bautès, N. (2013). Local and translocal systems of actors. In M. C. Saglio-Yatzimirsky, & F. Landy (Éds.), *Megacity slums. Social exclusion, space and urban policies in Brazil and India* (pp. 257–306). London: Imperial College Press.

Laslaz, L., Gauchon, C., Duval, M., & Héritier, S. (2014). *Les espaces protégés. Entre conflits et acceptation*. Paris: Belin.

Lézy, E. (2002) b. « Primitifs du passé » ou « primitifs du futur » essai pour une nouvelle géographie des peuples dits « primitifs » . In *Discours sur le primitif*, McIntosh-Varjabédian Fiona (Ed.), *Edition du Conseil Scientifique de l'Université Charles de Gaulle-Lille* (Vol. 3, pp. 191–212).

Lézy, E. (2012). Out of the park, le Parc National de Nairobi dans la géodynamique Maasai. In L. Bruno, & E. Lézy (dirs.), *Biodiversities. Les aires protégées urbaines, des laboratoires grandeur nature* (pp. 179–196). Paris: Editions Le Manuscrit.

Lézy-Bruno, L. (2014). Usages et conflits sous fond de protection de la nature en ville. Le cas du Parc national de Tijuca à Rio de Janeiro. In L. Laslaz et al. (Eds.), *Les espaces protégés. Entre conflits et acceptation* (pp. 209–227). Paris: Belin.

Magnaghi, A. (2014). *La biorégion urbaine, petit traité sur le territoire bien commun.* Paris: Association culturelle Eterotopia.

McNeely, J. A. (Ed.). (1993). *Parks for life: Report of the IVth world congress on national parks and protected areas.* Gland: IUCN.

MEMR. (2012). *Mapping wildlife dispersal areas and migratory routes/corridor (Part I): Southern Kenya Rangelands.* Nairobi: Ministry of Environment and Mineral Resources (MEMR).

Næss, Arne. (1973). The shallow and the deep, long-range ecology movement. *Inquiry, 16,* 95–100.

Nkedianye, D., Radeny, M., Kristjanson, P., & Herrero, M. (2009). Assessing returns to land and changing livelihood strategies in Kitengela. In K. Homewood, P. Kristjansen, & P. Chenevix Trench (Eds.), *Staying Maasai? Livelihoods, Conservation and Development in East African Rangelands*, (pp. 115–149). New York: Springer.

Noss, A. (1997). Challenges to nature conservation with community development in central African forests. *Oryx, 31*(3), 180–188.

Ogutu, J., Piepho, H., Dublin, H., Bhola, N., & Reid, R. (2009). Dynamics of Mara-Serengeti ungulates in relation to land use changes. *Journal of Zoology, 278,* 1–14.

Ogutu, J., Owen-Smith, N., Piepho, H., Said, M., Kifugo, S., Reid, R., et al. (2013). Changing wildlife populations in Nairobi National Park and adjoining Athi-Kaputiei plains: Collapse of the migratory wildebeest. *Open Conservation Biology Journal, 7,* 11–26.

Philips, A. (2003). Turning ideas on their head: The new paradigm for protected areas. *In the George Wright Forum, 20*(2), 8–32.

Pulsford, I., Lindenmayer, D., Wyborn, C., Lausche, B. Vasilijevic, M., & Graeme L. W. (2015). Connectivity conservation management. In L. W. Graeme et al. (Eds.), *Protected area governance and management* (pp. 851–889). Canberra: Australian National University Press.

Redclift, M. R. (1989). *Sustainable development: Exploring the contradictions*. London: Routledge.

Redford, K. H. (1991). The ecologically noble savage. *Cultural Survival Quarterly, 15*(1), 46–48.

Sandbrook, C. (2015). What is conservation? *Oryx, 49*(4), 565–566.

SANParks (2008). *Table mountain national park conservation development framework—Volume 2: Technical report*. Cape Town.

Sowman, M., & Wynberg, R. (Eds.). (2014). *Governance for justice and environmental sustainability: Lessons across natural resources sectors in sub-Saharan Africa*. London: Earthscan/Taylor and Francis.

Trzyna, T. (Ed.). (2005). *The urban imperative*. Sacramento: California Institute of Public Affairs.

Trzyna, T. et al. (2014). Urban protected areas: Profiles and best practice guidelines. *Best practice protected area guidelines series no. 22*. Gland, Switzerland: IUCN. http://cmsdata.iucn.org/downloads/bpg_urban_protected_areas.pdf.

UICN. (1997). *Congrès mondial de la nature: Résolutions et recommandations*. Gland: UICN.

United Nations (2014). *World urbanization prospects: The 2014 revision, highlights, population division*. New York.

Utting, P. (1994). Social change and political dimensions of environmental practice in central America. *Development and Change, 25*(1), 231–259.

Washington Convention (1940). *Convention on nature protection and wild life preservation in the Western Hemisphere*. http://www.ecolex.org/ecolex/ledge/view/RecordDetails?index=treaties&id=TRE-000085.

Wells, M., & Brandon, K. (1992). *People and parks: Linking protected area management with local communities*. Washington: The World Bank.

West, P. C., & Brechin, S. R. (Eds.). (1991). *Resident peoples and national parks: Social dilemmas and strategies in international conservation*. Tuscon: The University of Arizona Press.

Yeld, J., & Barker, M. (2004) *Mountains in the sea: Table Mountain to Cape Point. An interpretive guide to the Table Mountain national park* (p. 183). SAN PARKS.

Zérah, M. H., & Landy, F. (2013). Nature and urban citizenship redefined: The case of the national park in Mumbai. *Geoforum, 46*, 25–33.

Chapter 11
What Makes Urban National Parks "Urban"? Their Specifics Within the National Systems of Protection

Bernard Calas, Frédéric Landy and Theresa Mbatia

Through their construction, their use and where their users come from, most protected areas (PA) are of urban origin. As they are primarily frequented by people from cities, they are part of the fabric of the city and the metropolis. Fundamentally, if all protected areas are urban, how does the urbanity of Nairobi National Park (NNP) differ from that of other protected areas (PAs) in Kenya? How should this hypothetical difference in urbanity be assessed? The protocol of our approach consists of comparing NNP with the other PAs that are managed by the Kenya Wildlife Service (KWS) and which are further from Nairobi, identifying these differences and attempting to see whether they can be attributed to the proximity of a city. We shall then widen our perspective by more briefly examining other parks studied by the UNPEC project.

What criteria of urbanity should be used? The way PAs are compared with one another is much the same as the way other types of spaces are compared. They can be studied in a synchronic manner through their morphology, their uses, and how they are integrated into their respective surroundings. One can also take into account the dynamics of their differences and similarities by looking at the directions in which they are evolving.

We have chosen to focus on the actors, their lines and their choreography (Calas 2002), and to see how the approaches taken by these actors in parks that are adjacent to capital cities differ from the approaches they may take elsewhere. In a

B. Calas (✉)
Department of Geography, University of Bordeaux-Montaigne, Pessac, France
e-mail: Bernard.Calas@u-bordeaux-montaigne.fr

F. Landy
French Institute of Pondicherry, Pondicherry, India

F. Landy
University of Paris Nanterre, UMR LAVUE, Nanterre, France

T. Mbatia
Department of Geography and Environmental Studies, University of Nairobi, Nairobi, Kenya

F. Landy (ed.), *From Urban National Parks to Natured Cities in the Global South*,
https://doi.org/10.1007/978-981-10-8462-1_11

protected area, tourists, managers and the various flora and fauna—the wilderness—rub shoulders and, taking up their respective positions, act out a play of naturalistic tourism. We shall look at how this play differs from the one acted out elsewhere in the country, and then at the extent to which it is part of, if indeed it is part of, the fabric of the metropolis. By observing the people in the park: tourists, caretakers, managers, various users, etc., and by listening to their opinions on the differences between NNP and Kenya's other PAs, we can therefore take into account the uses of and the issues surrounding these areas in order to define what makes NNP unique. My hypothesis is that its uniqueness is related to its attachment to the city. Using the intersections between these situated and incomplete viewpoints, this work aims to explore a number of different factors and thus enable a comparison between various urban national parks in terms of how urban they are. Indeed, one assumption on which the entire UNPEC project relies is that these parks are unique because they extend into urban districts. However, this approach more or less stems from a tautology that can be dispelled by clarifying the differences between these urban PAs and other PAs, which are all part of the same national system of protection.

How does NNP differ from other parks in Kenya? We must consider whether these differences can simply be attributed to differences in geometric distance from cities: in other words, we need to know the extent of the role played by the spatial proximity of a capital city with a population of 4 million in the urbanisation—in the sense of the creation of urbanity—of a spatial structure such as a nature reserve. Who are the interested parties here? Do their identities reveal a certain level of urban cosmopolitanism? Do usage densities measured using the ratio of number of visitors/surface area of the park enable them to be classified? To what extent are the systems of the actors invested in the PA in question more or less composite?

How do these actors represent the park? As a piece of nature, as an empty space, as a landfill site, a place of recreation, a place of transit, a reserve? What kind of future do they foresee for the park? What projects are planned for the space and what type of ideology of development do these follow? A developmental type of development based upon the use of forcing or a sustainable type of development that treats the environment with care? In what way are the spatial structures produced part of the selective, exclusive, event-focused, club-based, productive, competitive city or do they contribute to the creation of a sustainable, inclusive, green and diverse city? In other words, to what extent is the creation of these spaces, through the actions of actor networks, involved in the development of an emerging metropolitan area?

This study will follow three stages.[1] First, taking the point of view of tourists, we shall see how the tourist experience at NNP differs from that at other Kenyan PAs (Fig. 11.1). Next, we shall take the point of view of managers and park rangers:

[1]The data needed for this research were collected on a specific mission carried out in July 2014, during which around 30 semi-directive interviews with those involved in nature conservation were conducted. However, the fundamental information needed for contextualisation is the fruit of scientific visits to Kenya and Nairobi over the course of over 30 years.

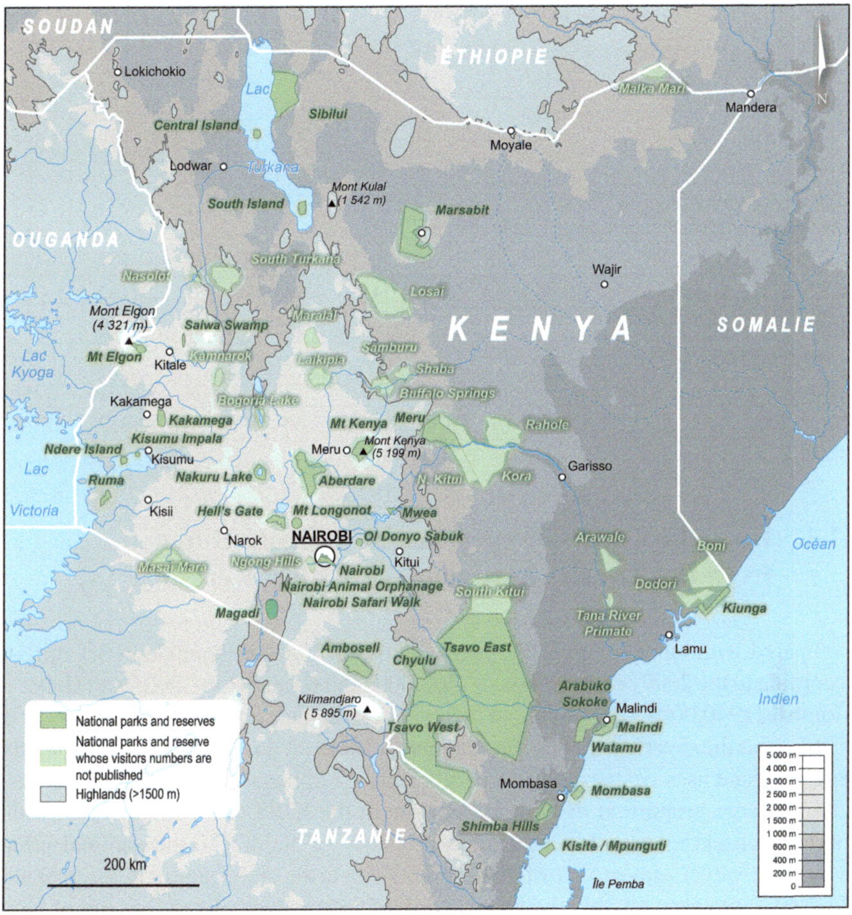

Fig. 11.1 The Kenyan conservationist system. *Source* Calas (2015). *Drawing* T. Auly. With permission from the journal *Dynamiques environnementales*

according to the KWS managers charged with implementing the conservation policy and the park rangers responsible for putting it into practice, what makes NNP different from other PAs? Finally, widening our observation to the metropolitan scale, we shall look at how NPP participates in the fabrication of the city and distinguishes itself from other PAs in Kenya. It differs so much so that a "metro-park", a type of hybrid space between a "natural park" and an "urban garden" is created and incarnates a kind of "naturbanity" which calls into question the abruptness of the divide between nature and city.

Table 11.1 NNP in relation to other major parks in the KWS

000 of visitors	2010	2011	2012	2013	2014
Nairobi National Park	120	135	151	154	131
Nairobi Safari Walk	201	176	147	136	133
Nairobi Orphanage	490	403	474	407	361
Nairobi Complex Total (117 km^2) Percentage of the national total of visitors (%)	811 29	714 26	772 30	697 29	625 28
Amboseli (390 km^2)	208	176	141	141	117
Nakuru (118 km^2)	241	245	253	262	226
Tsavo W (13,747 km^2)	138	98	70	68	51
Tsavo E (7,065 km^2)	264	267	176	153	103
Maasaï Mara	187	138	102	103	166
Kenya total	2,764	2,664	2,492	2,337	2,164

11.1 Does the Tourist Experience in Nairobi Differ from the Tourist Experience in Other Kenyan PAs?

Compared with other parks, urban parks—those in Nairobi (population 3.1 million according to the 2009 census), Nakuru (308,000) and Kisumu (388,000), and Haller in Mombasa (940,000)—are unique in their small size, their relatively high numbers of visitors and the diversity among those visitors (Table 11.1). The NNP complex—the park itself and its two annexes (Safari Walk and the Orphanage for Baby Elephants)—attract a large proportion of the total visitors to the KWS (28% in 2014); thus the tourism-conservation system reflects the primacy of Nairobi in the national urban system. However, almost 80% of these visitors stay at the orphanage or at Safari Walk. If we take into account only the park section of the complex, the proportion of NNP visitors in the total number of visitors to Kenya's PAs dwindles to 6%. This puts the urban primacy of NNP, in the strictest sense of the term, into perspective.

Furthermore, in terms of profitability, in 2005, the NNP complex made a profit of $1,247,000, compared with $2,857,000 for Nakuru, $2,700,000 for Tsavo East, $1,900,000 for Amboseli, $1,300,000 for Tsavo West and $800,000 for Aberdare National Park (World Resource Institute 2007). In terms of total value, NNP's gross profitability is not as high as other parks. However, relative to its area, its net profitability, much like other urban parks, is clear.[2]

[2]However, Nakuru NP overtakes Nairobi: Nakuru combines the high level of visits from schools and residents that is characteristic of urban parks with the significant number of non-resident visits (+50% of its total visitors) that is characteristic of bushland parks. The reason why Nakuru attracts more international visitors is that it has one distinctive feature that boosts its appeal in terms of animal tourism, one that has nothing to do with urbanity but which is not contradicted by it: flamingos. In 2015 NNP made a $2.7 million profit but we don't have the new figures for other national parks.

One of the most obvious aspects that distinguishes NNP from other Kenyan PAs is the greater diversity of its visitors in terms of origin and social group. The visitor statistics and price list for the KWS distinguish between three visitor types: Kenyans, non-citizen residents, and international tourists. Although the latter make up the overwhelming majority of visitors to protected bushland areas—95% of entries to Tsavo—they make up a significantly smaller portion of the visitors to Nairobi, where only 35% of visitors are international tourists, thus leaving more room for nationals and residents. This diversity of users shows that the diversity of the public is one variable of urbanity.

In India, Sanjay Gandhi National Park (SGNP) has its own defining features. It is also a small national park, which is characteristic of Indian parks located in the outskirts of cities (Bannerghatta near Bangalore—104 km^2) as well as fully urban parks (Guindy in Chennai—3 km^2!). But this feature is not as significant as it is in Africa, given that India's 103 national parks are generally quite small due to population density (393 km^2 on average nationally). It is also one of the most visited parks. Officially, it receives more than 1.2 million visitors per year. As in Nairobi, the proportion of visitors from abroad is very small—this feature is, however, slightly less striking as foreign tourism in India is less focused on national parks than foreign tourism in West Africa. Finally, this park is very "profitable" ($780,000 in entrance fees). Much like Tijuca National Park (TNP) or Table Mountain National Park (TMNP), one characteristic of urban national parks is their globally beneficial nature, enabling redistribution of revenue by the central services in order to supplement the budgets of other parks.

This diversity of origins is clearly the mark of a certain level of sociological diversity. Indeed, the civic categories mainly encompass sociological categories: Kenyan city residents and academics from the middle classes, expatriates and diplomats, international tourists. These sociological differences reflect the different roles that parks play in the Kenyan system of PAs: whilst a park like Sanjay Gandhi National Park (SGNP) is largely ignored by expatriates in India, NNP is one of the urban amenities offered to expatriates in Kenya's capital, a diplomatic and humanitarian hub in East Africa. One end-of-week activity is indeed to "go to the park". Furthermore, like most other urban parks (e.g., Nakuru, Kisumu, Mombasa or Hell's Gate), and like the three other parks mentioned in this book, NNP receives visits from entire schools. These groups of people even further distinguish urban parks and NNP from other PAs in the KWS.

This sociological diversity leads to a diversity in the ways nature and the reserve are used. According to the Senior Warden at NNP, who previously worked at Tsavo, one thing that makes the management of NNP more difficult is precisely this diversity of visitors. At NNP, "It is very difficult to tell your clients' intentions", whereas at Tsavo or Amboseli, users are more often the clients of tour operators who arrive with the sole intention of photographing the animals. In Nairobi, some visitors use the national park as an urban park, for fun, whilst others cross the park as a short cut on the way to work, and others even go with the intention of poaching. Similarly, the park in Mumbai attracts nature lovers as well as picnickers, (Mawdsley et al. 2009), pilgrims as well as delinquents. This diversity of use reflects the park's location in the city's outskirts and the diversity of urban populations.

Beyond this, the tourist experience in Nairobi is similar to that elsewhere. On a packed Japanese minibus, the tourist largely does the same things whether in bushland or in Nairobi: looking at the animals, and photographing them. The outcome of the experience remains the same since it involves looking wild animals in the eye, an experience that makes Nairobi much the same as other protected areas whilst distinguishing it from other urban green space, such as, in Nairobi, the Karura or Ngong Road forests, where the wild fauna is relatively hidden. But a number of characteristics distinguish what happens in Nairobi from what might take place in Meru, Tsavo or Maasai Mara.

First, in Nairobi, the actual experience of the safari is much shorter than that in other PAs: an afternoon, a morning, more or less; a full day in rare cases. Unlike the majority of other national parks, only a handful of visitors to NNP are overnight visitors who sleep in the park's one lodge. The urban park therefore reduces the safari experience to simply a game drive, whereas, in bushland parks, the experience is the epitome of an immersion that lasts longer, is less rushed, and is more encompassing. The suddenness of the dive into Nairobi's wilderness is temporal as well as spatial, since "the halo of contact metamorphism" (Calas 2006) between the protected space and the unprotected space, a halo that surrounds the great protected rural areas and which is delineated by various signs and markings, disappears into the city and is drowned in the city's morpho-functional diversity and its multitude of markers and signs. (The divide is just as spectacular in Mumbai, where a single avenue separates the park from the dense, noisy, concrete city. It is a little less spectacular in Rio and particularly Cape Town, mainly because the national parks there are more splintered, not in one piece and closed). In cities, the anticipation and projection that accompany the journey into PAs in this halo of transition, which acts as a veritable emotional buffer zone, are not punctuated by the signs and markers that would otherwise announce the arrival onto an exceptional heritage site. The violence of the transition between interior and exterior in Nairobi is jarring, much more so than in Tsavo, for example. Especially since, for the users of NNP, the game drive is just one of many activities available in the city, one that is often intended to simply fill free time, fill a void, or fill a gap in the schedule; it is a transitional activity, a stop on the way to the countryside or the beach. Conversely, for the majority of visitors, game drives in bushland safaris are the explicit reason for their often-intercontinental trip to these areas, the final destination, in a way, giving purpose to the entire trip. These differing temporalities are of fundamental importance in distinguishing the significance invested in the sightseeing programme.

Fig. 11.2 A violent landscape gradient, specificity of NNP. *Source* Photo by E. Calas

In protected areas, both at NNP and elsewhere, the purpose of the game drive is to catch animals off-guard and photograph them. However, the expected settings and backgrounds differ, and the explicit meaning of the images changes with them: the Eden-like, untouched look of the generic savannah of the "normal" parks contrasts with the apparent artificiality of the urban skyline when it is the setting for animal life: the quest for the illusion of timeless, untouched nature opposes the image of animals against a backdrop of skyscrapers (Fig. 11.2). A relative indifference to the décor as long as nothing "pollutes" the animal-focused gaze is characteristic of bushland PAs, but this contrasts with the attitude in Nairobi, where the gaze and the lens widen to the extent that they take in not only the countryside but also these human aspects of the urban backdrop.[3] In this respect, it is significant that the iconic skyline of NNP is urban; a visit to NNP is like a geographic

[3]This skyline apparently does not change the meaning of the pictures that tourists seek to capture in the animal parks, as the confirmation of the violent and irreducible difference/confrontation between the animal kingdom and human order. This ontological test is also geographical since it depends on the angle, the relative placement of the protagonists, the game of geometric and not ontological distances. The effects of the location are of fundamental importance, no doubt more so that is often the case elsewhere.

experience of a hybrid, an explicit quest to experience a border between two worlds, a liminal experience.[4] Nairobi's park thus acts as a decompression/compression chamber, a margin where the edge of the city is still visible whilst nature can already be seen.

The use in advertising of the image of wild animals against the backdrop of the urban skyline can be explained in part by the fact that NNP does not have a strong topographical identity and by the relative anomie of the landscape of the savannah, as well as by the commercial need to associate a park with an animal, a symbolic character, or a unique experience: here, that uniqueness is the fact that the reserve is attached to the city. A park is not fully urban unless its actors classify it as such: here, tourists and managers are on the same page.

The topographical identity of parks in Mumbai, Rio and even Cape Town is much stronger because they are mountainous. These parks therefore seem to have less of a need for this urban identity in order to exist than does Nairobi's park. But their specific features that are created by the proximity of the city are nonetheless noticeable, both in terms of leisure equipment (trains or cable cars) and in terms of visitor expectations: at Table Mountain, the tourist knows that s/he will not see any lions, as s/he might in Kruger, but perhaps only baboons; and only a few uneducated visitors expect to see tigers at SGNP.

11.2 How is NNP Different from Other Protected Areas in the KWS?

It is not so much different in its overall morphology, which obeys the practically universal morphology of the protected area: from this point of view, NNP is a PA like any other, sensitively following the same logic of territorial organisation and the same functional demands of the reserve. However, the on-site facilities and interventions are rather numerous at NNP: around 10 water reservoirs have been created in order to provide water and sustain the mammals in the dry season. Furthermore, the number of picnic areas, toilets, observation points and authorised vehicle disembarkation points, the density of paved roads and trails are all relatively high for the size of the park, in response to the needs of its higher visitor numbers. This relative over-equipment also demonstrates the characteristics of a population

[4]Whereas the iconic blueprint of certain parks, such as TMNP and TNP, focuses on the park's contour lines. However, this reversal is of no help in demonstrating urbanity, and instead highlights Nairobi's uniqueness compared with other urban parks covered in the UNPEC project sample.

that is generally less compliant with the behaviour codes that are universally allowed on safari, a population that is more fussy, with a tendency to use the park as a recreation area just as much as, if not more so than they use it as a natural park, and where the proportion of children is greater. The variability in the ways people use this relatively small space, which can also be seen in the three other parks mentioned in this book, is also marked by a greater tendency for fragmentation than is seen in other rural national parks. Of course, all parks experience this trend due to increasing diversity in the ways they are used (Calas 2015), the multiplication and fragmentation of the territories that manage them (the Maasai Mara, for example), and the emergence of concurrent uses. Nonetheless, this trend is exacerbated in urban parks, and particularly in Nairobi.

Second and more importantly, the KWS has made NNP different from other parks with its Main Gate. This is unique in the Kenyan conservation system. It is, in fact, both a classic gate and an enclave that backs onto a PA: this enclave contains the headquarters and central services of the KWS, the FoNNaP (Friends of Nairobi National Park—moved in 2016) offices, an orphanage, a zoo, a vast asphalt carpark, the KWS monument to the dead, three bar-restaurants, and a small shop. Such a density of functions is not seen in Mumbai even, where the entrance to the national park is rather majestic compared with other Indian national parks. This build-up of directional and specific facilities sets NNP apart from other PAs in the KWS. It is very much the flagship of the KWS, the capital park. This directional functionality is not insignificant, since it affects the relationship between the NNP and the hierarchy in the KWS, and even affects the relationship between the city and the PA itself.

Morphologically, this directional enclave is part of a belt of enclaves located between the heart of the park and the city, to the north of the protected zone: Wilson Airport, the Langata Barracks, the KWS studios, the David Sheldrick Foundation, official caretaker accommodation, and the New East Gate Campsite. To the north and the east, between the city and the park, these successive enclaves act as a veritable buffer zone that reinforces the electric fence and thickens the border between nature and the city. Even though they have been fenced off (Aberdare NP, Nakuru NP, Mont Kenya NP, Meru NP), none of the other PAs have had their fencing reinforced by such a system of specialised enclaves.

Beyond this specific feature, the eastern edge of the park—the one that borders the industrial zone between Langata Road and the southern tip of the reserve—provides another unique feature in that is the subject of a linear, composite, thrown-together, complex, negotiated urban construction project. This eastern edge in unparalleled in both its morphology and its genesis. From the Carnivore restaurant at Cheetah Gate in Athi River, it stretches 30 km and is a true ecological front (Guyot 2009 and Chap. 8) along which an ongoing battle plays out between a conservationist unit, which includes the KWS and the Kenya Association of Manufacturers, and a developmentalist front that is harder to identify. Spanning the

Fig. 11.3 East Gate in the GreenLine, July 2014. *Source* Photo by B. Calas

length of this 50-mile-wide GreenLine,[5] and because "NNP is under siege" (GreenLine flyer 2009), this project, once complete, will fight the threat of the city with a wall of "a million trees". It uses several species of trees (*Acacia xanthophloea*, *Acacia kirkii*, *Acacia nilotica*, *Olea europoea subsp. africana*, *Croton megalocarpus*, *Warburgia ugandenesis*, *Balanites aegyptiaca, Carissa edulis*) as well as *Acacia mellifera*, whose needles will be a deterrent to human intruders into the park. This GreenLine is in fact an urban construction as much as a conservationist measure in that it is part of a policy of urbanism that combines a cycle path, seven picnic areas, observation towers overlooking the savannah, and recreational facilities, all aiming to improve quality of life for residents of the housing schemes that are being built at the edge of the park (Fig. 11.3). This linear forest will apparently have multiple functions: to establish a thick boundary that is difficult to cross, to be a pleasant area for walkers, and to act as a visible and material sign of ownership to combat land grabbing.

The contemporary nature of this project with "clearance" operations in the corrugated metal slums that sit in the interstices of the industrial zone, between the edge of the park and Mombasa Road, highlights the connection between nature preservation issues and urban development issues. In proportional terms, it is a

[5]We shall retain the—perhaps significant—homonym with the Beirut green line that separated Christians from Muslims during the civil law (1975–1990) and which steered the "reconstruction" of the Lebanese capital.

reminder of the destruction of tens of thousands of houses in the slums within the park in Mumbai that has taken place since the end of the 1990s, a destruction that still has not been followed by tree replanting, giving rise to numerous suspicions about the greed of housing developers. In fact, trees are a spatial operator whose physicality increases awareness, among those involved, of the challenges that are created by the fact that the city and park are adjoined, the challenges that NNP represents in terms of Nature conservation and urban development, and the constraints that its protection places upon the urban ecosystem. This construction of a thick border between the PA and its surroundings is unique in the KWS (Fig. 11.3). No other parks are protected in this way: either the border is simply not a physical one (Tsavo, Samburu), or it is limited to two electric wires separated by a narrow path (Aberdares, Mount Kenya, the north of Meru National Park), or the public PA has managed to inspire the surrounding private operators to convert their land into conservancies and thus to build a very thick border between nature and man, as is the case around the Maasai Mara, for example.

"Service and industrial economies play an essential role in the country's economy but also have a significant impact on the environment. Yet the environment is a showcase for Kenya's economy. In Nairobi, the best symbol of nature is Nairobi National Park, in the way that the Eiffel Tower is the symbol of France. This is a good image for the economy, for investors and for foreigners who come to Nairobi": Julien Barbier (2011, p. 27), quoting the GreenLine communication manager, emphasises this relationship between nature conservation, urban policy, development policy and regional marketing policy, one component of which is the GreenLine project. It works in partnership with Better Globe Forestry (BGF), which carries out commercial tree planting. According to BGF, GreenLine combines "a safe commercial project and a vital environmental and human activity". Significantly, BGF also originated from the Kenya Association of Manufacturers.

This proximity of the city has an impact on the managers of NNP, particularly because the park acts as "a show room" (to use the words of the park director) for KWS' policies. It is shown to diplomats, politicians, journalists, development brokers, etc., on the initiative of not only the KWS hierarchy but also the local, and more importantly national, political authorities. The managers and caretakers of the park therefore play a vital role in the communications of the KWS: that of men/women of the land, witnesses to the concrete significance of policies and safety risks. Legitimate and legitimising, they are often asked for their opinion. In July 2014, the Senior Warden said that she had recently welcomed the Prime Minister of China, the President of Rwanda and the wife of the President of Burundi. The Deputy Park Warden thus highlighted the fact that one unique feature of NNP lies in its role in the KWS's communication strategy, due to its proximity to the authorities and the media. For the directors, the proximity of the park to these political players therefore considerably increases the demonstrative value of the park and, in turn, the pressure that is felt by its managers, whilst also highlighting the important role that nature and its conservation plays on Kenya's political-diplomatic-media stage. If a higher official on a visit expresses a desire to visit the park, the park staff must immediately prepare to receive him or her with all the pomp and circumstance that his or her rank demands. "You are forced to keep on your toes and you have only short notice". If an incident creates a conflict between the "communities" (in other words the residents of the areas on the south edge of the park, particularly Maasai farmers) and the animals, the press is often on the scene before any KWS officers (Chap. 7) whereas, by contrast, in Tsavo for example, it is the rangers who are tasked with contacting the media when an incident occurs. Under the eye of the media, the KWS authorities, the epistemic community, etc., Nairobi's park therefore truly acts as a showcase for the KWS. Its directional function requires an edifying showcase for nearby and available spectators, as well as rangers who are multi-skilled, efficient, polite, disciplined and urban. NNP plays an important role in the fight for nature conservation, combining soft power with hard power.

NNP's place in KWS' communication policy helps to make the threats of poaching and encroachment more unacceptable and inadmissible than elsewhere. This drives an exemplary policy of nature and animal protection, a policy that makes NNP the most controlled park in the KWS system. As part of this battle, NNP became a rhino sanctuary in 1998, and the rhinoceros has been its symbolic animal ever since.[6]

Though one might think that one indicator of the park's urbanity is its reduced vulnerability to poachers, a factor that is "traditionally" associated with bushland parks, interviews with managers at NNP refute this assumption completely. For around 30 years, NNP has played a key role in protecting the rhinoceros, which is an essential part of conservation's funding system. The park therefore has a high concentration of animals that have been gathered by the KWS and is a place where orphans raised by humans can be reintroduced into nature. This role explains the exceptional density of rhinos in the park. Yet an increase in demand in the Middle East and China has led to a resurgence in poaching. In Kenya, although this development primarily affects the enclosures of bushland parks (Meru National Park, for example), and private conservancies that house rhinos (Solio Ranch, Ol Pajeta), NNP is not necessarily spared. In fact, two NNP rangers are still scarred by an episode in 2013, when two rhinoceroses were killed. However, the KWS HQ and its additional units—Wilson Airport, where all KWS helicopters are kept, and the specialised canine unit and barracks—are close by, forming a group of "positive externalities" and a unique shield. So much so that, though the threat is extremely

[6]Symbolic species are classically used in nature conservation policies. In Kenya, this is first and foremost the rhinoceros, of which the numbers are known more or less exactly. According to the *KWS Annual Report 2015*, 59 rhinos were killed by poachers in 2013, 35 of which were in PAs managed by the KWS. For 2014, the respective figures were 59 rhinos killed including 18 in PAs. Among the most-targeted protected areas are Nakuru National Park, then Solio Ranch and Ol Jogi Conservancy.

Conversely, Mumbai's park seems especially less emblematic as it only houses leopards, which are much less prestigious animals than the "national animal" of India, the tiger. Curiously, at SGNP, in 2012, one could buy—desperate marketing tactic, perhaps?—polo shirts emblazoned with a tiger's head. The fact that the park is home to the highest density of leopards in the world was barely promoted (see Chap. 7).

high, the park's ability to respond to it is equally so. This explains the fact that only three rhinoceroses have died in NNP in five years.

However, NNP is not completely protected from visits by poachers: its location in the outskirts of the city, though an advantage in many ways, also poses specific problems for the managers of the KWS.[7] The tactics of poachers here are not the same as in more rural areas. The size of the Tsavo and Samburu PAs, their relative isolation, their relatively low population densities and the fact that animals and people in these areas roam much further, mean poachers are forced to make long, mechanised, logistically complicated hunting trips that are rather conspicuous; in Nairobi, their visits can be much more covert as they are sudden, brief and subtle. Like a normal visitor, a poacher simply has to enter the park through the main gate in a hired taxi, get out of the car, kill the animal when they spot it, cut off and collect its horn, and leave. "Our park is small, so in one hour, they can kill a rhino" (Senior Warden). This uniquely urban poaching technique means that NNP is not the safest park in Kenya. The sanctuary that one might expect its urban location to provide is not there.

Since 2013, the rangers of KWS have been using an innovative tactic in response to this unusual situation: the block system. The park is split into three sectors which are allocated to officers, and into ten blocks, each of which is constantly guarded by two or three rangers. Each of these blocks is monitored day and night from a tent equipped with a radio which, twice a day, morning and night, informs the NNP's central administration of the rhinos' state of health. Some of these blocks are equipped with huge watchtowers. Every week, this tight surveillance enables the distribution of the rhinos, all identified and named, to be mapped. This fixed system fortifies the classic system of patrols that is used in all PAs. Every morning, these patrols aim to locate the critical rhinos: the ones that have not been seen in the past 12 h and that need to be found. In Nairobi, these patrols are often conducted on foot, whereas elsewhere they use both vehicles and aircraft. This block system has also been adopted in Nakaru, where intrusions occur much more regularly; it is also being tested in some private conservancies that are targeted by poachers, such as Solio Ranch in Laikipia. This system obviously requires extra staff (190 rangers at NNP compared with 130 at Amboseli). In the words of the Senior Warden, "NNP is a very sensitive park", and this fact is both the explanation for and the result of this security context. Consequently, to be assigned to a "rhino area" is both a source of pride and a cause of stress and worry. But none of this really makes NNP particularly different to other PAs that have rhinoceros enclosures.

[7]The proximity of certain amenities and certain specific urban facilities, such as Jomo Kenyatta International Airport, are part of NNP's uniqueness. Indeed, the proximity of roads interferes with the management of the park. One the one hand, KWS' control of the spaces located under the arrival pathways helps to secure them, on the other hand, the air traffic related to JKIA prevents KWS rangers from using light aircraft in surveillance and means they can only use helicopters, which are less effective in the fight against poaching and more costly as they must be rented from the Kenyan Defence Forces.

Paradoxically, due to the quality of its surveillance, the relationship between man and nature at NNP has been greatly modified. "NNP is so controlled" "so well structured" says its director, that there are no longer (officially…) any illegal pasturelands or forest fires caused by pastoralists, meaning the ecosystem in the protected zone—built through a long-standing interaction between herders and nature—is changing, becoming less man-made, in a way. Paradoxically it is no longer as good at meeting the needs and habits of migrating herbivores—wildebeests, zebras, gazelles—even though the majority of conservation efforts focus on these animals! A large portion of the staff at KWS have come to regret the actions of the pastoralists and are regularly forced to light deliberate, controlled fires in an attempt to return to the previous, symbiotic system that has been disturbed by excessive conservation and protection efforts. Once again, a parallel can be drawn between this situation and the park in Mumbai, where both agriculture and animal rearing have been banned. The grass lands were then supposed to disappear and be replaced by forest—judged to be the more noble and appropriate environment for a national park. Now, SGNP's latest Management Plan is to rehabilitate these open spaces, as people have realised that without them herbivores cannot reproduce, causing problems in maintaining biodiversity in general and a food source for leopards in particular.

In Nairobi, although located outside municipal limits, although outside the limits of the agglomeration and beyond the expanse of buildings, the "wildlife dispersion corridor" of Kitengela, beyond the Mbagathi River to the south-west of the park, is evolving in the shadow of the city. It is the setting of a pressing urbanisation that is breaking up the space and impeding the free movement of animals. The KWS must therefore adapt its strategy accordingly. For the managers, this is where the future of NNP is at stake, in that the stability of NNP's ecosystem, strictly speaking, relies on the smooth passage and freedom of movement of herbivores. For this reason, the KWS is prioritising policies that focus on this dispersion area.

Several ideas shape this action to allow greater numbers of more and more urbanised people to live alongside animals. Even though NNP has not had an Environmental Management Plan since 2011 and, in terms of strategy, seems to function as an island, it in fact creates tight bonds with its pastoral fringes, the prominent members of which are involved in discussions and plans. NNP's Community Officer states that in open and semi-open parks, including Nairobi, the work of KWS agents indeed involves going to the forefront of neighbouring pastoral communities and interacting with all "stakeholders", including conservationist NGOs. In short, they help to build the "park beyond the park". He also does not fail to mention the intelligence work conducted around the parks, including Nakuru and Nairobi. These management practices thus dilute the urban features of Nairobi and Nakuru's parks, but the diversity of the interested parties surrounding both national parks increases the intensity of the social interactions that take place between KWS employees and the users of the parks.

For the KWS, is NNP unique in that it is subject to a specific human resources policy? In the upper echelons of the hierarchy, yes: the socio-systemic individuality of NNP seems to influence the KWS' assignment policy, posting very adaptable, flexible employees who are used to social stress and a sporadic work schedule. Since the early 2000s, the recruitment policy has been open to young graduates from universities, who can now reach the middle echelons of the hierarchy without having to climb up through the lower ranks, as was the case previously. For executives, NNP positions are important as they can act as career springboards, a foot in the door to jobs in central services.

However, at the lower rungs of the KWS hierarchy, is NNP unlike other parks? The ranger assignment policy in NNP does not seem to have any significance, either positive or negative. Only the needs of the service and staff availability explain the randomness of staff allocations, the criteria for which do not seem to be clearly/explicitly defined. When asked, the rangers at the NNP gates gave a semi-meaningful response when it came to how they saw their assignment to NNP. Some highlighted the facilities offered by the huge city, which make an assignment at NNP good and enviable news: e.g., schools, medical facilities, cultural facilities, etc. Others emphasised, conversely, the difficulties of city life, particularly the costliness of it. "Amboseli is cheaper than Nairobi, where things are too expensive for the family". Furthermore, the working conditions are tougher at NNP, where rangers are constantly on edge: "Here in NNP, nights and days, willing or not, You constantly are on duty". The danger of overworking is no myth, and the rangers proclaim it loud and clear: at NNP, the "pressure is a lot" (interview with the Community Officer). At every level of the hierarchy, the rangers emphasise the intensity of social interactions in the park, an intensity that sets apart their professional life in Nairobi from their previous experiences.

It is not unanimously agreed that the issue of safety plays a role in stress: whilst some cite the tension caused by the park's accessibility, others distinguish their experiences in Nairobi from their experiences in other parks such as Amboseli, where the Maasai come to kill lions, or Tsavo, where groups of armed poachers are a constant threat. It seems, however, that assignments to rhino areas (e.g., NNP Nakuru, Meru, Tsavo, Nairobi, etc.) is what makes the greatest difference between PAs in terms of safety-related stress.

NNP's consolidated budget is clearly affected by its attachment to the city. On the one hand, the KWS' central services do not hesitate to provide NNP with rather comfortable subsidies, taking into account the unique challenges posed by the park's location in the outskirts of the city and the urgency of any situations that require intervention; on the other hand, the ability to organise special events, such as concerts, bourgeois weddings, receptions for entrepreneurs, companies and charities, marathons and mountain bike races, all characteristic of the event-focused city, opens margins for financial operations that are unknown elsewhere. NNP's financial situation is therefore fairly healthy, in large part due to this urban situation.

Conversely, it is true that bushland PAs—parks as well as conservancies—benefit more from favours from large foundations, sponsors, global conservationist NGOs, and even large Anglo-Saxon universities, which are more sensitive to rural areas and the challenges faced in bushland PAs than to those faced at NNP. Furthermore, NNP's relatively good financial situation allows the KWS to transfer funds to other, less equipped PAs. This financial role does not help to facilitate a change in direction in its management. Some of the people involved in discussions about NNP's future confirm this, admitting that, of all possible scenarios, the most acceptable in their eyes is that which would perpetuate paying access by visitors in vehicles to a completely enclosed area (the megazoo option, see below).

11.3 How Does NNP Participate in the Fabrication of the Metropolis?

11.3.1 Firstly, It Is a Place of Conflict Between Several Different Visions of Nature and Development

On a daily basis, NNP is a place where two visions of nature conflict with one another: the "normal" vision where KWS and conservationists have hegemony, as expected in protected areas, and a plural and dissenting vision set up by users of the park, in their diversity and the diversity of the ways they use the park. Indeed, on the fringes of habitus and the law, certain users use the park as a green space more than a national park. When the rangers deplore the behaviour of users at picnic hotspots, incriminate drunkards, noisy people, disrespectful people and others such as reckless drivers, etc., aren't they seeking not only a change in individual behaviour but also a different vision of the way nature is used, one that works with the vision promoted by conservationists and which steers the habits of "normal" users? "Drunkards and WaIndis[8] have bad manners, complained a ranger. Some people even go so far as to entertain themselves by scaring off the lions!". Furthermore, some visitors no longer see NNP as a PA, but rather as a traffic-easing access route, using its tracks as shortcuts to avoid the traffic jams of the Mombasa and Langata roads during rush hour. These drivers have thus misappropriated the annual pass that was set up by KWS in order to increase customer loyalty. Every day, three industrialists from Kiserian cut across the park in this way, from Maasai Gate to East Gate. A few Langata residents also enter the park through the Main Gate on the way to work in the industrial zone, on the other side of the East Gate. This variability in the ways the park is used clearly reflects the city's diversity. However, implicitly, it highlights the difference in habitus compared with what is the accepted norm in PAs, and even becomes a sign of "naturbanity".

On another, longer timescale, as part of a prospective type of reflection, several scenarios show diverging representations of nature and its place in an emerging metropolis with a population growth rate of 5% per year. This medium-term prospective reflection covers four very open theoretical scenarios:

[8]Literally "the Indians": Kenyans of Indian origin.

- Maintenance of the current trends (the business as usual option), namely the protection of nature and its gradual separation from the Kitengela herbivore migration corridor under the pressure of encroaching development,
- Voluntary and abrupt closure of the southern border of the park using electric fencing of the same type as that which already separates the park from the city to the north (the megazoo option),
- The pure and simple construction of housing developments, paving over the park (the urban option),
- Securing of the Kitengela dispersion corridor located to the south-west of the park (the best option according to the rangers interviewed, since this would maintain the biological connection between the park and the herbivore migration area).

Each of these scenarios thus paints a very different future for NNP. The very variability of these futures is in itself a source of great uncertainty, and is a defining feature of urban PAs. The same diversity of possibilities is also seen in Mumbai, ranging from the worst (the disappearance of the park as it is eaten away by urbanisation and the rise of transport channels) to the best (ecological connectivity with Tungareshwar Wildlife Sanctuary and Aarey Colony). A 5 km road tunnel is planned: its construction and ventilation vents would be detrimental to the park, but its existence could in the long term reduce the threat to the very existence of the protected area. The future of most rural parks is thought of in a more classic way, on a longer timescale, in terms of renewal of the area rather than renewal of the human use and economic exploitation of it. The prospective opening up of paths of reflection regarding urban parks indicates the land's varied possibilities, a variability that is less likely or even highly unlikely in the countryside.

Alongside this, various development and transport infrastructure extension projects of national and even international interest contradict the KWS' desire to keep NNP's integrity intact. Plans for a ring road, a motorway bypass, pipelines, railroad expansion, a high-speed line and finally and a very high-voltage line are all huge projects that reflect Nairobi's logistical and transit functions and its place in the national development strategy (50% of Kenya's GDP). These projects would consolidate the city's competitiveness by modernising the infrastructure that connects the city to sub-regional logistical networks. For leaders and the elite of the national development, the relative emptiness of the park, at the gates to the city, is an opportunity that heightens their covetousness and the conflicts between them and the ecological-conservationist block. Of course, this contradiction between the two sides of development, the productivist side and the conservationist side, is not unique to cities. The lively debates that surround the future of most PAs, regardless of their distance from cities, and which sometimes even lead to physical confrontation, continually cite these tensions on which political ecology thrives. What is more strongly connected to a park's proximity to the city, however, is the concentration and interlacing of the tensions that surround such a small space, the media coverage that is given to them and the social mobilisation that they inspire and which, by focusing on or investing in this space, transform it into an

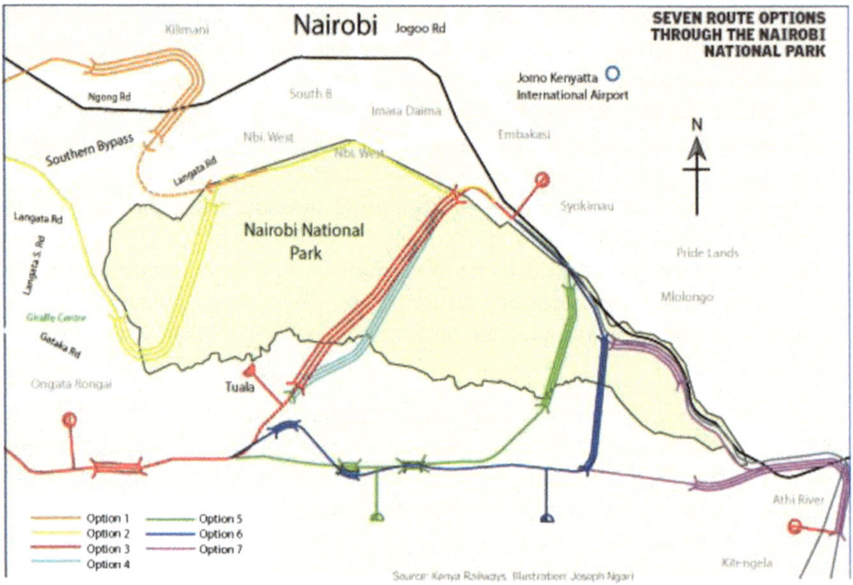

Fig. 11.4 The Standard Gauge Railway (SGR) project: The end of NNP? *Source* Habitat Planners (2016)

emblematic centre of either the environmentalist or developmentalist fight (see Chap. 9). A protest outside the Chinese Embassy was in fact held against the SGR railway project, financed by Chinese capital (Fig. 11.4). The park is truly a sounding board for the debates surrounding development.

Unsurprisingly, the very size of NNP (117 km^2, almost 20% of the municipality's surface area), a de facto land reserve, is highly coveted by certain metropolitan stakeholders, as is the case for the majority of low-density spaces (open spaces) or undervalued spaces (slums, for example) in the outskirts of large metropolises, and of course for the three other national parks covered here. Beyond the expected added value of a change in development strategy, bushland PAs are no different in that they are highly "coveted", but more often by farmers. In cities, quite varied types of actors are focused on the prospects of dividing, building upon, and paving over the landscape, radically transforming the way it is used.

Firstly, some of them are powerful: "some rich people have campaigned for development in the park" (NNP HQ, 27th July 2014) and indeed NNP is coveted and picked at by very active and highly-placed speculators. For example, the former President of the Republic, Moi, or the current Vice-President, Ruto, who respectively own Uchumi and Sunrise HS, and the Western Hotel that is located on the park's land. These land acquisitions are marginal, despite being highly covered in the media and serving as examples, even as legal precedents. More generally, the park's land is coveted by property developers. It is interesting to note that the ways these developers envisage the conversion of the park and that of the Kibera slums converge: poor people from the slums in the outskirts of the city, and wild animals from the reserve, should make way for production for the rich, with the savannah and the shacks being replaced by high-rise buildings, housing estates and malls.

However, alternative visions do exist and the park and its surroundings are sometimes seen as a space for potential housing developments for the poor. NGOs like Jami Bora, which is in favour of rehousing the residents of Kibera in decent accommodation, are critical to say the least when it comes to the securing of property in the park, and cultivate a classic rhetoric of social critique, recalling the colonial humiliation that is symbolised by the defence of the park.

11.3.2 Second, Because New Ways of Using the Park Contribute to Its Role in Metropolitan "Clubbisation", Whereby Residents of Certain Areas Group Together to Protect Their Own Interests

Access to many spaces in the parks of Cape Town and Rio is free, and the cost of entry to Mumbai's park is low (30 rupees, or around $0.20). But in Nairobi, a ticket for Kenyan residents costs 500 shillings, or almost $5.00: parks with an entry fee are more exclusive and thus participate in the creation of a selective and fragmented city, the antithesis of the inclusive and integrated city that is championed by supporters of sustainable urbanisation. In this respect, these special and very exclusive events only serve to reinforce this arrangement. In Kenya, where parks are, by definition, places where the well-off classes "intermingle", these special events organised at NNP reinforce the formation of an elite block based around distinct consumer practices in exclusive locations. However, if we compare the social choreography that plays out in bushland parks and in Nairobi, the "intramingling" that happens in bushland parks is predetermined, rigid, and difficult to overcome. It is an intramingling among the elite, on a global scale: the people who meet at Maasai Mara or Tsavo come from very different countries but all have more or less the same social standing. The "intramingling" that happens at NNP is more flexible, no doubt because interactions with others there are more short-lived, but also more intense and the visitors newer, more open to the diversity of middle-class people, creatives, and metropolitan superiors. This is above all a more metropolitan form of "intramingling", reflecting the cosmopolitanism of the city of Nairobi. The more recent and more commercialised ways of using NNP therefore contribute to "urban clubbisation" (Charmes 2011), in the same way that other selective spaces do: posh bars, trendy nightclubs, elitist golf courses, art institutions and clubs. These developments therefore exploit private access to NNP by extending it to other, non-touristic uses in order to build a space for "intramingling", to the benefit of the upper middle-class, a corollary of the demand for exclusivism in leisure activities (see Chap. 3).

However, in Nairobi, contrary to what Eric Charmes describes, it is not so much a case of a transformation of certain residential areas into residential clubs as it is one of an extension of the public space—access to which, it must be said, is already reserved

Fig. 11.5 Advertisement for a cultural event in NNP: culture for all? *Source* http://www.hangout.co.ke/artfest-art-in-the-park/

for a rather well-off population—and an exclusivism in the use of the space that starts with the transformation of these public spaces into quasi-private clubs for the happy few (Fig. 11.5). The spatial arrangement of the reserve lends itself well to this evolution, in that it already functions on the principle of reserved access.

This is why NNP is a contested space, an arena of complex conflict, especially since the proximity of the city increases the activism and the financial, human, diplomatic and media resources of those involved in the debate, such as conservationist NGOs. Therefore, compared with its counterparts in other parks, FoNNaP is a rich, strong, powerful NGO that is capable of mobilising people (see Chap. 9).

11.4 Conclusions

All in all, the urbanity of NNP compared with that of other parks in the KWS system is expressed through the greater diversity of its visitors and uses. The relative density of its use by its visitors, of which there are many in a small space, also reflects the proximity of the city. This impacts the way the park's management—the KWS—carries out its missions and integrates NNP into the mandate given to it by the State. Specifically, the KWS has had to set up an anti-poaching system—the block system—that is adapted to urban poaching practices, which differ significantly from bushland poaching practices. Beyond these differences that are revealed through observation of the players in the drama of conservationism, it is the extreme variability in the different futures that have been proposed for NNP, each by different actors, that makes this park different from other protected areas run by the KWS. Indeed, it is through the representations of these possible futures that one can see the real difference between rural PAs and urban ones: this difference, which substantiates the urbanity of NNP, seems to be scalar. In Nairobi, the contradiction between conservation and development is a national challenge, whereas in the countryside it is a "glocal" one. The urbanity of NNP is anchored in its national character, whereas in the countryside the national character of parks is erased by the generic character of the savannah, which is one of mankind's common resources, and the term "natural", which is both more localised and more globalised. In short, NNP seems to be urban because it is national, whereas other Kenyan PAs are "glocal", because they are natural. One is metropolitan, a "metropark", whilst the others fall into the category of commonplace infra-urbanity.

On a different scale, by isolating, in a truly artificial manner that is nonetheless common in geography, the various sections—the spaces that have not been built upon or cultivated—from their surroundings, i.e. the metropolis and/or the Nation—it is possible to put forward the idea that a continuum connects green spaces, from the small public garden to the strictly protected reserve. The "metropark" geotype is located along a continuum between urban park/forest and national park; the difference between the two poles lies in the fact that bushland parks are infra-urban whilst public gardens and metroparks are metropolitan. As a hybrid geotype, this metropark is a physical symbol of a type of naturbanity, one that fully constitutes metropolitan "humanimality" (Carnets de géographes 2013).

Bibliography

Barbier, V. (2011). *Les utilisations de la protection de la nature dans une métropole africaine en développement: Le Nairobi National Park (Nairobi Kenya)*, M.A. dissertation, University of Paris Nanterre.

Calas, B. (2002). La chorégraphie urbaine en Afrique orientale: L'hégémonie de la mondialisation revisitée. *Géographie et Cultures, 41,* 36–53.

Calas, B. (2003). Quel est le véritable patrimoine des parcs animaliers est-africains? In P. Cosaert (Ed.), *Patrimoines et développement dans les pays tropicaux* (pp. 333–342). Espaces tropicaux n°18, Dymset-Pessac.

Calas, B. (2006). De Biscarosse au Kilimandjaro: Évidences et vacuité des marqueurs spatiaux des espaces touristiques. In J. Lageiste (Ed.), *L'empreinte du tourisme. Contribution à l'identité du fait touristique* (pp. 321–342). Paris: L'Harmattan.

Calas, B. (2015). Activating the interactive landscape. Recent diversification trends in the Kenyan tourist industry. *Dynamiques environnementales, 35,* 204–224.

Carnets de géographes. (2013). «Humanimalité». http://www.carnetsdegeographes.org/mots_cles/Humanimalité.php.

Charmes, E. (2011). *La ville émiettée. Essai sur la clubbisation de la vie urbaine.* Paris: PUF.

Guyot, S. (2009). Fronts écologiques et éco-conquérants: Définitions et typologies. L'exemple des ONG environnementales en quête de Côte Sauvage (Afrique du Sud), *Cybergeo*, article 471, made available online on October 5, 2009.

Mawdsley, E., Mehra, D., & Beazley, K. (2009). Nature lovers, picnickers and bourgeois environmentalism. *Economic and Political Weekly, 44*(11), 49–59.

Peck, J., Theodore, N., & Brenner, N. (2009). Neoliberal urbanism: Models, moments, mutations. *SAIS Review, 29*(1), 49–66.

Sgard, A. (2008). Entre l'eau, l'arbre et le ciel, figures paysagères suédoises et construction de l'identité nationale. *Géographie et cultures, 66,* 121–138.

World Resource Institute. (2007). *Nature benefits in Kenya An Atlas of Ecosystems and Well-being*. Nairobi Central Bureau of Statistics.

Chapter 12
Environmental Education in Urban National Parks: A Political Tool for the Democratisation Process in Emerging Countries. Studies on Tijuca and Table Mountain National Parks

Nadia Belaidi, Rafael Soares Gonçalves and Glaucio Glei Maciel

The concept of environmental education has been through three major phases. At the international and institutional level, environmental education was raised in the context of the International Conference of the United Nations on the Human Environment at Stockholm in 1972. "Environmental education is the process of recognising values and clarifying concepts in order to develop skills and attitudes necessary to understand and appreciate the inter-relatedness amongst people, their culture and biological and physical surroundings. Environmental Education also entails practice in decision making and self-formulation of behaviour about issues concerning environmental quality" (IUCN 1971). This definition of environmental education, along with more recent ones, puts the emphasis on social change. It introduces an orientation towards "social criticism" into the educational process and encourages "learners" to investigate the social systems connected with environmental problems so as to gain a critical view of the human-nature relation. At the Stockholm Conference, it was stated that the development of environmental education is a fundamental element in dealing with the global environmental crisis.

Since the Earth Summit in Rio in 1992, however, questions bearing on "environment and development", following on from the publication of the Bruntland report, have tended towards the concept of "education for sustainable development" (UNESCO 1992). According to Agenda 21 "education is critical for promoting sustainable development and improving the capacity of the people to address environmental and developmental issues" (CNUED 1992, Chap. 36.3); this points up the close link between changes in the realm of environmental education and the

N. Belaidi (✉)
CNRS, MNHN - Dept Man and Environment, (UMR Eco-Anthropology and Ethnobiology), Paris, France
e-mail: nadia.belaidi@mnhn.fr

R. S. Gonçalves · G. G. Maciel
Department of Social Service, Pontifical Universidade Catholica, Rio de Janeiro, Brazil

F. Landy (ed.), *From Urban National Parks to Natured Cities in the Global South*,
https://doi.org/10.1007/978-981-10-8462-1_12

importance of the concept of sustainable development raised at the Rio Earth Summit. Education for sustainable development is aimed at a reorientation of training towards the sustainable use of resources on the basis of the belief that it is through the economy that all social and environmental problems may be solved (Rio Paradigm, V. par ex. Figuiere 2006). Though the role of education was emphasised afresh in response to problems of poverty and inequality in the world during the World Summit on Sustainable Development at Johannesburg in 2002, the turning point in Rio is acknowledged. The Declaration of Johannesburg explicitly focused on the doctrine of the free market as the engine for human development and on confidence in the power of techno-science to solve the problems (Hajer 1995; Young 2000).

Finally, following on from the seminar on "Educating for a sustainable future", organised by UNESCO and the South African Department of Education in 2002, the United Nations Assembly mandated UNESCO to coordinate the activities of the United Nations Decade of Education with a view to sustainable development (2005–2014). The object is to instigate operations to reform systems of education with the aim of reorienting programmes towards sustainable development. It was in this framework that "Education for a viable future" came into existence (UNESCO 2002). Yet the concept of a viable future is postulated on an essentially anthropocentric ethic: A new viable and lasting relationship between humanity and its habitat has to be envisaged, one that puts humanity in the centre of the scene without forgetting what is happening in the surroundings (UNESCO 1997, 17). The goal aimed at by everyone is "human security" (Ibid., p. 16). The alienation between subject and object, between human and nature, is very explicit in the proposal for a viable future. Confidence in technology is thus central to maintaining a "dynamic balance" of sustainable development (Ibid: 16–17). In this framework, environmental education is limited to being no more than an instrument, among many other "kinds of thematic education", for use in the interests of sustainable development; in this regard, environmental education is confined to a naturalistic or reactive approach to the environment, geared essentially towards the resolution of biophysical problems (Sauvé 2000).

Brazil as well as South Africa has actively participated in the international evolution of the idea of environmental education. Education for sustainable development came to the fore during the Rio Summit, and it is in South Africa that environmental education has become a tool of the global politico-economic programme of sustainable development. The economic weight taken by Brazil and South Africa legitimates their leaders in having their voices heard in international authorities. These countries have established themselves on the international scene, notably by heading environmental affairs, and promoting the ideas of development that their involvement supports.

The perspective on the way in which environmental education has been shaped by the recent process of democratisation of South African society (in 1994) and the earlier one of Brazil (in 1985/88) (part I), and adapted to their particular respective problematic, especially in their setting up of urban national parks (part II), is a useful illustration.

12.1 Environmental Education Established in the Process of Democratisation

Environmental education policies in Brazil were directly connected with the opening to democracy in the 1980s, but democracy was secured in the framework of an extremely liberal economic agenda in the following decade. Similarly, in South Africa the process of environmental education has been organised by and for the recent democratisation of society, especially in the appropriation of environmental justice, but with adaptation to the country's current issues in terms of redistribution.

12.1.1 The Weight of the Liberal Turnaround in the Establishment of Environmental Education in Brazil

Environmental education has been stable in Brazil from the 1980s, even though various programmes and projects had already been in existence since 1970. An important step for the institutionalisation of environmental education was taken with the help of the National Policy for the Environment. The obligation to include environmental education at all levels of instruction, including popular education, with the aim of readying it to participate actively in the defence of the environment, thereby bringing into being the capillarity required by this pedagogic practice was established in the legislative field in 1981 (PRONEA 2005, 22).

Article 225 of the Federal Constitution of 1988 stipulates that it is incumbent upon public powers to "promote environmental education at all levels of instruction, and of the public consciousness regarding the environment." In 1991, two authorities were created within the executive power to deal with the subject of environmental education: one was the Working Group for Environmental Education of the Ministry of Education and Brazilian Culture, which, in 1993, was turned into the General Coordination of Environmental Education,[1] and the other was the Division of Environmental Education of the Brazilian Institute of the Environment and Renewable Natural Resources (IBAMA). In 1992, the IBAMA

[1]The parameters of the programmes of national studies have included environmental education since 1996. The environmental dimension is "installed as an articulated and transverse study with different domains of knowledge, imbuing the practice of education and allowing for the creation of a global and comprehensive vision of the environmental question drawn from defined pedagogic projects", as affirmed by Loureiro et al. (2003, 14).

was constituted as the core of environmental education with delegations in all Brazilian states, in order to put into operation educative measures within the processes of the environmental management of units of conservation as part of the processes of the granting of environmental permits.

It was not, however, until 1997 that the National Programme of Environmental Education (PRONEA) was put into operation (MMA 1997), as the result of a partnership between the Ministries of education, environment, culture, and science and technology. This programme is primarily aimed at a consideration of environmental education with regard to environmental management (MMA 1997, 16).

The creation of centres for environmental education (NEAs) as part of the IBAMA delegations in the Brazilian states has added a supplementary level to the environmental education propositions. With the deployment of resources as of 1995, the environmental education of the IBAMA has become systematic, with annual planning meetings bringing together the NEAs and the General Coordination of Environmental Education with the intention of defining the directives of action and of creating the tools for the formulation, analysis and following up of NEA's projects. Finally, in 1997, Introduction to Education courses were put in place as part of the process of management, and hundreds of questionnaires were produced by IBAMA and other government organisations connected directly or indirectly to environmental questions (Loureiro et al. 2013).

The object of PRONEA was the consolidation of environmental education as a public policy, and that is the context in which the principles, directives, lines of action and organisational structure of environmental education were defined. The PRONEA is permanent and has to be recognised by all government departments. The perspective of its orientation is the sustainability of the environment towards the construction of a country for everyone: "Its actions are directed towards ensuring, in the domain of educational action, the interaction and balanced integration of multiple dimensions of environmental sustainability—ecological, social, ethical, cultural, economic, spatial and political—of the country's development, calling upon the involvement and social participation in the protection, recuperation and improvement of environmental conditions and the quality of life" (PRONEA 2003, 19).

Environmental education was definitively consolidated after the promulgation of law n°9.795/99, which instituted the national environmental education policy. Environmental education becomes an essential and permanent element in education and must be presented effectively at all levels and in all modalities of the education process (Art: 2). Educative activities and practices must be directed towards collective consciousness with regard to environmental questions, with the participation and the partnership of all sections of society, including government organisations. The environmental education ensured by these legal dispositions affirms the ideological nature of a method of building knowledge and stimulating political action for the transformation and conservation of life. Its goal is the democratisation of the debate on environmental questions, indicating that the educational process must be constructed in participative mode so as to ensure social justice to those who have been, and continue to be, deprived of their rights.

The process of environmental education is effective to the extent that it allows for the individual to be perceived as a social subject capable of understanding the complexity of society-nature relationships, and also of acting to prevent environmental dangers and damage caused by intervention in the natural and constructed environment, as the national environmental education policy prescribes. Environmental education is thus a basic tool in the operation of projects, particularly those conceived in the framework of urban protected areas. According to law n °9.795/99 (article 13, IV), public authorities must stimulate the raising of consciousness in society about the importance of the creation and management of conservation units.

Likewise, the national environmental education programme offers a constant transverse and interdisciplinary movement aimed at interiorising participation in the decisional process. As a principle, environmental education endeavours to improve the dialogue and address the adversarial debate in order to understand the conflicts in society, not as a pathological concern but, rather, as something inherent in social practice: "Environmental education must be carried out through a systematic approach, one capable of integrating the numerous contemporary environmental questions. This approach has to recognise the entirety of interrelations and the multiple dynamic determinations between natural, cultural, historical, social, economic and political domains. Over and above the systematic approach, environmental education demands a perspective of great complexity arising out of a world full of interactions at different levels of reality (e.g., objective, physical, abstract, cultural, affective, etc.)" PRONEA (2005, 34).

The educational directives of the IBAMA for the elaboration and operation of the government programme for environmental education (IBAMA 2005) asserts that education linked to a socio-environmental thematic is an education that uses the environmental management space as a teaching and learning space. Its aim is to stabilise, in practice, effective public management of the environment, seeking to ensure that the benefits of the environment are open to all.

The creation of the Chico Mendes Institute for the Conservation of Biodiversity (ICM-Bio) was announced, in 2007, as a strategy aimed at providing enhanced economic viability to the conservation units of Brazil. In 2009 ICM-Bio has introduced another approach, called the National Strategy for Communication and for Environmental Education. According to volume one of the collection (ICM-Bio 2015, 26), "communication and environmental education have a close interdependent relationship that demands that a conservation unit be able to create and articulate the activities in two domains so that each may complete and support the other". The reconsideration in the theoretical framework of environmental education associated with the management practices of the conservation units introduced at the time of the founding of ICM-Bio, seems to be angled more and more towards tourism and certainly towards the interests of big economic groups in this sector. As stated in volume five of environmental education recently published by ICM-Bio:

"the structuring of an effective system of conservation units calls for financial viability" (ICMBio 2015b, 18). ICM-Bio thereby introduces the principle of public and commercial usage as a new form of management of conservation units, which must be sustainable and able to raise funds. The augmentation of the number of conservation units in the National System of Conservation Units, has not, however, been accompanied by any significant increase in funding for their conservation.

The reconsideration of environmental education within ICM-Bio has had an impact upon the environmental education work of Tijuca National park, which has undergone major changes. The retirement of officers and the replacement of employees hired by the local environmental education centre combined with the park's change of direction—a new direction in closer connection with the concept of environmental management focused on the market—has led to a progressive diminishment of the dialogical and critical nature of environmental education inside the park.

In this framework, environmental education centres of federal conservation units in Brazil, and especially Tijuca National Park (TNP), take charge of environmental questions outside the framework of classical pedagogy and in contrast to the policies established for the conservation units.

12.1.2 The Weight of the Democratic Transition for the Establishment of Environmental Education in South Africa

After 1994, when the African National Congress (ANC) produced its Education and Training Programme, it featured the principle according to which scholastic programmes were to develop the understanding, the values and the competence necessary to achieve sustainable development (Irwin and Lotz-Sisitka 2010, 52–53) and an environment that ensures a healthy life.[2] Taken up in the White Paper of 1995 on education and training,[3] this principle allows environmental education to be designated as an "essential element" in the programmes at all levels, the goal being to create "environmentally literate and active citizens" (Department of Education 1995, 18). With the National Environmental Education Project for General Education and Training, proposed by the Ministry of Education in 2000, the environment is thus defined as a transversal subject that requires all teachers to

[2]Constitution of Republic of South Africa (1996), Chap. 2—Bill of Rights, art 24. Environment: "Everyone has the right a. to an environment that is not harmful to their health or well-being; and b. to have the environment protected, for the benefit of present and future generations, through reasonable legislative and other measures that a. prevent pollution and ecological degradation; b. promote conservation; and c. secure ecologically sustainable development and use of natural resources while promoting justifiable economic and social development".

[3]"*The curriculum will develop the understanding, values and skills necessary for sustainable development and on environment that ensures healthy living*", Department of Education (1995).

consider environmental questions in every field of apprenticeship. The processes of education for enduring development are henceforth integrated into all the areas of the official education programme. This means that each field must approach the problematic in the framework of its own teaching.

Institutional previsions relating to sustainable development have been adopted by South Africa and legitimise its educative activity. In the foundational texts on education for sustainable development, the same pedagogic principles are present and the same strategies for the establishment as those put forward for environmental education (The Tbilisi Declaration. Intergovernmental Conference on Environmental Education, UNESCO/UNEP 1977) but the significant change comes from the direction of educational action, operating henceforth on behalf of a development expected to be sustainable and articulate at the meeting point of the economic, societal and environmental spheres. Distanced from the axis of relationships between persons, the social group they belong to and the environment endowed with environmental education, education is nothing but a tool and the environment a resource in education for sustainable development (Sauvé 2007). Sustainable development offers, in fact, the sustainability of development itself, as the "purpose of humanity" (UNESCO-UNEP 1988, 1). This means that sustainability is seen as the supreme value on which all the other values, born of the needs (not discussed) of occidental type societies, current and future, must converge. That sustainability comes to be the foundation of an educational system where the suggested relationship with the world is confined within a managerial dynamic—as emphasised by the three interrelated spheres of economy, society and environment —often used to illustrate it, in which the economy is considered as an entity apart from and outside the social sphere. Human activities are thus interpreted solely in the frame of reference of "development" which, using the language of sustainability, is superimposed on diverse cultures and reduces the possibility of thinking about the realities in any other way (Sauvé 2000, 62).

The discourse has, however, been adapted to the particular problems of South Africa. The liberal view of the idea of development was the subject of a "differentiated appropriation" (Sauvé 2002, 4, 2006) by South Africa and is part of a distinct strategy. The connection between the different teachings provided with a view to its promotion is, in fact, established in the framework of the National Curriculum statement whose principles take note of the relation between human rights, inclusion, a healthy environment and social justice (Department of Education 2002), synthesised in the South African definition of environmental justice.

"Environmental justice is about social transformation directed towards meeting basic human needs and enhancing our quality of life—economic quality, health care, housing, human rights, environmental protection, and democracy. In linking environmental and social justice issues the environmental justice approach seeks to challenge the abuse of power which results in poor people having to suffer the effects of environmental damage caused by the greed of others. This includes workers and communities exposed to dangerous chemical pollution, and rural communities without firewood, grazing and water. In recognizing that environmental damage has the greatest impact upon poor people, EJNF seeks to ensure the right of those most affected to participate at all levels of environmental decision-making" (South African Environmental Justice Networking Forum 1997).[4]

In view of the environmental injustices experienced by South Africa (Kidd 1999), it would seem logical that a movement would arise to answer that. If, however, environmental justice in South Africa has, classically, had the objective of integrating environmental matters into the wider intellectual and institutional framework of fundamental rights and democratic responsibility (Wenz 1988; Low et al. 1998; Shrader-Frechette 2002) and of aiming at the widest possible definition of environment, it is very anthropocentric too and intentionally places populations, rather than flora and fauna, at the centre of a complex network of social, economic, political and environmental relations (McDonald 2002, 3). This is explained by the history of environmental policy in South Africa (Beinart and Coates 1995). Under the colonial and apartheid governments, thousands of black South Africans were forcibly expelled from their lands to make space for parks, and thousands of rands were given over to the preservation of the fauna and the protection of the wild flora in spite of the fact that the township populations were living without food, shelter or clean drinking water. The "whites only" policies in the national parks meant that black South Africans could not benefit from the country's natural heritage, and the draconian laws against poaching had the effect of shutting country dwellers off from essential resources.

Consequently, black South Africans (and anti-apartheid militants in general) initially paid little attention to environmental debates during the period of apartheid. The environment, seen either as a white question or as an explicit tool of oppression, was of little interest in the anti-apartheid struggle. The easing of apartheid legislation in the late 1980s and the liberalisation of South African political life at the start of the 1990s opened up a debate on the meaning and the causes and effects of the degradation of the environment and led to a reconsideration of environmental questions to include the work and the living spaces of black South Africans (McDonald 2002). From then on, environmental initiatives were speedily adopted as one of the post-apartheid democratic objectives. A wide range of unions, non-governmental organisations, associations and academics were quick to take up the new environmental discourse, and over a few years, to contest the environmental practices and policies of the past.

Then, at its election in 1994, the African National Congress (ANC), observing that poverty and the degradation of the environment were closely linked in South

[4]This Forum is a national organisation created to coordinate the activities of militants and environmental organisations concerned with social and environmental justice. It was formed following the conference "What Does It Mean to Be Green in the New South Africa?" organised by the Earthlife Conference Africa in 1992, see Hallowes (1993) .

Africa,[5] registered in its mandate for post-apartheid reconstruction and development the fight against environmental inequalities and injustices on the grounds that social, economic and political relations are equally part of the equation of the environment. One of the methods explicitly targeted is environmental education.[6]

According to the White Paper on the policy for the management of the environment in South Africa (Department of Education 1995), integrated and sustainable management of the environment, now and in the future, is the essential basis of sustainable development in all areas of human activity.[7] However, the definition of sustainable development in the White Paper requires that particular attention be given to meeting the needs of previously disadvantaged communities.[8] Consequently, if sustainable development is actually the basis of the South African policy for the environment, as defined at that time, it is aimed at respect for humanity's fundamental needs (e.g., nourishment, clothing, lodging and employment)—a crucial aspect for the South African people.

That clarification informs on the meaning given to environmental justice in South Africa. In this context, the government must integrate environmental considerations into justice and into social, political and economic development to meet the needs and rights of all communities, sectors and individuals.[9]

---

[5]2.10.1 "Apartheid legislation distorted access to natural resources, denying the majority of South Africans the use of land, water, fisheries, minerals, wildlife and clean air. South Africa's apartheid policies, combined with the underregulated activities of local and transnational corporations, contributed to the degradation of environmental resources, including soil, water and vegetation. They encouraged the misuse of fertilisers and pesticides. They placed workers' lives at severe risk because dangerous practices and substances were inadequately monitored (mining in South Africa remains an extremely dangerous job). Poverty and environmental degradation have been closely linked. In general, existing environmental policies allow inefficient and wasteful use of water, energy and raw materials, and high levels of air and water pollution" (ANC 1994, 38).

[6]2.10.6.3 "Environmental education programmes to rekindle our people's love for the land, to increase environmental consciousness amongst our youth, to coordinate environmental education with education policy at all levels, and to empower communities to act on environmental issues and to promote an environmental ethic" (Department of Education 1995).

[7]"Integrated and sustainable management of the environment, now and in the future, is the essential basis of sustainable development in all areas of human activity" (ibid., 16).

[8]"Sustainable Development: in the context of this policy sustainable development is defined as development which seeks to integrate environmental, social and economic concerns, now and in the future, and to keep within the carrying capacity of the environment. The focus is on ensuring that environmental sustainability, health and safety are not compromised, and that natural and cultural resources are not endangered. Sustainable development must ensure that the direction of investments, the orientation of technological developments and institutional mechanisms work together towards the goal of the sustainable use of environmental resources in a way and at a rate that will meet present and future needs. Sustainable development requires that particular attention be given to addressing the needs of previously disadvantaged communities" (ibid., 86—glossary).

[9]"Environmental Justice: To comply with the requirements of environmental justice, government must integrate environmental considerations with social, political and economic justice and development in addressing the needs and rights of all communities, sectors and individuals" (ibid., 22—principles).

In other words, environmental justice is not an end in itself but is, rather, part of a wider objective of social justice (Kidd 1999). It is only in these terms that the government has recognised environmental justice as a fundamental principle of the management of the environment in South Africa. In this framework, the State focuses upon natural spaces, especially on national parks, to make them symbols of a national project for sustainable development, and even environmental justice. Nature comes in here in a dialectic integrating global and local vision through the idea of environment which distinctly places these spaces in a "post-apartheid" logic—symbolic of reparation/reconciliation, but also brings them into the framework of an economic global and neoliberal policy (the programme GEAR: Growth, Employment and Redistribution). Although, during the presidency of Nelson Mandela, the defence of liberalism was carried out through the prism of the defence of fundamental rights and to the benefit of the reconciliation policy, with Thabo Mbeki and then Jacob Zuma, South Africa has adopted the process of liberalisation to promote its own economy. Liberalism, structured around free enterprise, democracy and respect for human rights, was seen in this light as a tool of development supporting reconciliation. Today, the promises of globalisation are counted upon—with a lack of anticipation of the social problems that globalisation may bring or exacerbate (Mielniczuk 2013).

12.2 Environmental Education, Inscribed in the Urban National Park: An Instrument of Social Management

Environmental education in both South Africa and Brazil appeared to be oriented towards action for social change. Assigning it a specific role would seem legitimate: it would look like a framework with the potential to help resolve social conflicts and conflicts of usage with the poorest, as well as promote citizenship. All the same, the objective of modernising the management of the use of national parks introduced by ICM-Bio in Rio especially, like the "urban marketing" of which Cape Town's Table Mountain National Park (TMNP) is the object, have turned it into an outright tool of social management.

12.2.1 At Rio, a Tool in the Process of Concession of the Corcovado/Paineiras Sector

Work on environmental education has been very active in the many *favelas* that surround the park, notably in the *favelas* of Guararapes, Cerro Corá, Vila Cândido and Prazeres. Founded on the principles of environmental education (and here on the theory of the "Pedagogy of the oppressed" (Belaidi et al. forthcoming), in TNP it has taken the form of touristic activities conducted mostly by *favela* residents (e.g., itinerant trade, parking attendants, taxi and minibus drivers). However, in conformity with the policies of revitalisation of the town of Rio de Janeiro, the managers of the TNP/ICM-Bio have occupied the spaces of the Paineiras de Corcovado sector and have gradually limited these activities in order to concede to tourism businesses. The dispute over control of the Corcovado-Paineiras sector, is not only a conflict between the institutions of federal and municipal government in the shared management of the TNP with the same object of commercialising that space by means of externalisation of concessions to satisfy the demands of tourism, but also acts to force out the inhabitants of those *favelas*.

In this context, ICM-Bio issued, in 2009, an Act of Reference calling for public bids for the concession to exploit and use the spaces of the Hotel Paineiras.[10] Earlier, there had been no competition because of the obligation to renovate the hotel's abandoned building so that it could lodge guests again. As of 2012, after the examination and revision of the Act of Reference, the offer was repeated less the obligation to resume lodging guests and the Consortium Paineiras/Corcovado won the contract. This is made up of the businesses Bel Tour, a traditional carrier operating in the tourist transportation sector since 1960; Esfeco, a company controlling, since 1979, the exploitation of the cable car that carries more than 900,000 tourists a year up to the statue of Christ the Redeemer; and the Cataratas do Iguaçu S.A company, a business offering touristic services in national parks such as Foz do Iguaçu and Fernando de Noronha.

The principal aim of the institutional occupation of the Corcovado/Paineiras sector was the maintenance of "order" in that space, so that necessary services could be provided to visitors. But the new system that went into operation following the concession has been to the advantage of particular social groups due to the favouring of management and the use of space. The concept of the organisation of territory established in the domain of the occupation of the Corcovado/Paineiras space depends on the idea that the space of the park is going through a "differentiated" phase and is associated with strategies of legitimisation of a specific project of the town. The idea of imposed "order" thus means, in actual fact, the criminalisation of poverty and an attempt to prevent access to the resources obtainable through tourism in the region. The process of putting the Paineiras-Corcovado sector in order has not taken into account access to resources bestowed by the TNP's installation of tourism. The criminalisation of activities carried out by the *favela* has prevented consideration of the importance of the function exercised by that population.

There is still some social resistance to this process involving those residents of *favelas* who continue to work in tourist services as members of van and taxi cooperatives, parking attendants, itinerant vendors and visitors' guides. Moreover, the forceful political mobilisation of the inhabitants has opened the way for the attributing of the total amount of the concession of the Corcovado/Paineiras sector

[10]To know more about the history of the Hotel Paineiras, see the site: www.light.com.br/web/institucional/cultura/ccl/memoria/hotel_paineiras/tehpaineiras.asp.

the sum of 4 million reais, divided into 20 instalments of 200,000 reais (about $60,000) to environmental education projects with a view to mitigating local repercussions caused by the touristic concession of the sector. The type of management of resources for environmental education chosen has been the subject of conflicts between managers and *favela* inhabitants, who want more extensive access to the modalities of utilisation of resources. Subsequent to numerous conflicts, the situation seems, at least for the moment, to have been resolved by a move in the direction of a more autonomous use by residents of this capital, with the support of the universities.[11] This amount, equivalent to 3.75%, of 96 million reais is anticipated from investments and/or 20% of profits is expected by ICM-Bio (20 million reais) per year for its services. The exploitation of the sector is a gold mine due to the statue of Christ the Redeemer, the most visited touristic site in the country.

In addition to the specific matter of the concession of the Paineiras-Corcovado sector, the current management of TNP involves the pathways of the park as far as possible where tourism, at the same time as providing resources, would facilitate a better control of the various areas of the park. The effort aimed at improving access for tourists is not necessarily at odds with the interests of the poorest members of the population. Some pathways start at the *favelas* and are commercially exploited by the inhabitants themselves. All the same, the priority given to tourists seems to be at the expense of the interests of other users. We may cite, for example, the meagre efforts made regarding the poorest groups, the blacks and the inhabitants of the *favelas*, who use the park for religious purposes, as much in the case of Afro-Brazilian worship as in that of Pentecostal groups (cf. Chap. 4). The dialogical effort of environmental education in the matter, as shown by the experiment of the Elos da Diversidade project,[12] may also ensure right of entry to the park to these groups to encourage constructive practices for the protection of nature and to involve the population very actively (Costa 2008).

Instead of discussing these practices, the current strategy of the TNP is to link up with the endeavours of the Ministry of Tourism and of the management of the environment in order to encourage the concessions of national parks and take advantage of the touristic demand in these fields. This collaboration chimes with the assumptions of strategic urban planning in favour of the commoditisation of the town

[11]Residents suggest that resources be managed by a university which would help in setting up environmental education proposals in the *favelas*. The universities will provide increased autonomy with respect to the PNT managers and will prevent these resources being used by mafia-type organisations, such as drug traffickers, in the *favelas*. Environmental education projects will be chosen after a study carried out by the university and the inhabitants. The call for bids is open till March 2017. See: http://www.tremdocorcovado.rio/ (accessed January, 2017).

[12]*Elos da Diversidade* is an environmental education project put into operation between 2007 and 2014 by the Secretariat for the Environment of the State of Rio de Janeiro in association with the Institute of Geography of the State University. The activities developed were aimed at identifying the religious practices of African and neo-Pentacostal origin in nature and to promote discussion of the maintenance of their rites in the conservation units while respecting the exigencies of environmental protection. See Corrêa and Costa (2013).

to the detriment of a managerial policy that would take the local residents' usage into account. Access to the area conceded in the park is very expensive and prohibitive for the poorer population. Both the cable car and the vans that carry tourists to the statue of Christ and to the new visitor centre in the renovated old building of the Hotel Paineiras are quite expensive too. The recovery of this building and the creation of the visitor centre has, nevertheless, introduced a new leisure centre for tourists, with a museum, restaurant, bistro and souvenir shop. It is as important to consider the impact the concession will have on the tour's supporting activities carried out by the surrounding population as to consider the new opportunities for employment for the inhabitants of the *favelas* in the tourist trade. What will be the game plans adopted by these inhabitants in the struggle for access to the park and to guarantee the rights and benefits of tourism support services in these sectors?

It is still too soon to put forward conclusions on this subject, but what can be asserted for now is that the process of commoditisation of the public good is manifest as much in urban as in environmental policies. It is most apparent in the management of urban parks such as TNP in favour of the tourism industry and to the detriment of more public usage, for the workers as well as for the tourists and residents (Maciel and Gonçalves 2016).

12.2.2 In Cape Town, a Tool of Urban Marketing

With its long-term goal of encouraging pro-environmental attitudes, of increasing environmental knowledge, of promoting citizenship and of encouraging stewardship activities, not only in situ but also in the "community" of origin,[13] environmental education has become a priority of the "People and Conservation" department, one of three executive departments of the South African National Parks (SANParks) organisation. The environmental education initiative run in TMNP is a pilot project (which tends nowadays to be generalised amongst the other national parks managed by SANParks) that is included in the objective of *social ecology* resumed by SANParks in the middle of the 1990s when, with the abolition of apartheid, it adopted the mission of promoting of a new idea of conservation in the national parks of South Africa. This notably evoked the "vision" elaborated for TMNP: "a park for all forever" that represents the park as a place accessible to all population groups of Cape Town, whatever their social conditions. It was seen as a place where the inhabitants of townships and suburbs would be able, through the link created or recreated with nature, to encounter or create a common identity, where today community life remains to be constructed. Moreover, through the benefits generated by the park, a redistribution of wealth is envisaged from which, this time, everyone will benefit.

[13]http://www.sanparks.org/parks/table_mountain/people/default.php (accessed October 6, 2015).

SANparks has developed three major environmental education programmes in the TMNP based upon the idea that *reconciliation ecology* is possible (Belaidi 2016). In this framework, not only is the park within the city, but the city is also within the national park.[14] The principal object of these programmes is to fuel a desegregation of the space towards self-esteem through awareness of the existence of a national natural heritage.

All these programmes have in common the granting of access to the park, an area that had been classified as white since 1957 and to which access is still limited by transport considerations, by entrance fees and also by habit. In order to promote appropriation of the place, "learners" have free entry to the park: the park bus carries them from their schools or homes in the townships to the education centre inside the park.

Another common aim of these programmes is to enhance the national character of the park. Here, the nation refers to the population (of Cape Town), and thus also to the black and coloured inhabitants—henceforth members of that nation. The stigmata of apartheid are still very much present and the feeling of belonging to the society has yet to come into being. The valorisation of the flora of the Peninsula is one of the tools in that it offers a powerful identity marker. The fynbos, calling to mind the "African" origin of the Cape's population, underlines the close relationship with, and the significance of, this vegetation for the precolonial populations—especially by way of an illustration of the use of nature by the San and the manner in which that use was then spread amongst other population groups (cf. Chap. 5). These systems of knowledge, incorporated into the cultural weave and the history of peoples (Odora-Hoppers 2001), are the basis of the (notably) social identity of the populations and SANParks too is betting that their diffusion through environmental education courses will be the way to bestow the foundations of a nation to the fragmented Cape Town society (Maila and Loubser 2003). This goes along too with the mission that is part of the Preamble to the Constitution of South Africa of 1996: the "need to create a new order in which all South Africans shall be entitled to a common South African citizenship".

Finally, these programmes are intended to have an impact on learners' environmental attitudes. The principal objectives are to inform and sensitise young people so that they may participate in the conservation of protected areas and relay these questions to the communities. In this framework, one of the programmes goes further with the explicit aim of training "ambassadors" of SANParks and of giving them basic competence in protected areas. This specific idea answers to another function of *empowerment* of the programme: it aims at identifying young people sufficiently interested in environmental subjects to involve themselves in supporting conservation. The park authorities tend to approve of, even to support, their access

[14]"The vision that Cape Town can be a city within a national park, rather than just a national park within a City. To achieve this the notion of 'reconciliation ecology' can be applied to Cape Town", Myrdal, Brett (park manager), *Park News*-Table Mountain National Park, Third Quarter, 2008, p. 1.

to higher education, still very unequal between blacks and whites, and then to employ them (mainly as rangers) in national parks.

The park is, at the same time, presented as a tool of redistribution/sharing of environmental spaces amongst the populations while ensuring protection of those areas, a base for the participation of populations, an instrument of repartition of wealth and self-affirmation within the society. It would seem, by this token, to be a vector of environmental justice. The fight against environmental injustice is also a guarantee of the integrity of the site. In fact, although ecological criteria are the basis of the designation of protected natural areas, their development is inextricably linked to the socioeconomic conditions of the local communities living around them (Niemela 1999). In Africa, it is in southern Africa that the greatest inequalities of urban revenue are found. South Africa is still detached from this panorama: even though the adoption of redistribution plans and policies have somewhat reduced inequalities in recent years (UN-Habitat 2008, 67), the economic and political model of concentration of resources continues. TMNP is encircled by populations living under diverse socioeconomic conditions (Chap. 3). On the western and eastern sides of the park are prosperous households (e.g., Bishops Court, Camps Bay, Sea Point, Constantia, Tokai, Newlands, etc.) while farther to the east of the park poverty prevails. The poverty of these populations is, depending certainly upon the actors, a serious threat to the perspective envisaged for this protected area (Chap. 4). It is crucial for park management that the people in or near the parks understand the connection between their own well-being and that of the protected area so that they may be conscious, or at least aware, of the values connected to the conservation of these areas and modify their behaviour accordingly, especially in the face of degradation, poaching and the exploitation of the areas.

This is a delicate position to maintain when dealing with people who have been excluded by law from these areas or who are still de facto. Recourse to environmental education goes some way to passing on the message. If it sticks to the function of modifying the behaviour of the populations so as to preserve those areas, it is also, and certainly, meant to ensure the economic potential of the spaces. As the social function attributed to the park is supported by dividends generated by its touristic exploitation, conservation has to be economically cost effective for it to be ensured. Redistribution of wealth cannot happen in the absence of this condition.

To the socioeconomic argument is attached a political one about the operation of environmental education initiatives in the park. TMNP seems to be the instrument of an urban marketing project orchestrated at the national level to the profit of Cape Town, in competition with the conservative parties in power in Cape Town just after the start of the democratic era (for an explanation of the vote in the Western Cape, cf Houssay-Holzschuch 1999, 236–244). TMNP is profitable to Cape Town to the extent that the emphasis on the values of nature tends to improve the image of the town. Cape Town then appeared as a “green” town, modern and ecologically responsible, quite distinct from the apartheid town known internationally for its profound social and racial inequalities. The “credit” for the transformation goes to the authority responsible for offering this leisure area: SANParks, the national authority that is the sole manager of a space totally integrated within the borders of the town. Before 1998, 14 institutions shared control of the public domain on the Cape Peninsula, but since the strategy of fusion of land ownership at the origin of the TMNP (Belaidi 2012, 271–275), the national authority is solely responsible for the management of this territory. The site thus seems particularly implicated in the power games between the two political parties in the Cape's mayoral race.[15]

The situation seems more and more hazardous for the ANC in its attempts to recover the Cape municipality, even to stay at national level in the future. The Democratic Alliance (DA) had been the official opposition party in Parliament since 2004. Considered at first as the party of white South Africans, this movement accentuated its multiracial electoral base at the general election of 2014. Thanks mainly to the coloured vote, the DA is the leading party of the Western Cape. The ANC therefore has a serious interest in marking the territory in a manner favourable to an electorate not yet won over.

12.3 Conclusion

It is now evident that environmental education is being used in the current context of the economic promotion of conservation units in Brazil. Much as the emphasis is still on their public use, that aspect is very evidently oriented towards business (Maciel 2015). Despite the fact that the principles of environmental justice continue to be evoked, what prevails in the logic of the public use of conservation units is the possibility of turning the protected areas as well as their tangible and intangible cultural heritage into merchandise, which directly impacts the praxis of environmental education and public management. The situation is the same in South Africa where, reduced to a simple education about the natural environment and associated with pro-environmental postures, environmental education has only moderately integrated ethical and critical competences. While the objective of environmental action is promoted, the action undertaken is rarely associated with a reflexive approach and remains instrumental. A discourse is revealed here that associates environmental education with a transfer of useful practices, and that considers education as a means of putting human potential to the service of economic growth.

With the importance given by the governments to environmental education, it may, however, be reasonable to anticipate that education might support the promotion of activities of reflection. But the pedagogical choice, and hence the returns in terms of reconciliation and emancipation, depends very much on the context in which it is adopted. Then, the instrumental conception of the conservation units is formulated by strategic planning that has dominated the management of major metropolises over

[15]On the idea of a "social bridge" attributed to biodiversity in the framework of municipality activities, see Davis (2005).

recent years, especially those of Rio de Janeiro and Cape Town. The towns and everything in them, such as the protected areas, are thus kept competitive. The logic of big city marketing, and business-oriented policies, affects not only urban policies but comes equally into conservation policies, and especially into the framework of efforts to implement tourism in protected areas (Maciel and Gonçalves 2016).

Bibliography

ANC. (1994). *The reconstruction and development programme*. Johannesburg: Umyanyano Publications.

Arantes, O., Vainer, C., & Maricato, E. (2000). *A Cidade do Pensamento Único. Desmanchando Consensos*. Petrópolis: Vozes.

Beinart, W., & Coates, P. (1995). *Environment and history: The taming of nature in the USA and South Africa*. London: Routledge.

Belaidi N., & Gonçalves Soares R. (Eds.), forthcoming, L'éducation à l'environnement: de l'émancipation politique à la "justice environnementale", *Droit et Cultures*.

Belaidi N. (2016). Ecologie & Réconciliation. Etude à partir des initiatives d'éducation à l'environnement sur Table Mountain National Park, Cape Town *in* R. Porteilla et al., *Afrique du Sud: 20 ans de démocratie contrastée?* (pp. 305–328). L'Harmattan.

Belaidi N. (2012). Le Patrimoine mondial pour créer une identité commune dans l'Afrique du Sud post-apartheid? Exemple de la ville du Cap *in* Bourdeau L., Garavari-Barbas M., Robinson M. (ed.), *Tourisme et patrimoine mondial* (pp. 269–284). Québec: Presses de l'Université Laval

Corrêa, M. A., & Costa, L. M. (Eds.). (2013). *A floresta: Educação, Cultura e justiça ambiental*. Rio de Janeiro: Garamond.

Costa L. M. (2008). *A floresta sagrada da Tijuca: Estudo de Caso de conflito envolvendo uso público religioso de Parque Nacional*, Mestrado em Psicossociologia de Comunidades e Ecologia Social. Universidade Federal do Rio de Janeiro.

Davis G. (2005). Biodiversity conservation as a social bridge in the urban context: Cape Town's sense of 'urban imperative' to protect its biodiversity and empower the people. In T. Trzyna (Ed.), *The urban imperative. Urban outreach strategies for protected area agencies* (pp. 17–21). California Institute of public affairs, Sacramento.

Environmental Justice Networking Forum. (1997). *Quarterly newsletter of the environmental justice networking forum, no 13*. Scottsville: EJNF.

Figuière C. (2006). Sud, développement et développement durable: l'apport de l'économie politique, *VertigO*, 7, 2, online.

Hajer M. A. (1995). *The politics of environmental discourse: Ecological modernization and the policy process*. Oxford University Press.

Hallowes, D. (Ed.). (1993). *Hidden faces: Environment, development, justice: South Africa and the Global Context*. Scottsville: Earthlife Africa.

Houssay-Holzschuch M. (1999). *Le Cap ville sud-africaine. Ville blanche, vies noires*. Paris: L'Harmattan.

Irwin P., & Lotz-Sisitka, H. (2010). A history of environmental education in South-Africa. In C. P. Loubser (Ed.), *Environmental education. Some South African perspectives* (pp. 47–51). Pretoria: Van Schaik Publishers.

Kidd, M. (1999). Environmental justice—A South Africa perspective. In J. Glazewski, & G. Bradfield (Eds.), *Environmental justice and the legal process* (pp. 142–160). Cape Town: Juta.

Loureiro, C. F. B, Azaziel, M., & Franca, N. (2003). *Educação ambiental e gestão participativa em unidades de conservação*. Rio de Janeiro: Ibase: Ibama.

Loureiro, C. F. B., Saisse, M. V., & Cunh, C. C. (2013). Histórico da educação ambiental no âmbito federal da gestão ambiental pública: um panorama da divisão do IBAMA à sua reconstrução no ICMBio, *Desenvolvimento e Meio Ambiente* (pp. 57–73).
Low, N., & Gleeson, B. (1998). *Justice, society and nature. An exploration of political ecology.* London: Routledge.
Maciel, G. G., & Gonçalves, R. S. (2016). Mercantilização da cidade do Rio de Janeiro e a concessão do Setor Corcovado/Paineiras do Parque Nacional da Tijuca. *Revista História e luta de classes, 21,* 44–55.
Maciel, G. G. (2015). *Mercantilização da cidade do Rio de Janeiro e suas implicações na gestão de unidades de conservação: um estudo sobre a concessão do Setor Paineras/Corcovado (Parque Nacional da Tijuca—RJ) e os efeitos sobre os moradores das favelas do Cerro Corá e do Guararapes.* Mestrado en Service Social, Rio de Janeiro: PUC-Rio.
Maila, L., & Loubser, C. (2003). Emancipatory indigenous knowledge systems: Implications for environmental education in South Africa. *South African Journal of Education, 23*(4), 276–280.
McDONALD, D. A. (Ed.). (2002). *Environmental justice in South Africa.* Cape Town: UCT Press.
Mielniczuk, F. (2013). Brics in the contemporary world: Changing identities, converging interests. *Third World Quarterly, 34*(6), 1075–1090.
Niemela, J. (1999). Ecology and urban planning. *Biodiversity and Conservation, 8*(1), 119–131.
Odora-Hoppers, C. (2001). *Indigenous knowledge and the integration of knowledge systems: Towards a conceptual and methodological framework.* Pretoria: HSRC.
Quintas da Silva, J. (2005). *Introdução à Gestão Ambiental Pública.* Ibama, Brasília-DF.
Sauvé, L. (2000). L'éducation relative à l'environnement entre modernité et postmodernité. Les propositions du développement durable et de l'avenir viable. In A. Jarnet, B. Jickling, L. Sauvé, A. Wals, & P. Clarkin (Eds.), *The future of environmental education in a postmodern world?* (pp. 60–64). Whitehorse, Yukon College.
Sauvé, L. (2002). L'éducation relative à l'environnement : Possibilités et contraintes, *Connexion, La revue d'éducation scientifique, technologique et environnementale de l'UNESCO,* XXV11, ½: 1–4.
Sauvé, L. (2006). L'organisation et la structuration du secteur de l'éducation en réponse au programme onusien du développement durable. In *Former et éduquer pour changer nos modes de vie,* Liaison Énergie-Francophonie, 72, 33–41.
Sauvé, L. (2007). L'équivoque du développement durable. *In Chemin de Traverse, 4,* 31–47.
Shrader-Frechette, K. (2002). *Environmental justice, creating equality, reclaiming democracy.* Oxford: Oxford University Press.
Wenz, P. S. (1988). *Environmental justice.* Albany: State University of New York Press.
Young, S. C. (2000). *The emergence of ecological modernization: Integrating the environment and the economy?* London: Routledge.

Documents

CNUED. (1992). Agenda 21, Rio de Janeiro, CNUED.
Constitution of Republic of South Africa. (1996).
Department of Education. (2002). *Revised national curriculum statement.* Pretoria.
Department of Education. (1995). *White paper on education and training.* Pretoria.
IBAMA. (2015). Orientações pedagógicas do IBAMA para elaboração e implementação de programas de educação ambiental no licenciamento de atividades de produção e escoamento de petróleo e gás natural. Brasília: IBAMA.
ICM-Bio. (2015). A unidade de conservação e o território: Reconhecendo o contexto socioambiental e geopolítico. Brasília: ICM-Bio.
ICM-Bio. (2013). *Gestão Estratégica,* in http://www.icmbio.gov.br/portal/gestaoestrategica.
ICM-Bio. (2015). Tornar-se visível: estratégia para promover articulações e captar recursos. Brasília: ICM-Bio.

ICM-Bio. (2012). Estruturação de Unidades de Conservação, in http://www.icmbio.gov.br/portal/visitacao1/estruturacao-de-unidades-de-conservacao.
IUCN. (1971). *Education and the environment*. Morges: IUCN Publication, New Series.
Ministry of Environmental Affairs and Tourism. (1998). White Paper on environmental management policy for South Africa, *Government Gazette*, 395, 18894, notice 794, 15 may 1998, Pretoria.
MMA (Ministério do Meio Ambiente). (2013). Governo investe em parques nacionais das cinco regiões do País, in http://www.brasil.gov.br/meio-ambiente/2013/11/governo-investe-em-parques-nacionais-das-cinco-regioes-do-pais.
MMA. (1997). Programa Nacional de Educação Ambiental (PRONEA). Brasília: MMA.
MMA. (2011). Quarto relatório nacional para a convenção sobre diversidade biológica: Brasil/ Ministério do Meio Ambiente. Brasília: MMA.
PRONEA. (2003 and 2005). Ministério do Meio Ambiente, Diretoria de Educação Ambiental; Ministério da Educação. Coordenação Geral de Educação Ambiental. Brasília: Ministério do Meio Ambiente.
UN-Habitat. (2008). State of the world's cities 2008/2009. London: Earthscan.
UNESCO-UNEP. (1988). Sustainable development *via* environmental education, *Connect Environmental education newsletter*, 13, 2.
UNESCO. (1997). Éduquer pour un avenir viable: Une vision transdisciplinaire pour l'action concertée. Conférence internationale Environnement et Société, Thessalonique. 8–12 Dec, 1997.
UNESCO. (2002). *Éducation pour un avenir viable: enseignements tirés d'une décennie de travaux depuis Rio de Janeiro jusqu'à Johannesburg*. Paris: UNESCO.
UNESCO. (1992). *Refonte de l'éducation pour un développement durable—Reshaping Education for Sustainable Development*, Dossiers Environnement et Développement. Paris: UNESCO.

Part IV
To Conclude

Chapter 13
Conclusion: National Parks Between Urbanisation and Globalisation

Frédéric Landy

As this volume draws to an end, How can we summarise our findings? Our main conclusion is not only that cities should think more about their national parks, in that they represent a fantastic asset for local urban management and global competition, but also and reciprocally, that parks should negotiate and coordinate more with the cities around them: even if they want to pursue a policy of "fortress" isolation, they still need cooperation from the urban sphere. The alternative can spell disaster, as was the case for a while with Sanjay Gandhi National Park (SGNP) in Mumbai.

Our study has been confined to urban national parks, parks within or else close to cities. The reader might think that this is a very unusual situation. Not at all. First, remember that there are almost a hundred urban national parks across the planet (e.g., Caracas, Los Angeles, Hong Kong, etc.), so there is nothing rural about many national parks, and more big cities than one might imagine have a national park. Moreover, we hope that we have convinced readers that studying urban national parks opens up questions that are in fact very general, for example what the (big) city is or ought to be, and what (the conservation of) nature is or ought to be. We have argued for the breaking down of conceptual, political and managerial barriers that may exist between park and city, in order to engender what we call "Naturbanity"—a hybrid neologism that expresses this need for symbiosis.[1] The fact that parks and cities must and will live

[1]This neologism has already been used by Gaulier (2013), an architect and landscape designer who argues for the creation or rehabilitation of urban landscapes with sustainability and conservation agendas.

F. Landy (✉)
University of Paris-Nanterre, Nanterre, France
e-mail: frederic.landy@ifpindia.org

F. Landy
French Institute of Pondicherry, Pondicherry, India

F. Landy (ed.), *From Urban National Parks to Natured Cities in the Global South*,
https://doi.org/10.1007/978-981-10-8462-1_13

together, for better or for worse, should first be considered *empirically*: challenges such as informal settlements encroaching on parks, or industrial pollution, may be offset, for example, by obtaining support from urban "civil society" for the funding of parks, or by making political capital from the awareness of the ecosystem services the park provides to the city. Through the organisation of several international conferences and intensive interactions, our project has helped to intensify and even create interactions between actors which had previously ignored each other. Our perspective must also be considered *philosophically*, since it blurs the "modern" dichotomy between nature and culture: animals and human beings can often jump the physical and ideological barriers separating parks from adjacent cities.

Box 13.1: Urban National Park or National Urban Park? (Monique Poulot)
Although urban national parks appear to be associated with the emergence of certain countries in the Global South, there is also a strong growth of this type of park in the Global North with some hybridisations resulting from meetings between park managers and dissemination of good practices. For the past two decades, the urban national park has thus engendered a discourse and a rhetoric that is on the point of becoming a model, with a strong tendency in the Global North in favour of a "*national urban* park" configuration.

Many urban national parks—often without the IUCN label—originated in the United States in the early 1970s, most notably as part of the parks-to-people initiative, to meet the growing demand of the urban poor and socially marginalised populations to access green spaces (Byrne and Wolch 2009). Two parks now seem emblematic of this movement because of the pre-eminent role played by their countries in the history of national parks and because of their park management skills: Djurgaden Park in Stockholm, Sweden, created in 1995, and Rouge Park in Toronto, Canada, created in 2012. Both parks were the result of lobbying carried out by associations against highway or industrial infrastructure projects. The proponents of each park had claimed that, in addition to protecting ecosystems, its fundamental benefit would be to provide proximity and accessibility to all users of the metropolis (residents and tourists) to a vast managed natural area that would become a national heritage.

Djurgaden, formerly an old royal park, now welcomes 15 million visitors annually. It is spread over an area of 27 km^2 and encompasses, in addition to natural spaces, many attractions such as the Grona Lund Amusement Park, the Vasa Museum and the Skansen Open-Air Museum. A true "mirror of Swedish society and Nordicity" (Girault 2017), its official name of Royal National City Park testifies to the transition from a national park to a national *urban* park. The "national" character manifests in several ways: in addition to the strong Nordic natural elements (e.g., marshes, fjords, etc.), the *mise en abyme* within the park of the Skansen Open-Air Museum, which showcases the whole of Sweden, reminds us of the rural heritage of the country and pays tribute to its past. The history museums also do the same, affirming in this way the national scope of the entire park. Finally, the presence of some famous places associated with the Sami—the Swedish indigenous peoples—also satisfies the national need for making reparations for past injustices. But all these elements find themselves now in an urban environment, enclaved within the city, with the national park and the city merging seamlessly one into the other. And it is up to the latter to ensure the protection, management and transmission of the former, in a rationale of openness and a desire to maintain a quasi-public space in the name of the urbanity that defines it.

As for Toronto's Rouge National Urban Park—note the specific order of the two adjectives in its official name—it relies on the same rhetoric around a threefold reconciliation that places the city at the centre: one between city and nature in an opposite of the first generation national parks; another between city and park with ongoing transactions between these two entities to shape—without stopping altogether—urbanisation in and around the park and to invent a model of shared governance between urban authorities and Park Canada; and the third between the urban population—city inhabitants or tourists—and the park, since the latter is above all designed to be an amenity.

This journey from national park to urban national park to national urban park must be viewed in the light of an urban expansion that leads Sieverts to refer to a "city of tomorrow … constituted by a concentration of compact urbanised areas and a set of fragments of included or encompassing landscapes which each fulfill specific urban functions" (Sieverts 2004, 26). It draws also from the changes of "regimes of historicity" in the terminology of historian François Hartog, the contemporary regime being that of "presentism". In this acceptation, the present has become the horizon without a future and without a past; it generates, on a day-to-day basis, the past and the future that it needs, day after day, and derives value from the here and now (Hartog 2016). The memory-nature of national parks that has to be protected against the trauma of destruction (Thoreau 1864; Muir 1901; Shama 1995) is replaced by the city-nature-heritage of the national urban parks, to be protected for the sake of the city and the urban inhabitants of today.

Byrne, J., Wolch, J. (2009). Nature, race and parks: Past research and future direction of geographic research. *Progress in Human geography, 33*(6), 743–766.

Girault, C. (2017). La *naturbanité,* une utopie réalisée par le Parc national urbain de Stockholm? *Nordiques, 33*, 61–78.

Hartog, F. (2016). *Regimes of Historicity. Presentism and Experiences of Time.* New York: Columbia University Press.

Muir J. (1901). *Our national parks.* Boston: Houghton, Mifflin.

Shama, S. (1995). *Landscape and memory.* New York: Alfred A. Knopf.

Sieverts, T. (2004). *Entre-ville: une lecture de la Zwischenstadt.* Marseille: Editions Parenthèses, coll. Eupalinos.

Thoreau, H. D. (1864). *The maine woods.* Boston: Ticknor and Fields.

It will be enlightening, in concluding this book, to end as we began: to return to the commonalities but also the factors of difference between our four parks, which we will tackle through a framework with two dimensions: "horizontal" (the city-park relation) and "vertical" (the local-global relation). The first part will focus on what our four sites have in common. The second part will deal with the question of flows and fences, a burning issue for our four locations: it is essential for parks that want to protect themselves from the city while gaining from its proximity, while in return the city benefits from environmental services but can also experience problems emanating from the park. Both park managers and city managers may perhaps be able to learn something from these pages.

Nonetheless, the differences between our sites are more than just details: the third part will therefore tackle these differences, and the factors that explain them, from disparities in size to disparities in policies. In the end, this will enable us to propose a number of typologies—classifications which, once discussed and explored, might subsequently lead to the implementation of differential adapted policies.

13.1 The Commonalities

These common points are questions, questionings, challenges that re-emerge in the same form in all four configurations. It is only the responses and solutions that sometimes differ. These questions can be classified into two groups, according to whether they correspond to "horizontal" or "vertical" tensions.

13.1.1 Horizontal Tensions: Urbanisation of the Park and Naturalisation of the City

Our Chap. 8 focused quite closely on this dimension, given that the tension (conflict or hybridisation?) between the two frontiers, the urban and the eco, primarily occurs in this plane. That chapter showed that making a stark opposition between city and park is too reductive, given that they are not sealed worlds. Even in India, a country where segmentation is the rule, this permeability is reflected institutionally in the fact that the Forest Department's Wildlife Division is authorised to operate outside protected areas in order to settle wildlife problems wherever they occur (notably in the city).

Nonetheless, the fact is that there are often sharp tensions between spaces that—whether we like it or not—remain distinct. Often, it is true, the reason for the creation of the protected area was to secure the city's water supply (Chap. 1): here, the protected area directly served the city and this urban dimension was almost its sole purpose. Everything changed, firstly with the nationalisation of protection, and secondly with the expansion of urban populations and spaces. A first point that our terrains have in common is therefore *urban pressure*. Population densities and sizes, which may be colossal in megacities (our four sites are home to almost 40 million people in total), industrialisation, real estate demand and pollution, are all factors that threaten the parks. Just in terms of communication routes, a growing presence in these cities, remember that Nairobi National Park (NNP) is bounded by the road and railway line linking Kenya's two principal cities; similarly, the Mumbai park is bounded by national highway 8, which links the country's economic capital to India's most industrialised state. Only Table Mountain National Park (TMNP) would seem to be somewhat more protected by its location at the end of a peninsula, at least in its southern part.

Urban pressure is also *social, not to say political, pressure*: pressure of poverty when slums are established near, or even in, the park (notably in Mumbai), but also pressure from the rich (cf. Chap. 3). "People have lawyers…" as the director of TMNP told us. "And whenever their dogs are found and fined in the park, their lawyers intervene!" In Nairobi as in Mumbai, the park services must sometimes intervene even when called out to deal with inoffensive animals, when the people doing the calling are socially or politically highly placed. Media reactions—whether mainstream or on social networks—are often disproportionate when incidents occur in these often highly "connected" big cities. Heads of state and other VIPs are very frequent visitors (except in Mumbai), which raises the international profile of the parks and thereby places heavy pressure on their managers (Chap. 11), as well as on urban and political leaders.

One unusual but very evocative situation is when political measures threaten the national park in pursuit of reparations that are in principle legitimate. Under India's Forest Rights Act, indigenous populations (or any farmer whose family has been there for generations) have the right to receive four hectares of land: this poses a big threat to Mumbai's national park, and consequently, to the Adivasis, some of whose villages in the park have up to now been implicitly tolerated; indeed, there is a suspicion that these "tribals" want to exploit this law to obtain fragments of the national park. The same problem is found in Rio de Janeiro, where collective ownership rights are recognised for the *quilombos* (communities previously formed by escaping slaves). A specificity of the State of Rio, a right of use can be granted to any farmer present for more than 50 years. In South Africa, finally, anyone expelled during apartheid was able to reclaim their land, even in a national park, though in reality they usually received financial compensation instead.

Finally, urban pressure falls on the *landscape*. Urban visitors want more tracks, or even roads, running through the parks. In Cape Town, the Cecilia and Tokai pine plantations had ceased to be profitable, and in 2000 the decision was made to cut them down to make way for the different kinds of endemic *fynbos* shrubs. This was an opportunity to generate biodiversity connections by linking lowland (Cape Flats Sand) *fynbos* to highland (Peninsula Granite) *fynbos*. However, controversy arose over this "tree felling madness": recreation needs shade, walkers wanted to keep their pines and objected to this environmentalist fundamentalism. In the end, a "middle path" was found for the local forests of Tokai: tree-lined alleys and shaded recreation areas would be retained, but now there is a biodiversity corridor linking the highland and lowland *fynbos*.[2]

However, we need to be careful not to see these processes as unilateral. The park also exercises pressure on the city, which in turn finds itself "naturalised". In Mumbai or in Rio, the existence of the park forces the authorities to develop rehousing plans for slum dwellers, even if this means constructing high buildings in previously non-constructible areas like Chandivali south of SGNP. Pressure is brought to close polluting industries. Indeed, environmental education, practiced because of the park, can itself be seen as a form of pressure exercised by the park on the city if it prompts a section of civil society to want a "greener" city (Chaps. 8 and 12).

The question then arises of the place of the park in urban policy, whether in terms of spatial management or urban branding. On the one hand, the city may see it as rewarding to emphasise the park in its policies. On the other hand, the task is far from easy: we have seen (cf. Introduction) how much adjectives count, and how it is difficult for a "national" park to be "local". While international rankings of "global cities" include indicators for environmental policies or the ratio of green space per inhabitant, not one of these rankings allows for the presence of a national park! Sustainable urban development rarely considers lions or leopards.

13.1.2 Vertical Tensions: Different Degrees of "Globalisation"

National parks (and also cities) are embedded in several scales, which overlap to varying degrees. These spaces of course have a local dimension, one of their official boundaries, but also a regional scale, especially when the country has a federal constitution and state or regional government plays an important role. Next, the national scale is important: obviously for a "national" park, but also for the city, and not only when it is the national capital: just as the park is part of the countrywide network of protected spaces, the city too is bound into an interrelation with the rest of the national urban network. And finally, there is the global scale: biodiversity objectives, pursuit of the status of a global/world city, etc., but also international partnerships between cities or between parks, like that between Tijuca and Table Mountain. Ultimately a "national" park needs to fulfil purposes that are not simply "national"—and this is a big part of the problem.

---

[2]Michael Slayen, "Tokai Plantation: Transition to an urban National Park", paper for UNPEC *BiodiverCities* Conference, *Rescaling Natural Parks and the City*, Cape Town, Apr 7–9, 2014.

We therefore have different degrees of "globalisation", understood as the expansion of a vertical dimension (from the local to the global), with both national parks and cities increasingly becoming global brands, often in parallel with the same process as the country's projection into the international arena, in order to attract investment, trade, tourists and conferences. This leads to the rescaling (Brenner 2009) of urban national parks as well as cities themselves: scales are not "pregiven", but "produced, contested, and therefore malleable arenas" (Brenner 2009, 63), "shaped by institutional arrangements, political strategies, and social forces" (ibid., 62). Parks and cities need to be able to play different scores between scales, and this malleability requires learning and resources.

However, not all our sites have achieved the same degree of globalisation. As the table in the introduction showed, Mumbai is proportionally much less globalised than Cape Town or Rio de Janeiro: the city attracts relatively few foreign tourists for international events, for example, and the majority of its activities and policies remain focused on national issues. Moreover, we should not be content to reason in terms of "sites", of a paired park-city identity—park and city do not necessarily share the same degree of globalisation. So the city of Mumbai, relatively introverted as it is, plays a much bigger role in the global arena than its park, which is primarily embedded in a local and regional scale.

The fact that a "national urban" or "national local" park is still considered an oxymoron shows the complexity of the interplay of scales. Of what advantage is it to Nairobians that "their" park is a strategic component of attempts by the Kenya Wildlife Service (KWS), and more fundamentally by international environmental organisations, to save the black rhino? For many inhabitants, the positive impacts are non-existent. Urban parks undoubtedly provide environmental services (e.g., microclimates, water reserves, leisure, etc.) but also many "disservices", primarily arising from the immobilisation of large land areas, which can be considered as "lost" to important urban imperatives such as housing or decongestion measures. Urban decision makers, but also civil society—city populations—are not necessarily ready to espouse the objectives and *raison d'être* of a "national" park. The legitimacy of national parks in urban locations is open to greater challenge than their rural equivalents, since the opposing actors and interests are potentially more numerous—and more powerful. It is therefore up to decision makers and environmental groups to turn this problem into an advantage: a national park in an urban environment also has more potential defenders than its rural counterpart. "If it is seen as a neighborhood park, it is loved locally, it increases commitment for protecting it".[3] The "pro-nature" lobby is undoubtedly disparate, ranging from deep ecologists to environmental civil servants to park users, but it can be very powerful, politically, financially and socially.

---

[3]Uma Adusumili, "The Mumbai debate", UNPEC BiodiverCities Conference, Rescaling Natural Parks and the City, Cape Town, Apr 7–9, 2014.

The relations between scales are also strongly affected by electoral political configurations. In Cape Town, or in Mumbai until 2014, the party of municipal government was not the same as that of provincial or federal state government. So Cape Town municipality accused the ANC of playing a double game: claiming to want to develop the country's third city, while at the same time putting obstacles in the way of the municipality's ruling Democratic Alliance (DA) party.

This political element is also clear if we recognise that the vertical dimension runs not only from the local to the global, but can also operate downwards, from the local to the micro-local. We then see that there is an indigenous temple in Mumbai's park, legacies of apartheid in the fragmentation of Cape Town's park or in the graveyards in Kruger Park, etc. Everywhere the history written by ordinary people has left its mark on national parks that cannot be left solely to biodiversity managers.

13.1.3 Connection Between the Horizontal and Vertical Dimensions

Our four sites can therefore be compared along two planes. On the one hand the horizontal plane, with highly variable levels of integration linking park and city, which in the best cases can make them not only compatible but also complementary —a situation that ecologists studying the relations between two species call "mutual symbiosis", and that in everyday language we now call "a win-win situation". And on the other hand, the vertical dimension, with differences of position in the globalisation process. Note that these two dimensions can influence each other. For example, the horizontal dimension of city-park integration may have an impact on the role of the urban agglomeration in the global arena: integrated functioning can give the city a green image, raising its international profile. Conversely, the vertical dimension may impact the city-park relationships: UNESCO guidelines, private investment by multinational companies, WWF and other powerful environmental NGOs, may have a big local impact. Yet our study found that this factor should not be overestimated, even in Cape Town and Rio, and that international factors are often used as scapegoats, baits or screens, as "lines to take" in official or unofficial discourses, rather than having a genuine impact. We should not overestimate the "greenwashing" of the global image of the city, or the role of global networks such as ICLEI (International Council for Local Environmental Initiatives), as against purely local priorities and factors. What is certain is that while our four cities are all to varying degrees globally connected, not all of them know how, or wish to, deploy this global scale. Mumbai's presence in international arenas is very low-key, whereas Cape Town has always been effective in mobilising multiple scales, and in attracting resources from outside the local sphere.

Bringing together the horizontal and vertical dimensions will be the aim of the end of this chapter, with recapitulating typologies. But perhaps we are already in a position to look at the very empirical consequences: the fences built are never impassable, just as the flows between spaces are rarely completely fluid.

13.2 Fences and Flows: Physical Barriers Versus Empirical Connections

Some protected urban areas are demarcated by fences and walls, while the boundaries of others are left entirely open. Yet people, wildlife (Chap. 7), knowledge, or values (Chap. 6) are often indifferent to the presence or absence of such barriers, and they flow (or not) in accordance with other logics. Rather than insisting on erecting impermeable barriers at any price, it would therefore be preferable in many cases to accept this porosity as inevitable, and turn it into an asset rather than a weakness. The dichotomy between fences and flows needs to be overcome in order to achieve the hybridisation referred to in Chap. 8. At first glance, the urban national park would seem to represent two opposing trends: the trend towards spatial closure typical of national parks, which are areas of protection *against* different dangers (e.g., indigenous populations, deterioration, agriculture, city, etc.), by contrast with the city, often considered as a space of co-presence, trade, movement and flows of all kinds. However, our approach has shown that the opposite can be true: segmentation, barriers, segregation in the city (and not only in Cape Town!); in the parks, flows—of flora, of animals, of visitors, of movement legal or illicit, and sometimes officially encouraged ("A park for all", announces SANparks). The opposition between fences and flows is therefore, at the very least, dialectical: too many flows can encourage fences (e.g., leopards and slums upset the managers of Mumbai park), whereas too many fences often prove untenable and ultimately give passage to larger flows, which might have been controlled by more open management.

13.2.1 *The Temptation of Enclosure*

The legacy of the "fortress-park" model remains hard to overcome for many national parks. It took a long time for governance models to change (Chap. 10), challenging forms of territorial exclusion and encouraging a more integrated governance that takes account of stakeholders and spaces other than those immediately affecting the park. Nonetheless, the very fact that permanent human presence is forbidden in much national park legislation encourages strong controls on the boundaries of the protected spaces, and even physical barriers. So in Mumbai, almost the whole of SGNP is officially surrounded by a wall, whereas two thirds of NNP is enclosed by a wall and/or an electric fence—only its southern boundary is open to allow wildlife migration.

The effect of urban pressure is further to legitimise this exclusion:

- when visitors are required to pay, access must be restricted to just a few points, otherwise the permeability of the boundaries in an urban environment will allow people to bypass the toll gates (which is harder in rural parks where there is often only a single, easily monitored, access road);
- above all, in response to encroachments of all kinds, the fence is there to give physical form to land boundaries that would otherwise exist only on paper. (And too bad if, as in Mumbai, these boundaries are disputed, given that some of them were drawn in the midst of legal chaos during the 1976 State of Emergency). The fence protects the park from the city;
- however, the fence also protects the city from the park, theoretically preventing the big mammals of Nairobi or Mumbai from getting out. Indeed, these animals are simultaneously a threat (potentially dangerous carnivores) and a treasure to be protected (the rhinoceroses of Nairobi). It is no accident that SGNP is India's only fenced national park, just as, in Kenya, the urban national parks (Nairobi, Nakuru) are all fenced (Chap. 11). In Cape Town, electric fences protect picnickers from the baboons. Tijuca National Park (TNP) and TMNP, on the other hand, have only very localised fencing: this is because they have hardly any large mammals, with the exception of the Cape baboons; but it is also perhaps (both cause and consequence?) because their mode of governance, which favours flows, is much more integrated and open than the mode that dominates in Mumbai and Nairobi.

13.2.2 And Yet, Flows of All Kinds!

Incoming and outgoing flows are of all types.

A. Fauna and flora:

  - Incoming: many alleged "invasive" plants penetrate the parks because of their location in very large cities, near seaports or airports. The very large numbers of people visiting urban national parks (4 million a year at Table Mountain) exacerbate the phenomenon. As does the proximity of millions of city dwellers: for example, the inhabitants of Cape Peninsula cultivate numerous exotic species, which then tend to spread.
  - Outgoing: the carnivores of Nairobi and Mumbai can terrorise the population when they encroach into the city, and it would seem that leopard territories extend outside these two national parks. In Cape Town, however, a much less frightening leopard, the endemic "western leopard toad", is preserved through satellite tracking; its migration routes are protected by roadblocks and flashing lights; a committee instigated by the South African National Biodiversity Institute encourages the population to support the campaign, for example by growing plants that attract the toads.

B. Human flows:

- Legal: tourists or local inhabitants enter on foot, by car (compulsory in Nairobi), train (Rio), or cable car (Cape Town). In Mumbai, local people can get annual passes (for morning yoga), a possibility also available in Nairobi to executives who drive through the park to avoid the morning traffic jams.
- Illegal: poaching can be a source of significant rewards (a Nairobi rhinoceros is worth hundreds of thousands of dollars). More subtle, but perhaps equally destructive, is the effect of local strollers secretly entering essentially open areas of Mumbai's park. After nightfall, Maasai pastoralists sometimes pasture their herds in NNP—this is illegal, but many managers close their eyes to it in order to maintain good relationships with Maasai leaders, and some ecologists believe that this helps to regenerate the grasslands.

C. Mental and institutional flows:
Whereas Mumbai and Nairobi are characterised by a fairly clear institutional divide between city and park, in Rio and in Cape Town there is a degree of hybridisation: for example, Tijuca consultative committee brings together stakeholders from different origins. In Cape Town, this hybridisation is both spatial (with the existence of numerous reserves, some municipal, some private conservancies, inside the urban area itself) and institutional: it is not SANParks, but the City of Cape Town itself that possesses a Biodiversity Department, a Parks Department, a Watercourses Department, skills that act as bridges encouraging coordination with the administration of the national park.

13.2.3 So What Should Be Done?

Fences are of limited use in preventing the movement of flora and even of fauna. Leopards in Mumbai climb 3-m-high walls, baboons dig under the fences of Cape Town's park, lions escape though the conduits of Nairobi's park. And sometimes the power fails in the electric fences.

So three positions are to be recommended.

A. Of course, fences should be retained if necessary, if for no other reason than to provide pay points in the cases of Nairobi and Mumbai.
B. But these fences must also be as smart as possible. The case of the GreenLine project, in Nairobi, is a good example. Forming a second boundary inside the park, it undoubtedly consumes some of the area available for wildlife: however, the resulting corridor, with its plantings of trees, open for use by some of Nairobi's population for picnics or jogging, is a form of social fencing that could enhance citizen acceptance while reinforcing the barrier: a win-win in terms both of efficient wildlife protection and equity in social progress. It would also be desirable to do more to involve the Maasai and the Adivasis in the

surveillance and management of the parks in Nairobi and Mumbai, and for these indigenous populations to be seen as contributors to conservation policy (to control poaching or fires by their presence) rather than as predators. More generally, with all the local park defense groups (FoNNaP in Nairobi, Mumbaikars for SGNP in Mumbai, etc.), "civil society" as a whole could contribute to this social fencing. Ultimately, supported by better environmental education for the whole population, "environmentality" (Agrawal 2005), a shared environmental awareness, could come to represent both the most effective and equitable fencing.

C. Nonetheless, the permeability of fences—the impossibility of making them secure against all inward and outward flows—needs to be accepted. Just as the management of the park needs to recognise the presence of the city, rather than pretending that it does not exist, it is better that this permeability should be taken into account and put to use as far as possible, rather than denied.

   - one example is the development of the leopard awareness campaigns conducted in Mumbai by environmental activists, aimed both at slum dwellers and the middle classes (the Adivasis need them less, because they know how to behave with wild animals). The purpose is to protect human beings, whatever their social and residential status. However, the purpose is also indirectly to protect the animals, given that fewer leopard attacks mean a lower risk of reprisal.

Another goal is to strengthen spatial connectivities:

- connections between park and city: there should be regular buses running from the city to the park. The city should be required to maintain environmental standards in an intermediate zone around the park: given how difficult it is to define this buffer zone, why not define it empirically, for example in Mumbai as the area where leopards penetrate? Leopards are "good cartographers" (Landy 2017), because they know how to represent the real fluidity of urbanisation, beyond the rigidity of maps and urban planning statistics.
- ecological corridors: the real or potential connections between "natural" "green" areas are crucial for species reproduction. In some cases, decision makers have shown long-term and systematic commitment to this (Mosaica Carioca in Rio and Source to Sea river corridors in Cape Town). Sometimes, however, the task is difficult because of the difference in status between protected areas (national park vs. municipal parks), and on many sites urban pressure makes it very difficult. In particular, the development of transit routes, admittedly often very inadequate in the past, is introducing more and more fractures: the Tungareshwar Wildlife Sanctuary is becoming increasingly isolated from SGNP by railway lines, and railway lines are going to cut NNP in half.

It is something of a paradox that ecologically justified barriers are sometimes transgressed, whereas others that make no sense end up being maintained. Permeability is not necessarily a disadvantage, but often an asset, though one that is difficult to protect.

13.3 The Factors that Differentiate Landscapes, Management Systems and Trajectories

Beyond the questions and issues common to our four sites, what are their differences? Each of our park-city pairings represents a *configuration*, a spatial system. This configuration consists of the park itself, i.e. its size, its biodiversity, etc., but also of its environment: the geographical environment (horizontal and vertical, as we have seen), obviously marked by the presence of the city, but also the historical, political, etc., environment. These configurations might seem to correspond to the "socio-spatial formations" previously defined by di Méo (1985), if it were not for the limitations in the dimensions of identity and territory: which Mumbaites really identify with their national park? Are Nairobians really attached to NNP?

We believe that in order to complete this book and to understand the reason for the diversity in the cases analysed, we need to go back to these configurations, in which some components more than others explain the disparities encountered, whether in the landscapes, the processes, or the policies. Let us look successively at the factors of differentiation in the national contexts of each of the countries, then at those relating to each park, and lastly, those relating to each city.

13.3.1 Differences in the National Contexts

- **The national political and legislative context:**

This may vary in its concern for environmental protection, and the approach to this protection may vary in its degree of social inclusivity. In this, for example, we see a contrast between South Africa and India. In Cape Town, and in South Africa in general, decision makers often place emphasis on both environmental and social policies. Since 1994, social and racial inclusion has been a stated priority, and strong environmental awareness has led to the implementation of a fairly strict policy of environmental assessment. Of course, it can be said that this is not always reflected in the facts, which in reality leave a great deal of wriggle room for social elites and economic and financial priorities. It is often a matter of lip service (Chap. 12). Fair enough. But it should be noted that even this lip service is not present in Mumbai, where SGNP is not a "park for all", where national integration is no more than a relic of Nehru's time, where the system of Public Interest Litigations is primarily used by a social elite, on ostensibly environmental grounds, to institute legal proceedings designed to exclude poor populations, and where environmental regulations are overtly bypassed by the most powerful economic actors.

– **Social divisions: living standards, cultural levels and "modes of identification"**

> As for the love of nature, there is not much of it in the slums. What people there need is money
> (Albert Cohen, *Belle du seigneur*, Paris: Gallimard, 1968, p. 307).

The societies of the four countries studied are characterised by sharp divisions into social hierarchies where disparities of wealth combine with those of race, ethnicity, or caste. The societies of the so-called "emerging" economies are increasingly unequal, with well-off social classes developing alongside persistent high levels of poverty. Although this process of emergence would now seem to be slowing (Brazil, Kenya), a social elite continues to maintain much of its privileged status. These social inequalities are reflected in often quite disparate practices and ideologies relating to "nature" (Chaps. 3–5). The "national park" is a concept that was largely invented in the United States and often introduced by colonial powers: the reason for its creation and maintenance is generally well understood by largely westernised social elites, much less so by poor populations.

Chapters 3 to 5 nevertheless showed that we need to be wary of strict social determinism. In India, the "modernity" that presided over the creation of national parks around the world and reinforced the separation between nature and culture, continues to be widely shared by the lower middle classes and by Forest Department officials. Visitors to Mumbai's park are often more attracted by the toy train or the tiger safari bus that runs through a small animal park, than by the forest itself, a wild place that either frightens them or repels them as a symbol of poverty and backwardness. Conversely, certain westernised ecologists from the upper social classes often unknowingly share with the indigenous populations of the park (whom they often reject both politically and socially) the same belief in an ecosystem of which human beings are just one part.

That is why, according to Guha and Martinez-Alier (1996), the fact that indigenous or disadvantaged populations protest against protected areas inspired by Western perspectives, in no way means that they are not environmentalists. This is a *political* conflict for the control of natural resources. For such populations, "these are not wild areas that they threaten allegedly by their presence, but their world" (Hache 2012, 26). That having been said, while Guha and Martinez are right to write about and rehabilitate the (other) environmentalisms of the South, they are probably wrong to contrast them with the North. When Guha, in an admittedly fairly old article (1989), contrasts American deep ecology with the ecology of the Chipko movement in India, for example, he underestimates the power of what were to become the Indian environmental movements originating in well-off, often "anti-poor" civil society, which were ultimately fairly close to wilderness ecology. With the country's economic emergence, full-belly ecologists have become fairly widespread in India as well.

A path that we have not been able to follow for a long time is that of the links between nature and culture, according to what Descola (2013) calls modes of identification. Could these sociocultural differences between various relationships with nature reflect more deeply the "ontologies" of each civilisation, the differences between natives and park managers being in fact explained by the fact that the latter are very Westernised? They would then have adhered to a "naturalistic" mode of identification, to use the typology of Descola, which tends to oppose humans and nonhumans, and thus to make the protected area an enclosed space devoted to "nature". Whereas the animistic indigenous people would oppose the very notion of a protected area because they do not distinguish between humans and nonhumans. The hypothesis is seductive. However, for our case study it is not necessary to invoke Descola to go "beyond nature and culture", the title of his book, since a national park is a well-maintained, humanised space with roads, tourism, animals introduced or captured, plantations, etc., which can sometimes bring it closer to a garden and to the sphere of "culture". It should also be noted that one of the four countries studied in this book, India, is in the typology of Descola placed in another mode of identification, that of "analogism": according to this mode, both humans and nonhumans are characterised by certain traits. For Hindus and most other Indians, correspondences do exist between elements (e.g., water, fire, air, etc.), and human faculties, soils, celestial signs, etc.: this also tends to break the nature-culture dichotomy of modernity.

How do we situate the respective weight of all these ontologies in contemporary India? This is something that went beyond the capacities of the UNPEC project, and the research team was divided on this point. To take only the case of Mumbai, some of us thought that if there were so many faults in the park fortress, if the leopards could go out and the slums come in, it was because the managers of the national park were not in a "naturalistic" but an "analogical" mode of identification. They had "never been modern", to paraphrase Latour (1993). For other researchers, however, these loopholes were not due to an ontology associating humans and nonhumans: the foresters are trained in a very "naturalist" school of thought, and for them the protected areas remain the places ideal for conserving nature in a variety of isolates. If the faults in the fortress exist, it is simply due to very material reasons: enormous housing crisis, greedy politicians, corrupt officials, etc. Who is right, who is wrong? An exciting issue for a new research project.

In any case, we need to avoid any risk of cultural racism, but also essentialist caricatures of the values of the peoples of the South, or of "Indians", or of "indigenous peoples", etc. Hence we should adopt an empirical approach that takes into account the history and the geography of the park-city pairing, and in particular the past or present power relations between social groups. Supposedly "culturally determined" practices relating to access to the park are in fact often explained by the history of that park—reflecting colonial domination to one degree or another—and by urban social history, with its varying levels of segregation and inequality (Byrne 2012). To give a simple example, the reason why the inhabitants of Khayelitsha township rarely visit TMNP is primarily that they live a long way from it, that transport is expensive, that they would need a car, and that they would have to get home before dark for fear—so they imagine—of being murdered! And the reasons why the native populations of Mumbai's park are such avid worshippers of the tiger god Waghoba are perhaps as much political as cultural: "What these communities lack in political and material resources in modern discourses, they make up with their non-modern systems to negotiate with leopards directly. Thus, the persistence of Waghoba could be interpreted as a survival strategy in the context of the weaker position occupied by these communities vis-à-vis the state" (Ghosal 2013, 120).

- **The place of the park in sometimes antagonistic scales of political power:**

The electoral victories of the BJP party in India, at the levels of Mumbai municipality, the State of Maharashtra and the federal Government, should in principle make the relations between levels of power less conflictual, given that there can be said to be an "alignment of political planets". This is not the case in Cape Town. Not only are city and province held by opposition parties to the federal authority, but also TMNP has only recently become profitable (Rodary, personal communication): earnings from it go into the revenues of SANParks in Pretoria, as part of the policy of redistributing revenues from the profitable parks—Kruger and TMNP—to the money-losing parks, but they do not go to the local institutions. The tensions between the City of Cape Town and SANParks thus coincide with political opposition between the DA and ANC parties and spatial opposition between the municipal and national levels. Similarly, it remains to be seen whether, for Nairobi, the growing power of the counties will not upset the status quo (Chap. 9), since Kajiado county manages the whole of the dispersal area outside the park, with the result that the park depends on it ecologically, while being administratively located in the county of Nairobi.

13.3.2 Differences Relating to the City

- *Urban size and pressures:*

Mumbai agglomeration (population 20 million) is five times larger than Nairobi or Cape Town. One might therefore expect greater challenges in the Indian case because of the absolute population figure. Yet this remains to be demonstrated, especially in the case of NNP, which is currently under such threat. A conurbation can expand in directions other than its national park, and a large population does not necessarily imply large numbers of potential visitors if the park is away from the centre or attracts few outside visitors. On the other hand, more than the absolute number of inhabitants, the speed and nature of urban growth are undoubtedly decisive. Cape Town grew by 2.6% between 2001 and 2011, compared with 12.1% for Mumbai Urban Agglomeration: not only is the percentage higher in Mumbai, but the type of growth there is particularly damaging, given the catastrophic state of housing and infrastructure, and the scarcity of effective regulation.

One point that our cities share is that the sometimes desperate attempts to establish buffer zones around their national parks have come too late to establish a sort of protective veneer. Urban growth has already put paid to this. This is regulation coming "from the park", imposed on the city, within the framework of what we have called an "ecological front" (Chap. 8). The question is, "how to apply regulations in the city that emanate from the protected area? In urban conditions, this means that there must be complementarity between the plan for the management of the protected area, the urban master plan and the other documents relating to land rights, despite the lengthy process of cultural change and of its translation into legal and legislative texts" (Bruno et al. 2016). So cities may be justified in their frequent claim that such regulation is impossible to apply as it stands. That having been said, significant differences exist: in Cape Town, it has not been possible to establish a formal buffer zone, but regulations, such as fire prevention measures, are enforcing an area that could give rise to one. In Mumbai, some maps show a buffer zone inside SGNP itself, but no such zone actually exists. Another buffer zone, this time outside the boundaries of the park, should stretch 10 km beyond (as with all the other national parks in India), but in Mumbai's case, it would include the airport, the two ports, and the sea itself. In Rio, the buffer zone also extends 10 km and therefore faces the same problem as in Mumbai, because of the proximity of the sea; nonetheless, it is more of a reality, although its details have not been legally established and there are differences between federal laws and the laws of the State of Rio. A second buffer zone here is set by altitude: construction is not permitted on areas above 100 m, which protects TNP. Finally, in Nairobi, the erstwhile buffer zones were handed over (notably in the time of President Moi) to the industries closest to government. The county council is trying to establish minimum plot sizes, but this is at least as much to maintain pastures for the Maasai herds as to protect the national park.

– *Importance of socio-spatial differences:*

By contrast, it would appear that our four cities are on a roughly equal footing in this respect, all marked by a legacy of social segregation which may be further reinforced today in certain urban policies. Socioeconomic differences are reflected in spatial differences between neighbourhoods, and in return these spatial differences tend to reinforce socioeconomic differences, with the effects of distance from the centre or of stigmatisation (Saglio-Yatzimirsky and Landy 2014). It is therefore only at close-grained neighbourhood scale that sharp differences can be found, not between our four cities but between the sub-spaces that border each park. A rich white neighbourhood will often make different use of the park than a poor neighbourhood. The question is why: we have seen how cultural and educational factors play a less significant role than cost factors or physical distance.

This opposition between rich and poor needs to be overcome, and this is what our Chaps. 3 and 4 have sought to do. Nonetheless, many urban plans, and indeed urban policies, show how this polarity has remained anchored in attitudes and discourses: either with the aim of combating it (official rhetoric in Cape Town in favour of better access to the park for the poor and for Africans), or as a component of "anti-poor" rhetoric, as in India: are there not still more than 100,000 people living in slums within Mumbai's park? We saw in the Cape Town case how the well-off white populations use the environmental alibi to practice socio-spatial exclusion. In Delhi, the terms "aesthetic governmentality" or even "aesthetic criminalization" (Ghertner 2015) have been used to describe the principle that allows poor neighbourhoods in central areas to be demolished, whatever the rights of their populations in a democratic state under the rule of law, as soon as those neighbourhoods become visible and "cause a stain". This is true of all our cities, caught up as they are in a process of "worlding" (Roy and Ong 2011), of access to the status of global cities through beautification and greenification.

13.3.3 Differences Relating to the Park

- The role of history:

– *Age of the park*: it might be assumed that the oldest parks would retain the legacy of an authoritarian vision of what a national park should be, in other words more exclusionary and conservationist, whereas more recent parks would be more open, having been created in postcolonial times when belief in the conservation fortress had faded. A comparison between the oldest park (Nairobi 1946) and Cape Town (1998) might seem to confirm this hypothesis. However, a strict distinction is difficult to maintain, in particular because the creation of the "national" park is one thing, the start of the protection process is another: Tijuca, for example, became a national park in 1961, but the area had already been protected for a century, and the Table Mountain area was also protected well before 1998 (Chap. 1). More subtly, it might be imagined that a park that historically has always been managed by national services (Nairobi, Mumbai) would differ from one like Table Mountain, long managed by the local authorities. In Cape Town, the municipality's decision in 1998 to devolve the management of a number of areas of the park to a single operator, entrenched the SANParks national office as operator of TMNP. This procedure reflected a process of municipal disengagement. Today, however, the city bitterly regrets the decision, given the significant revenues it forgoes, as well as its loss of control over fundamental urban issues that the national park represents. It therefore pursues its involvement through multiple initiatives and collaborates much more with the park than the municipalities of Mumbai or Nairobi, for which the park was always national government territory.

- This explains, at least partially, the disparity in such *joint management institutions* that exist. "A park for all … by all"?[4] This is far from the case in Cape Town, where a forum ran between 2003 and 2011 without much success. Nonetheless, both the municipality and civil society are quite active in joint management efforts, as they are in Rio (Chap. 10), whereas the management committee in Nairobi has only just been created, and there is no such entity in Mumbai.

• The size of the park (Fig. 13.1)

TNP is the smallest of our parks, at 39 km^2, six times smaller than TMNP (250 km^2, excluding the 1,000 km^2 maritime area). Is this an advantage or disadvantage?

In theory, a large size is an advantage: it makes it easier to resist urban pressure, to produce buffer zones that are real, if not necessarily planned. What would remain of the southern part of SGNP, with its slum encroachments, if the park were

Fig. 13.1 The four national parks seen at the same scale. *Source* Map by J. Robert

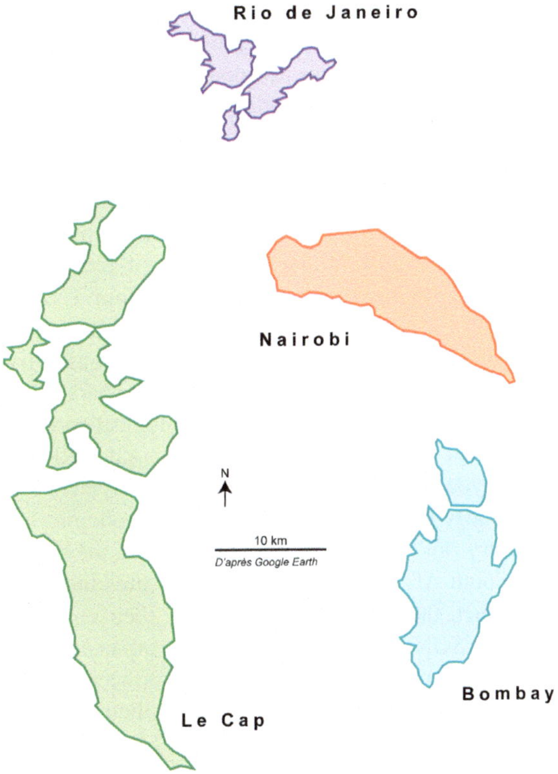

[4]J. Swanepoel, S. Didier, "A park for all… by all?", BiodiverCities Conference, Cape Town, April 7, 2014.

smaller? Size is also a political asset: in principle, it brings greater economic influence, a higher profile, and more local people can be mobilised to defend it. Yet all these arguments can be reversed: a large surface area means more urban contact, therefore greater fragility, with resources that are too dispersed and boundaries too wide for the park authorities to defend. A small but famous park can attract more revenue but also greater national or international visibility Tijuca than a larger but little-known park (SGNP). Remember that Kruger Park is 40 times larger than TMNP, but brings in only, four times as much revenue.

The other physical determinisms are equally disputable, as we will see.

- The shape of the park:

NNP is exceptional in that it is a single block and always has been. In contrast, TMNP and SGNP are patchworks that have developed gradually, with varying degrees of difficulty and with no real continuity, for reasons that are either physical (a narrow sea sound in Mumbai) or historical (private plots still outside municipal control in Cape Town). Tijuca is an intermediate case: it was formed in a single step, but is divided into three separate sections. In principle, the patchwork situation is the least comfortable, since it results (SGNP) in differences in land status or even arguments over legitimacy and appropriation. The Tijuca archipelago might also seem fragile, given that its ecological continuity must be constructed on the ground in the absence of any legal standing. But whereas the situation of NNP seemed somewhat favourable at the time we began our research (2012)—a single, homogeneous park, well enclosed and protected by a fence or savanna—subsequent years have shown its fragility, culminating in the plan for a railway that will divide it in half. So it does not appear that there is any rule that a park created in a single block is more robust than others.

A more interesting question is that of ecological continuity: we have seen that this continuity has to be constructed within TNP and in TMNP themselves, connecting protected patches. It could not be an issue that arises as such within NNP, and hardly in SGNP where the park has a coastal river and a road running through it. Nonetheless, it is clear that none of our parks, even TMNP—the largest of them—can survive as isolated entities. Ecological corridors thus need to be formed at the scale of the whole urban region and beyond. TNP must be linked with Pedra Branca and beyond, SGNP with Tungareshwar Sanctuary, etc. At this scale, therefore, differences in population density and economic infrastructures emerge clearly: a "country full" such as India contrasts with the relatively low densities of a country like South Africa, to the advantage of the latter.

In fact, the parks differ greatly in their environment. TNP is primarily urban; likewise SGNP, surrounded as it is by urban expansion, but also because of its own expansion up to the 1980s: only its northern tip, cut off from the rest, gives on to open spaces. NNP and TMNP differ in their configurations, with only some of their edges touching the city. An advantage? Probably, but also probably a weakness too, since TMNP is encircled by the sea (also a protected area) and is therefore unable to "retreat", whereas the savanna south of Nairobi, so crucial for the presence of animals within the park, is facing development.

- The topography:

All our parks are characterised by altitude, or at least by slopes. All, as it happens, except the one at the greatest risk, NNP, which, if it had been high up, would have been more difficult to put a railway through. The land relief acts as a foil, and the highway or railway corridor to Gujarat are only a threat to the edges of SGNP. Plans for a tunnel across it are under discussion: if they are implemented, the park will be affected by the construction process, and by the ventilation and safety shafts. But it will suffer less than NNP, with the Single Gauge Railway running from one side of it to the other.

- Park vegetation:

Although there are pinewoods in TMNP and open forest in the west of NNP, two broad types of vegetation can be identified: parks can have a relatively open landscape, with essentially grassland (Nairobi) or bush vegetation (Cape Town), or they can be dominated by a largely closed landscape, characterised by forestland (Mumbai, Rio). It is very risky to ascribe an advantage to one or the other in terms of urban integration, since landscape representations vary greatly between cultures and identities. Perhaps it would go to forest landscapes, which often attract more visitors (especially in hot climates): it is not the savanna in NNP that brings in the tourists, but the animals that live on it. And only cognoscenti with an appreciation of the endemic biodiversity of the *fynbos* are likely to make it the purpose of their visit: otherwise, people go to TMNP to admire the sceneries, or else to picnic—if possible in the shade of a exotic tree.

However, the question needs to be considered ecologically, and not only from the perspective of the city. Landscapes can be equally open but differ widely in their values: there is nothing biologically very remarkable about the Nairobi savanna, whereas the Cape Town *fynbos* is unique in the world. Protective measures need to reflect this, which is why in Nairobi they want to protect the park's surviving fragment of forest, in the wet zone, whereas most ecologists in Cape Town want to cut down the pines.

- The park's functions:

We will return to the differences in function between parks that are primarily dedicated to the protection of biodiversity (Mumbai, Nairobi) and those where the priority is visitor access for purposes of environmental education, leisure or tourist revenues (Cape Town, Rio). For the moment, we will simply emphasise a very simple but essential criterion of differentiation: whether or not the park is home to large mammals (what both public and rangers commonly call wildlife, forgetting small fauna and all flora). This can be an attraction to encourage visitors (in fact, the only function in Nairobi) and have a positive impact on park revenues as well as on the human, material and financial resources provided by the national management department. However, it would seem that this richness is more of a curse. These animals tend to leave the boundaries of the park, whether voluntarily (e.g., because they know nothing of administrative boundaries, because there are holes in the fences, etc.) or involuntarily (e.g., poaching). The city has to be protected from these sometimes dangerous animals (Chap. 7) just as these animals have to be protected from the city. The ever-growing worldwide rhetoric on the need for cities that are "sustainable" and "green", etc. always refers to green spaces, vegetation and small animals (e.g., insects, amphibians, perhaps foxes, etc.), but much less frequently to the lion or zebra of Nairobi.

The embarrassment is understandable. What is to be done with these animals in the city? Is not endemic vegetation such as the Cape *fynbos* easier to incorporate into urban policy than baboons? So we should not be too quick to criticise the management of the NNP and SGNP, or to call it archaic because of their failure to coordinate their policies with the city, or because they operate on a closed rather than an open mode in respect of landscapes or institutional approaches alike. It is not an easy task, perhaps even impossible. Could Tijuca be as open to tourism and urban encroachment if it contained rhinoceros or leopard? In the same way, one can understand why the parks of Nairobi and Mumbai still appear fundamentally "modern", in the sense that they prioritise surface and continuity over linearity and discontinuity: at one point, NNP wanted to extend and incorporate a adjacent piece of unfenced land partly belonging to the government ("Sheep and Goat Land"), in order to expand its protective buffer, whereas the parks of Rio and Cape Town were quicker to understand the advantage of biodiversity corridors as a means to avoid the need to control huge continuous areas. But this was much easier for them, given that vegetation reproduces easily in situations of discontinuity (seeds carried by the wind), while big mammals need vast continuous land areas in which to feed and move.

13.4 What Park Typology, for What Policies?

To finish, how can we introduce order among all these similarities and differences? Could we not classify our parks according to different criteria, and thereby implement different policies from one type to another, but the same policies within each type? In the previous pages, we outlined different classifications, ranging from the simplest (surface area of the park) to the most complex (the degree of coordination with urban "civil society"). We have nevertheless seen that it is far from easy to draw practical lessons from this, given that, for example, a small size can prove an advantage or a problem, depending on the regional configuration. Nonetheless, broad-brush as it is, the synthesis in Table 13.1 shows how Nairobi, and in particular Mumbai, are unfavourably placed in terms of the possibilities of integration between park and city.

In order to go further, below we reproduce another table (Table 13.2) that was proposed in the introduction to this book, in which our four parks are classified along two axes. The horizontal axis shows the park's degree of integration with its urban and regional environment, notably through the formal or informal institutions that are used to coordinate the objectives and policies of the park and the city. The

Table 13.1 Configurations favourable or unfavourable to naturbanity (park-city integration)

	Cape Town	Rio de Janeiro	Nairobi	Mumbai
National context				
National political and legislative context	++	+	+	
Sociocultural divisions		+		
Scales of power		+	+	++
The city				
Urban size and pressure	++	+	+	
Socio-spatial differences				
The park				
Impact of history	+	+		
Size	Criteria	Not	Relevant	Here
Shape	Criteria	Not	Relevant	Here
Land relief	++	++	+	
vegetation		+		+
Functions (wildlife)	++	++		
Total number of +	9	10	4	3

In this table, one or two + signify the presence of a factor favourable to this integration. No + signifies an unfavourable factor

Table 13.2 National parks between globalization and urban integration

	High city-park integration	Low city-park integration
High globalisation of the park	*Rio, and to a lesser extent Cape Town*	*Nairobi*
Low globalisation of the park		*Mumbai*

In this table, one or two + signify the presence of a factor favourable to this integration. No + signifies an unfavourable factor

vertical axis shows what we call the degree of globalisation, i.e. the importance of supranational issues. TNP and TMNP operate as logos for their respective cities, or even countries, whereas SGNP has very little international recognition, a fact of indifference to both the park's managers and the Mumbai authorities.

The case of NNP case stands out. Like SGNP, it coordinates little with the city, concerned as it is to accomplish two major functions: to protect biodiversity (in particular rhinos), and to attract tourists, preferably wealthy tourists. On the other hand, it seems much more advanced than idem in terms of globalisation: many of the tourists who come are foreign, and Nairobi's important role in hosting international bodies and conferences is reflected through the park (visits by heads of state, see Chap. 11). Admittedly, it is questionable whether this globalisation is as "high" as in the parks of Rio and Cape Town: the park is not the first choice of foreigners, the prestigious visitors are not used for enhancing the global image of the park. Yet, in this respect, NNP has resemblances with TNP and TMNP, although it does not share their urban trademark function.

We could sum up the situation by saying that there are two types of urban national park here.

1. *The fortress park* (Nairobi, Mumbai), showing little integration with the city, heirs to a conservation ideology founded on separation, boundaries, and the dichotomy between nature and culture. The management model differs little from the country's other "rural" parks. These parks tend to withdraw into themselves in an attempt at self-protection, in this case protection from the city. However, we would be wrong to attribute this solely to the legacy of their origins, or to the sometimes paramilitary aspects of the KWS or the Indian Forest Department: as we have said, one essential factor is that these parks are home to big mammals, which encourages isolation.
2. *The integrated or multidimensional park* (Rio, Cape Town), connected both horizontally (to the city) and vertically (to the global scale): conservation remains an important objective, but it is clearly secondary in Tijuca, where the priority is pragmatically on environmental education for visitors and the influx of revenues that can be used to fund conservation in the remote regions of Brazil. Here, the objectives of development, ecosystem services and citizen quality of life are compatible with those of conservation. Nature must not only be preserved, but must be "brought back into" cities (Rademacher and Sivaramakrishnan 2017, 11). Bruno et al. (2016) refer to these parks as "emerging parks", because they are located in so-called emerging economies. Just as emerging countries tend to align with the countries of the Global North, the models of these parks are often inspired by, or at least constructed with, parks in the Global North (Chap. 11): they are generally more open to the outside, contributing to the construction of the "sustainable city", where a concern for urban quality of life that leaves more room for "nature" combines with reflection by city dwellers themselves regarding the space that should be left for plant and animal species in the urban environment (Blanc 2004).

Like any typology, ours has its limitations, given its lack of nuance. Is NNP really a fortress, when a programme of payment for ecosystem services provided by the Maasai cattle herders has been set up to encourage them not to close or divide up their pastures so as to allow wildlife to move around? Mumbai is a long way from such a step. (However, this programme was not launched by the KWS but by an American NGO, and it has since stalled. The integration of the Maasai cattle herders into the park's operation remains a work in progress.)

Is SGNP a fortress, when it receives more than a million paying visitors a year, not to mention an unquantifiable number of local walkers, some legal, some less so? If we ignore the institutional aspect, SGNP seems highly "integrated" into the city, heavily used by urbanites, given that for the inhabitants of the neighbouring districts, in particular the middle classes, it is an area much used in the early morning for walking or yoga. However, SGNP employs very few Adivasis, and even fewer illegal local slum dwellers. By contrast, Cape Town Park is a fairly large source of employment for the inhabitants of the poor townships bordering the protected area, and the park seeks to be a tool of social advancement by "*providing benefits beyond boundaries*": the national Expanded Public Works Programme (EPWP) provides the funding for jobs such as planting or weeding, and SANParks acts as the implementing agency (Swanepoel 2013). Of course, jobs of this kind are highly precarious, but it is still hard to imagine that, in India, the National Rural Employment Guarantee Scheme (NREGS), which slightly resembles the EPWP, will one day hire poor people to work in the national park.

In fact, institutional and official integration needs to be distinguished from informal integration. The latter is a reality in all the parks, even when illegal. First of all, economic integration: TMNP has its flower harvesters and fish poachers, Nairobi its clandestine ox herds, not to mention the poaching also found in Mumbai (in addition to illegal wood selling). Then spatial integration: the encroachments are most limited in Rio; in Nairobi they are confined to the savanna in the dispersal area outside the park's official boundaries; there are few in Cape Town (on the squatters of Hangberg, read Swanepoel 2013); obviously, it is in Mumbai that this violent and suicidal spatial integration between park and city is at its most extreme, with tens of thousands of people still living in slums in the park.

What are the right policies for our two types of park? One thing is sure, there needs to be action on both fronts: mitigation of challenges, and proactiveness to avoid these challenges and to take advantage of the assets.

1. *For the fortress parks*, the imperative is greater openness. As was said earlier (part 2), park management needs to recognise the presence of the city rather than maintaining the pretence that it does not exist; similarly, it is preferable that the permeability of boundaries and fences be accepted and exploited rather than denied. Municipalities need to develop genuine plans for "urban nature", to recognise the ecosystem services that the park provides, while for its part the park needs to acknowledge the services that the city can deliver (setting a physical or conceptual buffer zone by "hybridisation" (see Chap. 8), financial revenues, political support from the municipality or civil society, etc.).
   For Mumbai, therefore, the advice might be as follows:

   - that the Forest Department which manages the park should follow the Nairobi model in creating a community wing, focusing on the Adivasis but also on all city dwellers who are in contact with the park. The number of intervention units responding to leopard sightings should be increased, together with education in the schools, in residences, and also in the slums.
   - with a view to conflict mitigation as well as proactiveness, environmental education for the population should be improved, but alongside better education in population for the rangers: social sciences are important disciplines for managing a park. City dwellers will let the (rare!) leopards go about their business without harassing them and making them dangerous, while the

Forest Department will gain a better understanding of the desires and concerns of both the urban locals and the Adivasis.

- better use of indigenous knowledge: just as in Nairobi, the policy of recruiting Maasais into the rangers could be continued. Adivasis could be engaged in the management of SGNP (apart from the menial tasks that they already do). Beyond the income that these jobs would generate, this would reflect a highly political recognition of the role of the indigenous populations in the national parks, perhaps culminating in the granting of certain rights. Could not grazing be officially permitted in NNP at certain times of the year, which—if properly managed—would improve the quality of the grassland for the wild herbivores, while at the same time providing access to compensation for livestock killed by carnivores? This would mitigate conflicts with the herders, but would also be a proactive way of tackling the risks of fire and poaching through the surveillance they could exercise: the park's future would be guaranteed by greater ecological resilience and improved social acceptance (Laslaz et al. 2014).
- The policies needed are not confined to the "horizontal" dimension, bringing together park and city within a single urban region, but lie also in the vertical dimension of globalisation. SGNP remains ignorant of certain success stories amongst comparable parks abroad, and could organise more international visits for its officials and rangers. It could also organise visits from more foreign colleagues and experts for a two-way exchange of experience. The Municipal Corporation of Greater Mumbai, and perhaps at a higher level, the Mumbai Metropolitan Regional Development Authority, could get more involved in international networks of "green" cities and regions in spite of their limited collaboration (Pethe et al. 2011): there they could find support and suggestions on ways to integrate the national park and make it both a flagship for international visibility and an asset for local quality of life and urban sustainability.
- All this would need to be done gradually, and cautiously. "If the Forest Department is on leave for two weeks the SGNP shall not exist!", a consultant working for SGNP exclaimed at the BiodiverCities Conference (Cape Town, April 7, 2014). We are not arguing for openness at all costs, but a gradual and considered opening of windows, along with efforts to ensure that the remaining exclusionary and coercive measures have a human face. So in Mumbai, the greatest strictness needs to be shown in dealing with the slums in the park (and with the politicians who have allowed them to develop), but their inhabitants need to be rehoused before their homes are demolished—a requirement in fact stated in the High Court judgement of 1997. In addition, no doubt, the wall needs to be increased in height, repaired where there are gaps, and topped at least partially by an electric fence. However, the plan announced in 2017 to build a wall around the peri-urban enclave of Yeoor seems extremely debatable, since it will preclude the movement of species! "A wall will prevent animals from crossing the road and prevent them being run over by speeding vehicles. This will also help avoid man-animal

conflicts. Another advantage of this wall is that it will clearly define forest and human areas, thus preventing encroachment of the forest land", explained a range forest officer to the *Hindustan Times* (April 27, 2017). Here however, as in Nairobi, there is a risk of turning the national park into a mere zoo, where the animals would be fed by people since species movement and food chain functioning would become impossible.

2. *For the integrated and multidimensional parks*, the aim, one might say, should be to keep up the good work. As noted in Chap. 10, these parks are interested in France's new institutional national park system, set up in 2006, as it could introduce a certain freedom of management, though of course the model should not be imitated servilely without regard for the specific conditions of the different countries. Alongside the "core zone" (*zone coeur*), that is the park itself, it introduces a buffer zone outside the park, called the *zone d'adhesion* (adherence area): the localities (*communes*) around the park choose whether or not to sign the park charter, which defines rights and obligations. This adherence area can extend very far: on the island of Porquerolles, located in Port-Cros National Park on the French Riviera, tourists who climb up onto an escarpment slightly over 100 m high, may be surprised to learn that the adherence zone includes *communes* located on the other side of Hyères Bay, more than 10 km away from the island on the distant horizon of the mainland. The idea is to achieve a twofold solidarity: first a solidarity of landscape, in the visual sense of the term (to maintain the quality of the view from the promontory); then ecological solidarity (e.g., the island's bats go to feed on the mainland salt marshes, the Posidonia seagrasses provide food for mainland species, etc.). Commitment to the charter is voluntary, so it demands a degree of environmental awareness amongst local politicians and populations, and possibly environmental education, for the restrictions introduced by the document's regulations to be accepted. However, it also requires these populations to have a clear vision of all the *advantages* in signing: the "zone d'adhésion" label can boost the local economy by attracting more tourists, and the environmental constraints should not be seen as park diktats, but as possibilities for improving the day-to-day lives of local people (e.g., better waste management, industrial standards, etc.). The "environmental front" described in Chap. 8 should not be seen as an encroachment of the national park into an urban jurisdiction that lies outside its purview, but as a tool of hybridisation in which all areas and social groups have something to gain.

13.5 Conclusion: Naturbanity, A "Wild Urbanisation"

For readers who find the neologism "naturbanity" off-putting, another description might be *wild urbanisation*: not "wild" in the sense of uncontrolled—urban sprawl in pursuit of real estate profits and short-term economic objectives—but wild in the sense of acceptance of a nature that is not completely tamed, which will be given a "place" in the city (cf. the dual meaning of the title *Places of Nature in Ecologies of Urbanism*—Rademacher and Sivaramakrishnan 2017). Many efforts have already been made in Cape Town to allow access to this nature (wastelands, weeds), but there is nothing easy about it: remember that in France, people sometimes complain of the presence of trees in their streets, because of the dead leaves they drop. Any form of nature can be seen as a nuisance. And if vegetation is not so easy to accept, what about animals, particularly carnivores! (Chap. 7; cf. Gullo et al. 1998, on how the pumas of Los Angeles challenge urban development modes).

Urban national parks, strange geographical objects up to now trapped between a conservation role and an urban location, can play a major part in such a paradigm. Instead of being squeezed between somewhat contradictory objectives (e.g., conservation of nature or urban recreation? environmental education or profit motives?, etc.), they can turn their problems into assets, and make their hybrid nature (in all senses of the word) the spearhead of a new policy and a new philosophy, reconciling nature and the city—even better, making them mutually supportive.

As Rosenzweig (2003) argues, *reservation* ecology, still dominant today as a way of protecting natural species, is threatened by human expansion and its direct or indirect effects. One solution is the attempt to practice *restoration* ecology, by reintroducing biodiversity into degraded areas. Here again, however, the potential is limited. What is needed, therefore, is the development of a *reconciliation ecology*, "the science of inventing, establishing, and maintaining new habitats to conserve species diversity in places where people live, work or play" (Rosenzweig 2003, 15). This notion can be fairly criticised as utopian or excessively focused on species and not enough on ecosystems. It can, however, be retained to give urban national parks —and all the other protected areas in urban environments—the central role that they deserve, but which is not always attributed to them. Indeed, our national parks are like a spatial metaphor for this aspiration towards reconciliation ecology. They can be divided into three more or less concentric spaces. In the core area, or its equivalent, reservation ecology is practiced, sometimes in the form of totally reserved spaces. Their edges are more the domain of restoration ecology (replacement of pines by *fynbos* in Cape Town, reforestation of demolished slums in Mumbai). And it is the city, outside the park, that could become the place of reconciliation ecology, where nature and culture are reconciled, and where biodiversity can spread and reproduce as best it can—often not so well. The compromise would be preferable to an urbanisation that entirely neglects the imperatives of nature. Instead of the city being seen as a hard, mineral structure, it should be seen as a complex of thousands of gardens, of balconies, of yards, of flowerpots, of open spaces, of public, private or shared spaces, each of which can accommodate, within its own capacity, a medium for the movement of species, fauna and flora, from the proteas of Cape Town to—why not—the lions of Nairobi. National parks are in any case cramped: they need to be networked not only with other protected areas, but also with certain urban subspaces which, pooled together, could function as green corridors to support the national park.

There is nothing naively angelic in our position. Let us not idealise all the Maasai or the Adivasis, or the leopards, or the poachers, the "real estate sharks" or predatory urbanites. And let us not practice the cult of the "local", with its need to be protected from higher echelons. Is it to play devil's advocate to argue that if the national interest (e.g., protection of biodiversity, more tourists, more revenue, etc.) takes precedence over the special interests of the inhabitants of a *favela*, it can be ethically and politically legitimate to favour the national interest? In fact, many conflicts seem to be much less conflicts between spaces (park vs. city) than between special interests and the public interest. Of course, this begs the question of defining the public interest, a concept often used to mask special interests.

So we do not underestimate the difficulties. How can endemic biodiversity be protected while still opening up the parks to the city and to visitors? Under what conditions can city dwellers accept the incursion of wild carnivores into their neighbourhoods? None of these questions is easy to answer. There is no obvious connection between policies to protect the leopard toads of Cape Town and the leopards of Mumbai. But the questions at least need to be asked, whereas too often relations between the world of the national park and the world of the city are characterised by mutual ignorance or mutual antagonism.

References

Agrawal, A. (2005). *Environmentality: Technologies of government and the making of subjects*. Durham: Duke University Press.

Blanc, N. (2004). De l'écologique dans la ville. *Ethnologie française, 34*(4), 601–607.

Brenner, N. (2009). Restructuring, rescaling and the urban question. *Critical Planning*, pp. 61–79.

Bruno, L., Landy, F., Lézy, E. (2016). Des parcs nationaux et leurs métropoles: L'évolution des rapports à Rio de Janeiro, Mumbai, Nairobi et au Cap. In S. Barles & N. Blanc (Eds.), *Ecologies urbaines*. Paris: Economica.

Byrne, J. (2012). When green is White: The cultural politics of race, nature and social exclusion in a Los Angeles urban national park. *Geoforum, 43*(3), 595–611.

Descola, P. (2013). *Beyond nature and culture*. Chicago: University of Chicago Press.

Di Méo, G. (1985). Les formations socio-spatiales ou la dimension infra-régionale en géographie. *Annales de Géographie, 94*(526), 661–689.

Gaulier, J. M. (2013). *Landscape transformations: Urbicus*. Hong-Kong: Design Media Publishing.

Ghertner, D. A. (2015). *Rule by aesthetics: World-class city making in Delhi*. New York: Oxford University Press.

Ghosal, S. (2013). *Intimate beasts: Exploring relationships between humans and large carnivores in western India*. Ås: Norwegian University of Life Sciences. (PhD).

Guha, R. (1989). Radical environmentalism and wilderness préservation: A third world critique. *Environmental Ethics, 11*(1), 71–83.

Guha, R., & Martinez-Alier, J. (1996). Introduction. In R. Guha & J. Martinez-Alier (Eds.), *Varieties of environmentalism: Essays north and south* (pp. xi–xx). London: Earthscan.

Gullo A., et al. (1998). The cougar's tale. In J. Wolch & J. Emel (Eds.), *Animal geographies. Place, politics and identity in the nature-culture borderlands*. New York: Verso.

Hache, E. (Ed.). (2012). *Ecologie politique. Cosmos, communautés, milieux*. Paris: Editions Amsterdam.

Landy, F. (2017). Urban leopards are good cartographers: Human-nonhuman and spatial conflicts at Sanjay Gandhi National Park, Mumbai. In A. Rademacher & K. Sivaramakrishnan (Eds.), *Places of nature in ecologies of urbanism* (pp. 67–86). Hong Kong: Hong Kong University Press.

Laslaz, L., Gauchon, C., Duval, M., & Héritier, S. (Eds.). (2014). *Les espaces protégés et territoires. Entre conflits et acceptation*. Paris: Belin.

Latour, B. (1993). *We have never been modern*. Cambridge: Harvard University Press.

Pethe, A., Gandhi, S., & Tandel, V. (2011). Assessing the Mumbai metropolitan region: A governance perspective. *Economic and Political Weekly, 46*(26–27), 187–195.

Rademacher, A., & Sivaramakrishnan, K. (Eds.). (2017). *Places of nature in ecologies of urbanism*. Hong Kong: Hong Kong University Press.

Rosenzweig, M. L. (2003). *Win-win ecology: How Earth's species can survive in the midst of human enterprise*. Oxford: Oxford University Press.

Roy, A., & Ong, A. (Eds.). (2011). *Worlding cities: Asian experiments and the art of being global*. Oxford: Blackwell.

Saglio-Yatzimirsky, M. C., Landy, F. (Eds.). (2014). *Megacity slums. Social Exclusion, Space and Urban Policies in Brazil and India*. London: Imperial College Press.

Swanepoel, J. (2013). *Custodians of the Cape Peninsula: A historical and contemporary ethnography of urban conservation in Cape Town*. Master's dissertation, University of Stellenbosch.

Chapter 14
Afterword: Seen Through the Eyes of Researchers, Are Practitioners Partners, Research Objects or Hurdles?

Glen Hyman, Frédéric Landy and Louise Bruno-Lézy

Practitioners have no time to read scientific literature. They need half a page with bullet points.

An Indian consultant representing Sanjay Gandhi National Park, UNPEC BiodiverCities Conference, Cape Town, April 7, 2014.

In 2010, during the first international BiodiverCities Conference in Paris, the representatives of the city of Nairobi and Nairobi National Park opened their presentations by saying that their first meeting took place in the plane, on the way to Paris. Up until then, contacts between the city and the park had been limited to a telephone call, when a lion entered the city and the Kenya Wildlife Service had to come and fetch it. Two administrations, two jobs and no communication between them.

Ensuring that such meetings took place, promoting dialogue between cities and national parks as well as developing exchanges between practitioners and researchers, were some of the ambitions of the UNPEC project, within the framework of action research not limited to publishing academic articles, but seeking—together with practitioners—to offer common discussions and new avenues to manage urban national parks—these very unusual objects—in the best way possible. Let us admit it: we have not been very successful. With some exceptions, managers have not turned out to be real *partners*; in most cases, they remained research *objects*; sometimes, they even appeared to be *obstacle* to research work

G. Hyman (✉)
Center for the Sociology of Organizations Sciences Po, Paris, France
e-mail: glenhyman@gmail.com

F. Landy
University of Paris-Nanterre, Nanterre, France

F. Landy
French Institute of Pondicherry, Pondicherry, India

L. Bruno-Lézy
Essonne District Council/Institut Libertas, Île-de-France, France

F. Landy (ed.), *From Urban National Parks to Natured Cities in the Global South*,
https://doi.org/10.1007/978-981-10-8462-1_14

that sought to avoid constraints linked to the control of research permits, deemed too restrictive or difficult to obtain, and at the same time to avoid the politically correct narratives imposed upon practitioners within the framework of their duties. Let us not forget that, in our four cases, we are dealing with civil servants who are subjected to professional confidentiality.

This afterword is proposing a critical analysis of the successes and failures linked to our potentially over-ambitious objective of building bridges between two worlds underlain by very different principles and operations which, sometimes, appear incompatible. As shown in the first section of this text, the cycle of BiodiverCities conferences, which was inaugurated in 2010 by the Urban Protected Areas Network, through the project entitled SETUP and extended by UNPEC, has been an interesting tool contributing to exchanges between researchers and practitioners, and between practitioners themselves.[1] However, it was not enough to build long-lasting links or bring out common projects. Practitioners have often been considered as objects of research, or even perceived as hindering research, which is the argument of the second section of this chapter. Why this—if not total, at least dominant—failure? To what extent are researchers responsible? Why did practitioners not show more interest in collaboration? Yet, according to them, urban parks (and *a fortiori* their relationship to the city) suffer from a lack of research which, paradoxically, was due to their proximity to the city (indeed, these parks were considered as being either too specific or not "wild" enough to be prioritised for studies). Was it the format or the temporality of the research that did not manage to interest managers? Or are the human and social sciences still not recognised as sufficiently important in the management of national parks with their primary biodiversity conservation objectives, and where human presence is still taboo? All these questions are dealt with in the third section.

Finding out the reasons for this failure could save other projects from the same pitfalls. This can also, as shown in the conclusion, shed light on certain research characteristics which often make research incompatible with the operating method of park and city managers. Finally, it will also open up suggestions for future solutions.

1

- BiodiverCities Paris 2010: "Enjeux and stratégies de gestion durable des aires protégées urbaines and périurbaines au Nord and au Sud", in partnership with Sciences Po-Paris and ONF Agence Fontainebleau.
- BiodiverCities Rio de Janeiro 2012: "Les aires protégées urbaines: enjeux, acteurs, espaces", in partnership with the Tijuca National Park, the City of Rio de Janeiro and the Pontifical University of Rio.
- BiodiverCities Le Cap 2014: "Rescaling Natural Parks and the City", in partnership with the Table Mountain National Park, the City of Cape Town and the French Institute of South Africa.
- BiodiverCities Marseille 2015: "Parks naturels: construire une nature urbaine?", in partnership with the National Park of Port-Cros, the National Park of Calanques and the cities of Hyères and Marseille.

14.1 Walkways, not Bridges: Conferences on BiodiverCities

The project entitled UNPEC took over the first research project which, since 2009, had aimed at developing an international network of urban protected areas (http://www.upa-network.org/). The objective was already to promote collaboration between research teams, natural area managers, local and regional authorities as well as civil society representatives around issues relating to the management of protected areas faced with urban dynamics. Institutional partnerships established around the organisation of each BiodiverCities conference, have contributed to exchanges between organisers, researchers and practitioners from each site hosting the event. The conference which lasts three days (two days for the conference and one day for a field trip), is an opportunity for the hosting site to show participants its initiatives in the field. These outings have been an opportunity for researchers and practitioners to exchange ideas on site management, as well as to identify common problems and solutions which could be adapted to other sites. However, the advantages of the links created or reinforced during these occasions, did not really benefit exchanges during the conference discussions because these outings took place at the end of each event, instead of starting them, which would have contributed to breaking the ice between individuals and institutions.

Most managers were delighted to have been put in touch with their counterparts in another country, thanks to our invitations.[2] Even though, for example, Cape Town had already opened up internationally while Rio de Janeiro had not done so to the same extent, it came as a surprise or even a relief for some when they discovered that the problems in "their" urban national park could be found in other countries. "You realise that in another city, a park has more pressure than yours".[3] During the SETUP project (2009) which organised a meeting in Sanjay Gandhi National Park (SGNP), the foresters of Mumbai were surprised by a presentation given by researchers on the park in Rio, pointing out the existence of other parks that were *also* subjected to the pressure of the city, with the threat of slums as well as wealthy residences, the religious use of the park and the issue of invasive species! All these subjects were becoming the common fate of protected areas in poor or emergent countries. Not only did this prove that practitioners were not automatically "guilty" of bad management, it also showed that they could still hope to find ideas for solutions through

[2]The National Park of Réunion Island (France), seeking to exchange experiences on urban issues, learnt in 2016 from Cape Town practitioners that they could engage with their French counterparts from Calanques and Port-Cros, who had taken part in the fourth BiodiverCities encounter in Marseille.

[3]The Director of the Nairobi National Park, UNPEC BiodiverCities Conference, Cape Town, April 7, 2014.

the experience of others. "We are never recognised at home.... But when we rely on international experience, people pay more attention to what we are saying!"[4]

As such, managers were not just seekers of the comparative matrix built by UNPEC which, thanks to indicators of surface area, budget and number of visitors among others, made it possible to look at parks quantitatively. They were also seekers of experiences and success stories that they were often able to learn about, in addition to those they learned about thanks to pre-existing relationships (as between the parks of Tijuca and Table Mountain in particular). The balance sheet varies a lot depending on where we stand. In Cape Town for example, UNPEC did not bear much fruit as far as practitioners were concerned, because the latter were already promoting efficient coordination between city and park, as well as internationally. However, in Mumbai, things appeared far more positive. What happened between the time of the first meeting with the director, and the follow up promises after UNPEC, for studying the leopard issue in particular! We went from complete mistrust to cautious trust. For a French researcher, the mere fact that he had been invited to talk one afternoon in February 2013, at a training session for rangers, was a considerable achievement and recognition of the project's legitimacy. Of course, UNPEC is far from being the only entity to have opened up SGNP: civil society institutions such as Mumbaikars for SGNP or Krishna Tiwari's group, but also the arrival of younger, more open and better trained directors, have led the park to engage with the city and vice versa. However, it is very likely that the project has had a big hand in bringing both sides together.

14.2 Relations Between Practitioners and Researchers: Partners, Research Objects or Hurdles?

The initial ambition, which consisted of creating bridges between the researchers and managers of cities and parks, was undoubtedly well intentioned. In addition to producing scientific literature, together with the managers we would also be co-producing knowledge, in a form that was accessible to them or even other non-academic actors. In return, this co-production of knowledge could enrich our scientific research. Managers were thus, a priori, fully fledged partners: not only were they contact persons who could give us authorisations, or informants we could interview, but they were also colleagues contributing, just as we did, to boundary work that would better link knowledge with action (Clark et al. 2016).

Yet, our research questions were not built with the practitioners—only part of our answers. And from their side, they rarely came to see us to ask us for our opinions on problems they encountered. This is perhaps what explains, for example, our failure in anticipating the doom that was to strike Nairobi National Park in the

[4]The Head of Environmental Management, City of Cape Town, at UNPEC *BiodiverCities* Conference, Cape Town, April 7, 2014.

last year of the project. In 2012, when UNPEC was launched, the park appeared relatively well protected, threatened by the city but only on the margins, and with initiatives at the rural interface in the south, which led one to believe that the fauna's migration zone would be preserved. And yet, today, not only does the park appear more threatened than ever, but it could be ended altogether because of the Standard Gauge Railway project. As researchers, our perception was rather optimistic at first, and we certainly did not see it coming. Was it because the managers were themselves surprised by decisions and processes over which they had no control or did not expect? Or is it rather that they did not want to tell us about it?

14.2.1 Practitioners as Research Objects in the End

The names of these practitioners were known to most from the beginning of the project, seeing that members of our team had met most of them during previous projects, at one time or another. However, they were not contacted to contribute to the draft of the research project submitted to the French National Research Agency. Moreover, no specific methodology was elaborated for this partnership. Only the intention was clear—although some researchers still wanted to conduct a more traditional form of research, leaving other UNPEC members the task of liaising with the practitioners.

The latter, it is true, had been invited on many occasions to take part in UNPEC workshops and conferences. As already pointed out, these events were the only proper meeting points during the project. But, let us confess it, the fact that practitioners were invited mostly gave us an opportunity to listen critically to what they had to say as well as really work with them, while considering them as research material as much as partners. Although the collaboration rhetoric could certainly appear to be sincere on both sides, there was always uneasiness reflected by, for example, our hesitation during the preparatory meetings for these encounters, in deciding the format of the conferences and the status practitioners should be given. We most often had practitioners speak between themselves, and researchers speak between themselves, without seeking to mix both worlds. Yet, interesting round tables were organised. During the BiodiverCities Conference in Rio (2012), one afternoon was dedicated to the "role of research in the management of urban protected areas". However, the needs pointed out by Tijuca in particular, concerned life science themes more than social science ones. In Cape Town in 2014, round tables gathered practitioners only, with one researcher acting as moderator. Perhaps a longer conference format would have been better, a conference during which, apart from the days when it was open to the public, other days dedicated to closed workshops between researchers and practitioners would have been an opportunity for a more in-depth and confident dialogue on issues raised during the conference, and on actions observed in the field. This requires a workshop methodology that breaks away from the usual format of scientific conferences, during which presentations succeed one another in the form of monologues offering often very interesting information, but not bringing about discussions and exchanges.

In the end, quantum physics could be applied to research on urban national parks. We know Heisenberg's uncertainty principle—in the way that it has been popularised at least: just as there is a limit to the precision with which it is possible to know two physical properties of the same particle simultaneously, in our case it became very difficult to work with practitioners who would have been partners and research objects at the same time. Some of us have tried to consider them as both, partners and research objects—to somehow measure the particle's speed and location at the same time… but at the risk of losing in precision on both fronts. Interviewing and observing practitioners leads to their being affected by the very fact that they are being observed. How can one introduce research objects into a research team? How can one encourage them to think about and be critical of their approach and position? The UNPEC team was too attached to the "purity" of the research object to engage in a real partnership. Considering practitioners as objects prevented us from giving them important responsibilities, and that was the price we had to pay, we often thought, for keeping our freedom of access to certain subjects and stakeholders.

14.2.2 Practitioners Turned into Hurdles

It is true that, in return, with remarkable differences in the case of Rio and Cape Town, managers barely offered to take more of a part in the UNPEC project from within. But is not this due to the fact that we did not give them a clear chance to do so? Some of us had concerns about discussing certain matters with them: were we to explain the political and social ins and outs of our project, would they still agree to work with us? Worse still, would they not close the doors of the park or the city to us? One of us refused to present a paper at our conference in Cape Town, for fear of the political reactions his talk on the Cape Peninsula would generate. Another of our researchers kept remembering his first encounter with the director of the Mumbai's park, which prevented him from opening up to the successors of the park's director. Indeed, in 2007, that director had ordered him not to study "social issues" related to the park, and to limit his studies to fauna and flora. That director had compared the park's management, which was undermined by political pressure and corruption, having accepted to house up to 500,000 slum dwellers, to the pressure G. W. Bush underwent in relation to the war in Iraq. Serious suspicions of corruption weighed on him in any case. In addition, the Indian Forest Department is very uncommunicative, making it difficult to obtain research permits in particular, if required (e.g., gathering signatures at the level of the regional State Government, the Central Government, or even the Wildlife Institute of Dehradun, etc.). In Mumbai, practitioners were considered more as potential obstacles than aids.

Nonetheless, permits were requested by UNPEC and obtained—although we did request the minimum in this regard, even when the new park officials turned out to be far more open to research and social issues than their predecessors. To us, these authorisations were a token of trust and, at the same time, a source of information. They gave us an opportunity to look into the past, by enquiring about authorisations given to former projects, thereby circumventing the forgetfulness of researchers rotating and practitioners being transferred.

14.3 Why Did This Failure Occur?

Two possibilities can explain this failure: we had something to offer in our basket and they took nothing from it, because they were not interested or because we did not know how to promote ourselves. Or we had nothing to offer in the basket. The truth lies probably in both assumptions. Most practitioners and researchers felt that closer collaboration would be a waste of time. Perhaps they were right to suppose it but with such a state of mind, nothing great could be expected. If our methodology had been better, would we have followed it no matter what? It was hard to build; we did not realise exactly how hard it would be to try; and we did not try hard enough.

14.3.1 Different Concerns

We have already mentioned our concerns when it comes to applying for research permits that cost too much[5] or take too long to obtain for reasons such as the bureaucracy and format not being adapted to social sciences among others. In addition, we were also concerned about causing any harm to our informants. F. Landy did not dare talk about the fact that he had slept in an Adivasi village, inside the park of Mumbai, seeing that these "tribal groups", according to the director, were to be expelled one day or other. He also did not mention some of the residents' illegal practices he had seen. Practitioners are always officially on the right side of the law, but not automatically of what is legitimate. They most often represent a police force, and researchers need then to wonder about ethics, in considering what should be mentioned and what should not. They are often and cautiously alert in order to avoid saying one word too many that would endanger certain fragile actors, not to mention the progress of the research. This is the particularity of these practitioners: while sometimes they are weak in the face of powerful urban dynamics (e.g., major devastating development projects, real estate greediness, etc.), they can turn out to be strong or even brutal towards subordinate actors.

[5]The total budget of the UNPEC project was €250,000 and was to be divided over four and a half years between 25 people.

14.3.2 The Low Status of Social Sciences in the Culture of Conservation

We did not know how to promote what we had to offer. Researchers look at practitioners critically, while practitioners look at researchers doubtfully (as far as being able to find rapid and efficient answers is concerned). The objective of a national park is to protect biodiversity; it is not to integrate society which, in the missions of managers, is at best recent and remains secondary. Researchers in the life sciences can bring useful answers to managing species and environments, even if the long term aim of this type of research does not correspond exactly to the necessary adaptability of the manager's action. Yet, human beings are not quite recognised, in the world of conservation, as a fully fledged element in the management of protected areas. That is why humanities and social sciences remain peripheral in research on protected areas.

14.3.3 Two Different Cultures

The world of researchers and that of practitioners correspond to these autonomous "fields" defined by Bourdieu (1980). The argument is valid for both sides: when we are part of a field, it is difficult to see what can be of interest to the other field and create an interface, a "boundary" (Swiss Academy 2014); it is far easier to see immediately what separates them. They do not speak the same language and are not formatted to communicate between each other. Without even mentioning the potential superiority or inferiority complexes of one or the other, even though there is goodwill to forge links, we need to recognise our reciprocal asymmetries and create a process for long term co-learning. Learning something directly from the mouth of a Chinese person is marvellous, but whether one learned Chinese or not beforehand is another matter.

During a conference organised in Paris in 2010 by SETUP, a practitioner confided in us that "As far as we are concerned, we turn complex matters into pragmatic matters, while you, the researchers, take concrete matters to transform them into abstract and complex matters"! Indeed, the approach used by researchers is perhaps often too conceptual and too difficult to translate for those who make everyday decisions. Concepts are not always directly useful a priori, and researchers would need to use only those that are relevant to practitioners wisely—but they are not trained for that. The outings organised during the conferences held in Rio, Cape Town and Marseille were undoubtedly more interesting to practitioners than the actual conferences: in the field, they understood the technical and practical sides of what they were visiting—but without for all that automatically connecting it with the theoretical content that we wanted to provide. On the other hand, some practitioners came to our conferences with often Powerpoint presentations for a very general audience, filled with photos of flowers and birds, and their trust lacked so much that they tended to cant. These conferences were definitely not the right platforms to mention certain sensitive subjects.

In order to support the action-science boundary, we probably did not know how to best use what we could call boundary agents (at least two of us had professional dealings with policymakers) and boundary institutions, i.e. the research wings that exist within the administration of national parks to issue permits. In Nairobi, the KWS has a Biodiversity Research & Monitoring wing. In the case of Rio, in 2007, IBAMA created SISBIO (the Biodiversity Authorization and Information System), which can deliver automatic research approvals, following applications made online: according to an official, 91% of applications are processed in less than 45 days. Some institutions are used to dealing with researchers and can offer them housing in the park. On the other hand, this is more difficult in Mumbai, where research services are almost non-existent: they are contracted out to ancient institutions, such as the Bombay Natural History Society or, more recently, to civil society organisations such as Mumbaikars for SGNP (MfSGNP). There are a few houses in the park but, rumour has it that until recently, they were mainly used for parties organised for the personalities of the city.

The fact that we hardly sought to apply for research permits can then be understood from a different perspective: not only were we concerned that we would not obtain these permits and we needed to save on our budget money, but in our view making researchers pay for such permits was not legitimate. Although paying for the permit could have been seen as a commitment on our part to support the park's management, most of us, on the contrary, felt that it was logical that researchers should not pay for a permit, and should benefit from free entrances or at least reduced fees to the park, considering that we were supposed to be partners to the managers of the park. Who was right, who was wrong?

14.3.4 The Difficulties of Building a Network

We established a *list* of names of people or institutions rather than a real *network*. With a few exceptions: as already mentioned, during the meeting held in Paris in 2010, the representatives of the KWS and the city of Nairobi had never met in Kenya before; but in the end things did not work out, because they were transferred. In Marseille, in 2015, the governor of the county of Kajiado and the deputy director of the KWS also met for the first time, thanks to UNPEC, but will they be able to maintain their relationship?

It is not an easy task, truly. Not only did we want to build bridges between researchers and practitioners, but also between park managers, between city managers, and between park managers and city managers. Bridges between city and park managers existed already in Cape Town and, to a lesser extent, in Rio. But in Nairobi they were only outlined, while in Mumbai, where both worlds have been traditionally ignoring each other, everything had to be started from scratch. It was only during the last conference, in 2015, that we managed to have the director of the

Mumbai's park participate. Up until now, either the then directors clearly did not want to attend (one of them missed his visa deadline "because it was his birthday on that day"), or their administration did not want to let them attend (one of the directors could not even obtain a passport). For a long time the park did not have an official representation at our conferences, or was represented by a consultant who had his ways with the administration. As for the city of Mumbai, most officials we contacted declined our invitation to take part in conferences overseas, for the simple reason that the national park was part of the Forest Department, and was not under the jurisdiction of the city. In the end, only one elected representative and two officers (including one consultant) working for the Mumbai Metropolitan Region Development Area, were able to take part, although the latter did not get much support from their administration.

14.3.5 The Structural Problems of Financing Research

The UNPEC project was financed by the French National Research Agency (ANR), which is supposed to promote more basic than applied research. As such, the project had no incentive to involve practitioners. To the contrary, the greater their involvement, the more complicated the management. On the one hand, the ANR refuses to finance foreign institutions directly: paying invoices is possible, but not *ex ante* transfers to partners. On the other hand, requesting ANR funds to finance a project involving too many practitioners would have been suicidal, because it would have led to the non-selection of the project. The actual fundability of the project would have been endangered by real action research. And since the project is drafted, budgeted and finalised at the time when financing begins, it is too late afterwards, or at least too difficult, to bring in new partners in the team. This temporality issue was translated, for example, into the fact that UNPEC published a small brochure without any input from practitioners (since we wanted to use this brochure to introduce them to the project that was already off the ground), a brochure which practitioners did not appropriate, and distributed even less, even though their institution was (briefly) mentioned in it. Also, the project did not take the time to produce a new version of the brochure, this time in partnership with the practitioners.

Time was also against us in other areas. Indeed, how can we build relationships for the long term when managers rotate and are transferred? Even the UNPEC team has changed partially over four and a half years, which represent a period too short to build strong links, and too long to keep the same interlocutors.

Ultimately, time was lacking on both sides. Green shoots grew, but we lacked the time to water and fertilise them. A workshop was almost organised in Mumbai by the practitioners themselves at the end of the project, with a view to elaborating a guide to good practice guide, but the defection of one Indian civil servant ruined everything. It is true, however, that very few UNPEC members were prepared to volunteer for this elaboration.

Researchers had their monthly salaries already guaranteed. In the end, UNPEC was only a part-time research job, part of a schedule that was already very busy with lecturing for the academics, the bureaucracy of universities, and other projects which the members of UNPEC had joined, for fear of being short of research credits. Symmetrically, the practitioners often seemed overworked. How do you interest the director of Table Mountain National Park in the notion of "naturbanity", when his telephone is constantly ringing due to a problem encountered by a fire-fighting helicopter?

Action research is still too taboo in France, at least in the social sciences. Our ambition was scientifically valid but practically meaningless.

14.4 Difficult Work Ahead: Avenues for Building a Long-Lasting Partnership

But let us not end too pessimistically. On the one hand, practitioners have admitted the value of our encounters and exchanges, and many have expressed their wish for more. By crossing the results of research works conducted since 2009 on urban national parks, with those of the feedback from field actor experiences, a joint request was made to prolong UNPEC with an operational project of cooperation between new urban parks, with the support of their cities. This came about because partners in the countries of the Global North are interested in the experiences of the South and vice versa. Within the framework of the UPA Network, and in partnership with the Urban Conservation Strategies Specialist Group of the IUCN WCPA, the idea is to develop a programme of cooperation and exchange between the teams of urban protected areas, based on the principle of mutual capacity building, and aimed at implementing joint projects. The idea is to share successful experiences in urban protected areas that produce co-benefits for protected areas and cities, through the implementation of working groups and training courses. The priority themes would be inspired by those identified from UNPEC and BiodiverCities. The working groups and joint projects would benefit from the support of academic partners responsible for evaluating actions.

On the other hand, a few lessons can be learnt from UNPEC, these being fully-fledged results (even if they are "negative"). In the light of these results, what could be suggested to the new teams wanting to engage in similar action research? It will involve a "struggle", which is what happens when one wants to reinforce a "biodiversity science-policy nexus" (Spierenburg 2012). Such a struggle will at first concern the structuring of research policy. This is where it should be decided to develop popularisation actions as well as action research, so that researchers do not become obsessed with producing academic research in peer-reviewed journals with a view to improving their CVs. More prosaically, a requirement would be that any research project begins slowly, so as to build a team progressively on a solid and non-academic basis, and with limited financing, before embarking on the adventure.

This is what calls for projects such as the Initial Support for Exploratory Projects of the Centre National de la Recherche Scientifique (CNRS) (PEPS) are all about. This would also require the opening up of social sciences, by encouraging researchers to work with policymakers from the very first drafts of answers to calls for projects.

Another reorganisational need would consist in breaking away from the usual research model emanating from the countries of the North, and France in particular, concerning the countries of the South: namely, researchers who reside in France and who go on short field trips (sometimes less than two weeks) in the country under study. This model is barely valid for "standard" research in the social sciences, and is not valid at all when one wants to forge strong links with practitioners, and *a fortiori* with all local stakeholders. This is what explains, *a contrario*, the success of colleagues working on other projects, who developed a better partnership in the south of the NNP: Western as well as Kenyan researchers posted at the International Livestock Research Institute (ILRI), on assignment for several years in Nairobi, or on secondment from Colorado State University, had time to become involved in the creation of a management plan for the area, together with the Maasai populations as well as the KWS (Reid et al. 2014). In a way, the election of a former researcher from the ILRI as Governor of Kajiado, can be considered as a side achievement of the project! For this partnership to succeed, the two factors required were time, and the fact that the lifeblood of the team remained physically located and focused on one region. The UNPEC team members were, to the contrary, scattered not just over the four sites under study (with each team member making a detailed study of only one site, rarely two), but also on other projects (some team members were working on up to four projects at the same time!) in other countries and on other issues.

Perhaps the "struggle" is also taking place at the level of the group of researchers in particular. As suggested by the guidelines of the Swiss Commission for Research Partnerships with Developing Countries (Swiss Academy of Sciences 2014), it is better to be frank from the very beginning and to lay all the cards on the table, so as to start on a positive footing, with no hidden meanings or misunderstandings. However, let us specify that the problem does not only concern a good understanding between researchers and practitioners. There were dissensions within the UNPEC group from the very beginning, concerning the role practitioners should play. Many researchers, subsequently, did not wish to collaborate in the drafting of the policy briefs, arguing that they were not qualified to do so. A decision had to be taken in this regard by the coordinator, either with or without the practitioners. Both options would have been legitimate. However, the decision was never taken.

This being said, satisfying the first and second principles of the Swiss Academy, "Set the agenda together", and "Interact with stakeholders", is easier said than done, because both can be contradictory: which practitioners would accept to deal openly with *all* stakeholders, including the slum dwellers residing in the national park, the local mafias or the real estate lobby?

References

Bourdieu, P. (1980). *Questions de sociologie*. Paris: Minuit.

Clark, W. C., Tomich, T. P., Van Noordwijk, M., Guston, D., Catacutan, D., Dickson, N. M., et al. (2016). Boundary work for sustainable development: Natural resource management at the Consultative Group on International Agricultural Research. *PNAS, 113*(17), 4615–4622.

Reid, R. S., Kaelo, D., Nkedianye, D. K., Kristjanson, P., Said, M. Y., Galvin, K. A., & Gambill, I. (2014). The Mara-Serengeti ecosystem and greater Maasailand: Building the role of local leaders, Institutions, and Communities. In J. N. Levitt (Ed.) *Conservation catalysts. The academy as nature's agent.* Lincoln Institute of Land Policy.

Spierenburg, M. (2012). Getting the message across. Biodiversity science and policy interfaces—A review. *Gaia, 21*(2), 125–134.

Swiss Academy of Sciences. (2014). *A guide for transboundary research partnerships* (2nd ed.). Bern: KFPE, online.

Printed by Printforce, the Netherlands